Introductory Business Statistics

Introductory Business Statistics

VOLUME 1

Melissa Cardon

Kruger Brentt Publishers
2025

Kruger Brentt Publishers UK. LTD.
Company Number 9728962

Regd. Office: 68 St Margarets Road, Edgware, Middlesex HA8 9UU

ISBN: 978-1-78715-396-7 (Vol. 1)

Disclaimer:

Every possible effort has been made to ensure that the information contained in this book is accurate at the time of going to press, and the publisher and author cannot accept responsibility for any errors or omissions, however caused. No responsibility for loss or damage occasioned to any person acting, or refraining from action, as a result of the material in this publication can be accepted by the editor, the publisher or the author. The Publisher is not associated with any product or vendor mentioned in the book. The contents of this work are intended to further general scientific research, understanding and discussion only. Readers should consult with a specialist where appropriate.

Every effort has been made to trace the owners of copyright material used in this book, if any. The author and the publisher will be grateful for any omission brought to their notice for acknowledgement in the future editions of the book.

For information on all our publications visit our website at http://krugerbrentt.com/

PREFACE

"Introductory Business Statistics" embarks on a foundational exploration of statistical concepts and techniques essential for making informed business decisions, offering readers a comprehensive understanding of the principles, methods, and applications of statistics in the business context. This book serves as an indispensable resource for students, professionals, and entrepreneurs seeking to harness the power of statistics to analyze data, draw meaningful insights, and drive evidence-based decision-making in business settings.

Statistics plays a crucial role in modern business operations by providing tools and techniques for summarizing, analyzing, and interpreting data to inform strategic planning, resource allocation, marketing strategies, financial forecasting, and performance evaluation. In an era characterized by big data and data-driven decision-making, statistical literacy has become an indispensable skill for success in various fields, including finance, marketing, operations management, and entrepreneurship.

The primary objective of this book is to provide readers with a comprehensive introduction to business statistics, covering key topics such as data collection and presentation, descriptive statistics, probability theory, statistical inference, hypothesis testing, regression analysis, and correlation analysis. Each chapter offers insights, theoretical frameworks, practical examples, and hands-on exercises to help readers develop proficiency in applying statistical methods to real-world business problems.

Contributions from leading experts in the field explore various aspects of business statistics, including techniques for analyzing categorical and numerical data, conducting market research, forecasting sales and demand, evaluating investment opportunities, and assessing risk and uncertainty. Special emphasis is placed on the use of statistical software packages such as Excel, SPSS, and R for data analysis and visualization, enabling readers to gain practical experience in using statistical tools to solve business problems.

As businesses increasingly rely on data-driven insights to gain a competitive edge in today's fast-paced and dynamic marketplaces, the ability to understand and interpret statistical information has become a critical skill for success. By mastering the principles and techniques of business statistics, individuals can make better-informed decisions, identify opportunities for growth and innovation, and drive sustainable business performance in an ever-changing world.

We extend our sincere gratitude to the contributors whose expertise and dedication have enriched this book. It is our hope that "Introductory Business Statistics" will serve as a valuable resource and reference guide for readers seeking to enhance their statistical literacy and analytical skills, empowering them to thrive in the modern business landscape.

Melissa Cardon

CONTENTS

VOLUME-2

1 Chapter | FUNDAMENTAL CONCEPTS OF SAMPLING AND DATA

Figure 1.1 We encounter statistics in our daily lives more often than we probably realize and from many different sources, like the news. (credit: David Sim)

INTRODUCTION: AN OVERVIEW

You may be contemplating the inquiry, "When and where will I apply statistics?" Statistical information is prevalent in newspapers, television, and the Internet. Statistics exist about crime, sports, education, politics, and real estate. Generally, when one reads a newspaper article or views a television news show, one is presented with representative information. This information enables you to determine the accuracy of a statement, assertion, or "fact." Statistical approaches can assist in formulating the most informed conjecture. As you will always encounter statistical data at some stage in your life, it is essential to acquire strategies for assessing this information critically. Consider purchasing a residence or overseeing a financial plan. Contemplate your selected vocation. Disciplines such as economics, business, psychology, education, biology, law, computer science, police science, and early childhood development necessitate the completion of a minimum of one statistics course. This chapter encompasses the fundamental concepts and terminology of probability and statistics. You will rapidly comprehend the interrelationship between statistics and probability. You will also learn how data is collected and how to differentiate between "good" and "bad" data.

1.1 DEFINITIONS OF STATISTICS, PROBABILITY, AND ESSENTIAL TERMINOLOGY

The science of **statistics** deals with the collection, analysis, interpretation, and presentation of **data**. We see and use data in our everyday lives.

In this course, you will learn how to organize and summarize data. Organizing and summarizing data is called **descriptive statistics.** Two ways to summarize data are by graphing and by using numbers (for example, finding an average). After you have studied probability and probability distributions, you will use formal methods for drawing conclusions from "good" data. The formal methods are called **inferential statistics.** Statistical inference uses probability to determine how confident we can be that our conclusions are correct.

Effective interpretation of data (inference) is based on good procedures for producing data and thoughtful examination of the data. You will encounter what will seem to be too many mathematical formulas for interpreting data. The goal of statistics is not to perform numerous calculations using the formulas, but to gain an understanding of your data. The calculations can be done using a calculator or a computer. The understanding must come from you. If you can thoroughly grasp the basics of statistics, you can be more confident in the decisions you make in life.

Probability

Probability is a mathematical tool used to study randomness. It deals with the chance (the likelihood) of an event occurring. For example, if you toss a fair coin four times, the outcomes may not be two heads and two tails. However, if you toss the same coin 4,000 times, the outcomes will be close to half heads and half tails. The expected theoretical probability of heads in any one toss is $\frac{1}{2}$ or 0.5. Even though the outcomes of a few repetitions are uncertain, there is a regular pattern of outcomes when there are many repetitions. After reading about the English statistician Karl Pearson who tossed a coin 24,000 times with a result of 12,012 heads, one of the authors tossed a coin 2,000 times. The results were 996 heads. The fraction $\frac{996}{2000}$ is equal to 0.498 which is very close to 0.5, the expected probability.

The theory of probability began with the study of games of chance such as poker. Predictions take the form of probabilities. To predict the likelihood of an earthquake, of rain, or whether you will get an A in this course, we use probabilities. Doctors use probability to determine the chance of a vaccination causing the disease the vaccination is supposed to prevent. A stockbroker uses probability to determine the rate of return on a client's investments. You might use probability to decide to buy a lottery ticket or not. In your study of statistics, you will use the power of mathematics through probability calculations to analyze and interpret your data.

Key Terms

In statistics, we generally want to study a population. You can think of a population as a collection of persons, things, or objects under study. To study the population, we select a sample. The idea of sampling is to select a portion (or subset) of the larger population and study that portion (the sample) to gain information about the population. Data are the result of sampling from a population.

Because it takes a lot of time and money to examine an entire population, sampling is a very practical technique. If you wished to compute the overall grade point average at your school, it would make sense to select a sample of students who attend the school. The data collected from the sample would be the students' grade point averages. In presidential elections, opinion poll samples of 1,000–2,000 people are taken. The opinion poll is supposed to represent the views of the people in the entire country. Manufacturers of canned carbonated drinks take samples to determine if a 16 ounce can contains 16 ounces of carbonated drink.

From the sample data, we can calculate a statistic. A statistic is a number that represents a property of the sample. For example, if we consider one math class to be a sample of the population of all math classes, then the average number of points earned by students in that one math class at the end of the term is an example of a statistic. The statistic is an estimate of a population parameter, in this case the mean. A parameter is a numerical characteristic of the whole population that can be estimated by a statistic. Since we considered all math classes to be the population, then the average number of points earned per student over all the math classes is an example of a parameter.

One of the main concerns in the field of statistics is how accurately a statistic estimates a parameter. The accuracy really depends on how well the sample represents the population. The sample must contain the characteristics of the population in order to be a representative sample. We are interested in both the sample statistic and the population parameter in inferential statistics. In a later chapter, we will use the sample statistic to test the validity of the established population parameter.

A variable, or random variable, usually notated by capital letters such as X and Y, is a characteristic or measurement that can be determined for each member of a population. Variables may be numerical or categorical. Numerical variables take on values with equal units such as weight in pounds and time in hours. Categorical variables place the person or thing into a category. If we let X equal the number of points earned by one math student at the end of a term, then X is a numerical variable. If we let Y be a person's party affiliation, then some examples of Y include Republican, Democrat, and Independent. Y is a categorical variable. We could do some math with values of X (calculate the average number of points earned, for example), but it makes no sense to do math with values of Y (calculating an average party affiliation makes no sense).

Data are the actual values of the variable. They may be numbers or they may be words. Datum is a single value.

Two words that come up often in statistics are mean and proportion. If you were to take three exams in your math classes and obtain scores of 86, 75, and 92, you would calculate your mean score by adding the three exam scores and dividing by three (your mean score would be 84.3 to one decimal place). If, in your math class, there are 40 students and 22 are men and 18 are women, then the proportion of men students is $\frac{22}{40}$ and the proportion of women students is $\frac{18}{40}$. Mean and proportion are discussed in more detail in later chapters.

NOTE

The terms "mean" and "average" are frequently utilized synonymously. The replacement of one word with another is a widespread practice. The precise phrase is "arithmetic mean," while "average" refers to a central position. In practice, among non-statisticians, "average" is frequently synonymous with "arithmetic mean."

EXAMPLE 1.1

Determine what the key terms refer to in the following study. We want to know the average (mean) amount of money first year college students spend at ABC College on school supplies that do not include books. We randomly surveyed 100 first year students at the college. Three of those students spent $150, $200, and $225, respectively.

Solution 1.1

The population is all first year students attending ABC College this term.

The sample could be all students enrolled in one section of a beginning statistics course at ABC College (although this sample may not represent the entire population).

The parameter is the average (mean) amount of money spent (excluding books) by first year college students at ABC College this term: the population mean.

The statistic is the average (mean) amount of money spent (excluding books) by first year college students in the sample.

The variable could be the amount of money spent (excluding books) by one first year student. Let X = the amount of money spent (excluding books) by one first year student attending ABC College.

The data are the dollar amounts spent by the first year students. Examples of the data are $150, $200, and $225.

Try It 1.1

Determine what the key terms refer to in the following study. We want to know the average (mean) amount of money spent on school uniforms each year by families with children at Knoll Academy. We randomly survey 100 families with children in the school. Three of the families spent $65, $75, and $95, respectively.

EXAMPLE 1.2

Determine what the key terms refer to in the following study.

A study was conducted at a local college to analyze the average cumulative GPA's of students who graduated last year. Fill in the letter of the phrase that best describes each of the items below.

1. Population ____ 2. Statistic ____ 3. Parameter ____ 4. Sample ____ 5. Variable ____ 6. Data ____

a. all students who attended the college last year
b. the cumulative GPA of one student who graduated from the college last year
c. 3.65, 2.80, 1.50, 3.90
d. a group of students who graduated from the college last year, randomly selected
e. the average cumulative GPA of students who graduated from the college last year
f. all students who graduated from the college last year
g. the average cumulative GPA of students in the study who graduated from the college last year

Solution 1.2

1. f; 2. g; 3. e; 4. d; 5. b; 6. c

EXAMPLE 1.3

Determine what the key terms refer to in the following study.

As part of a study designed to test the safety of automobiles, the National Transportation Safety Board collected and reviewed data about the effects of an automobile crash on test dummies. Here is the criterion they used:

Table 1.1

Speed at which cars crashed	Location of "drive" (i.e. dummies)
35 miles/hour	Front Seat

Cars with dummies in the front seats were crashed into a wall at a speed of 35 miles per hour. We want to know the proportion of dummies in the driver's seat that would have had head injuries, if they had been actual drivers. We start with a simple random sample of 75 cars.

Solution 1.3

The population is all cars containing dummies in the front seat.

The sample is the 75 cars, selected by a simple random sample.

The parameter is the proportion of driver dummies (if they had been real people) who would have suffered head injuries in the population.

The statistic is proportion of driver dummies (if they had been real people) who would have suffered head injuries in the sample.

The variable X = the number of driver dummies (if they had been real people) who would have suffered head injuries.

The data are either: yes, had head injury, or no, did not.

EXAMPLE 1.4

Determine what the key terms refer to in the following study.

An insurance company would like to determine the proportion of all medical doctors who have been involved in one or more malpractice lawsuits. The company selects 500 doctors at random from a professional directory and determines the number in the sample who have been involved in a malpractice lawsuit.

Solution 1.4

The population is all medical doctors listed in the professional directory.

The parameter is the proportion of medical doctors who have been involved in one or more malpractice suits in the population.

The sample is the 500 doctors selected at random from the professional directory.

The statistic is the proportion of medical doctors who have been involved in one or more malpractice suits in the sample.

The variable X = the number of medical doctors who have been involved in one or more malpractice suits.

The data are either: yes, was involved in one or more malpractice lawsuits, or no, was not.

1.2 DATA, SAMPLING, AND VARIATION IN DATA AND SAMPLING

Data can originate from a population or a sample. Lowercase letters such as x or y are typically employed to denote data values. The majority of data may be classified into the following categories.

- Qualitative
- Quantitative

Qualitative data are the result of categorizing or describing attributes of a population. Qualitative data are also often called categorical data. Hair color, blood type, ethnic group, the car a person drives, and the street a person lives on are examples of qualitative (categorical) data. Qualitative (categorical) data are generally described by words or letters. For instance, hair color might be black, dark brown, light brown, blonde, gray, or red. Blood type might be AB+, O-, or B+. Researchers often prefer to use quantitative data over qualitative (categorical) data because it lends itself more easily to mathematical analysis. For example, it does not make sense to find an average hair color or blood type.

Quantitative data are always numbers. Quantitative data are the result of counting or measuring attributes of a population. Amount of money, pulse rate, weight, number of people living in your town, and number of students who take statistics are examples of quantitative data. Quantitative data may be either discrete or continuous.

All data that are the result of counting are called quantitative discrete data. These data take on only certain numerical values. If you count the number of phone calls you receive for each day of the week, you might get values such as zero, one, two, or three.

Data that are not only made up of counting numbers, but that may include fractions, decimals, or irrational numbers, are called quantitative continuous data. Continuous data are often the results of measurements like lengths, weights, or times. A list of the lengths in minutes for all the phone calls that you make in a week, with numbers like 2.4, 7.5, or 11.0, would be quantitative continuous data.

EXAMPLE 1.5 Data Sample of Quantitative Discrete Data

The data are the number of books students carry in their backpacks. You sample five students. Two students carry three books, one student carries four books, one student carries two books, and one student carries one book. The numbers of books (three, four, two, and one) are the quantitative discrete data.

TRY IT 1.5

The data are the number of machines in a gym. You sample five gyms. One gym has 12 machines, one gym has 15 machines, one gym has ten machines, one gym has 22 machines, and the other gym has 20 machines. What type of data is this?

EXAMPLE 1.6 Data Sample of Quantitative Continuous Data

The data are the weights of backpacks with books in them. You sample the same five students. The weights (in pounds) of their backpacks are 6.2, 7, 6.8, 9.1, 4.3. Notice that backpacks carrying three books can have different weights. Weights are quantitative continuous data.

TRY IT 1.6

The data are the areas of lawns in square feet. You sample five houses. The areas of the lawns are 144 sq. feet, 160 sq. feet, 190 sq. feet, 180 sq. feet, and 210 sq. feet. What type of data is this?

EXAMPLE 1.7

You go to the supermarket and purchase three cans of soup (19 ounces tomato bisque, 14.1 ounces lentil, and 19 ounces Italian wedding), two packages of nuts (walnuts and peanuts), four different kinds of vegetable (broccoli, cauliflower, spinach, and carrots), and two desserts (16 ounces pistachio ice cream and 32 ounces chocolate chip cookies).

Name data sets that are quantitative discrete, quantitative continuous, and qualitative(categorical).

Solution 1.7

One Possible Solution:

- The three cans of soup, two packages of nuts, four kinds of vegetables and two desserts are quantitative discrete data because you count them.
- The weights of the soups (19 ounces, 14.1 ounces, 19 ounces) are quantitative continuous data because you measure weights as precisely as possible.
- Types of soups, nuts, vegetables and desserts are qualitative(categorical) data because they are categorical.

EXAMPLE 1.8

The data are the colors of backpacks. Again, you sample the same five students. One student has a red backpack, two students have black backpacks, one student has a green backpack, and one student has a gray backpack. The colors red, black, black, green, and gray are qualitative(categorical) data.

TRY IT 1.8

The data are the colors of houses. You sample five houses. The colors of the houses are white, yellow, white, red, and white. What type of data is this?

NOTE

You may collect data as numbers and report it categorically. For example, the quiz scores for each student are recorded throughout the term. At the end of the term, the quiz scores are reported as A, B, C, D, or F.

EXAMPLE 1.9

Work collaboratively to determine the correct data type (quantitative or qualitative). Indicate whether quantitative data are continuous or discrete. Hint: Data that are discrete often start with the words "the number of."

a. the number of pairs of shoes you own
b. the type of car you drive
c. the distance from your home to the nearest grocery store
d. the number of classes you take per school year
e. the type of calculator you use
f. weights of sumo wrestlers
g. number of correct answers on a quiz
h. IQ scores (This may cause some discussion.)

Solution 1.9

Items a, d, and g are quantitative discrete; items c, f, and h are quantitative continuous; items b and e are qualitative, or categorical.

TRY IT 1.9

Determine the correct data type (quantitative or qualitative) for the number of cars in a parking lot. Indicate whether quantitative data are continuous or discrete.

EXAMPLE 1.10

A statistics professor collects information about the classification of her students as freshmen, sophomores, juniors, or seniors. The data she collects are summarized in the pie chart Figure 1.2. What type of data does this graph show?

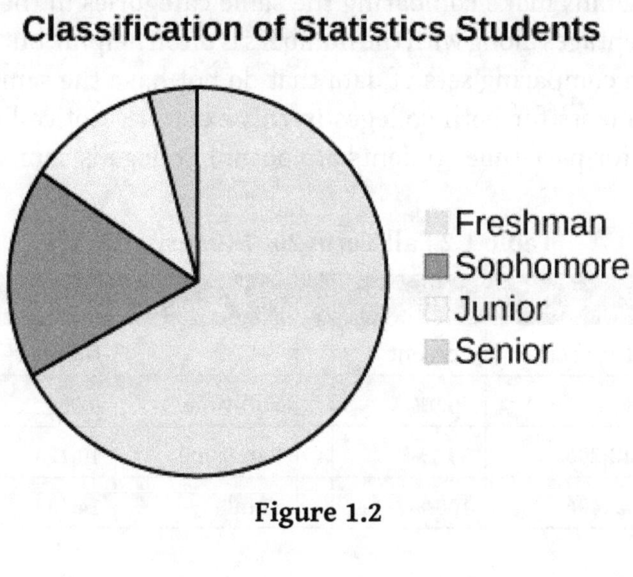

Figure 1.2

Solution 1.10

This pie chart shows the students in each year, which is qualitative (or categorical) data.

TRY IT 1.10

The registrar at State University maintains records of the credit hours completed by students each semester. The data he gathers are encapsulated in the histogram. The class intervals are 10 to under 13, 13 to under 16, 16 to under 19, 19 to under 22, and 22 to under 25.

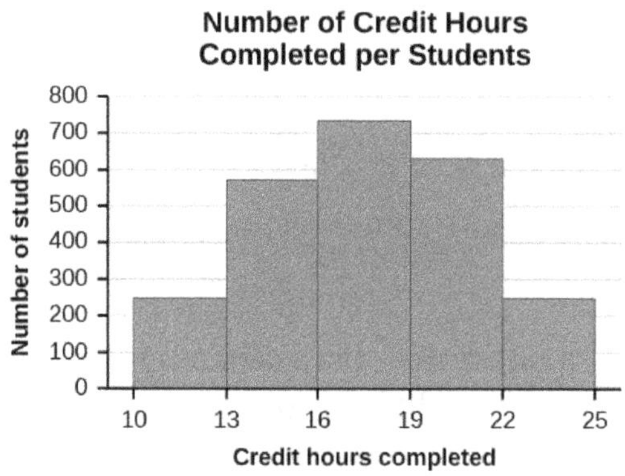

Figure 1.3

What type of data does this graph show?

Qualitative Data Discussion

Below are tables comparing the number of part-time and full-time students at De Anza College and Foothill College enrolled for the spring 2010 quarter. The tables display counts (frequencies) and percentages or proportions (relative frequencies). The percent columns make comparing the same categories in the colleges easier. Displaying percentages along with the numbers is often helpful, but it is particularly important when comparing sets of data that do not have the same totals, such as the total enrollments for both colleges in this example. Notice how much larger the percentage for part-time students at Foothill College is compared to De Anza College.

Table 1.2 Fall Term 2007 (Census day)

De Anza College				Foothill College		
	Number	Percent			Number	Percent
Full-time	9,200	40.9%		Full-time	4,059	28.6%
Part-time	13,296	59.1%		Part-time	10,124	71.4%
Total	22,496	100%		Total	14,183	100%

Tables are a good way of organizing and displaying data. But graphs can be even more helpful in understanding the data. There are no strict rules concerning which graphs to use. Two graphs that are used to display qualitative(categorical) data are pie charts and bar graphs.

In a pie chart, categories of data are represented by wedges in a circle and are proportional in size to the percent of individuals in each category.

In a bar graph, the length of the bar for each category is proportional to the number or percent of individuals in each category. Bars may be vertical or horizontal.

A Pareto chart consists of bars that are sorted into order by category size (largest to smallest).

Look at Figure 1.4 and Figure 1.5 and determine which graph (pie or bar) you think displays the comparisons better.

It is a good idea to look at a variety of graphs to see which is the most helpful in displaying the data. We might make different choices of what we think is the "best" graph depending on the data and the context. Our choice also depends on what we are using the data for.

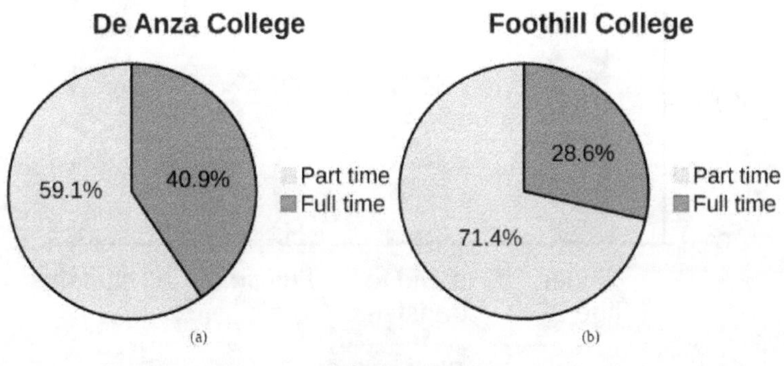

Figure 1.4

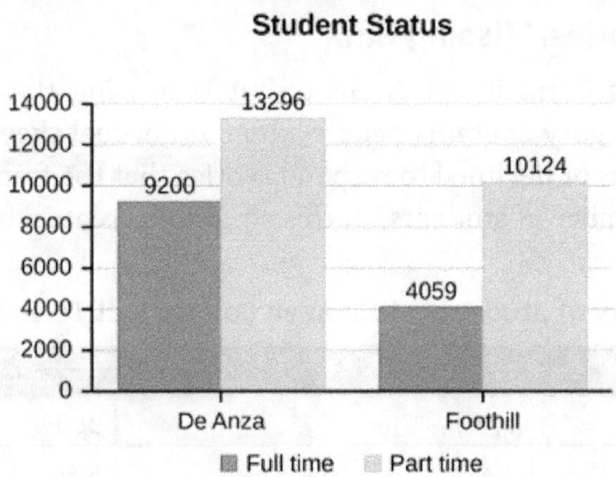

Figure 1.5

Percentages That Add to More (or Less) Than 100%

Sometimes percentages add up to be more than 100% (or less than 100%). In the graph, the percentages add to more than 100% because students can be in more than one category. A bar graph is appropriate to compare the relative size of the categories. A pie chart cannot be used. It also could not be used if the percentages added to less than 100%.

Table 1.3 De Anza College Spring 2010

Characteristic/category	Percent
Full-time students	40.9%
Students who intend to transfer to a 4-year educational institution	48.6%
Students under age 25	61.0%
TOTAL	150.5%

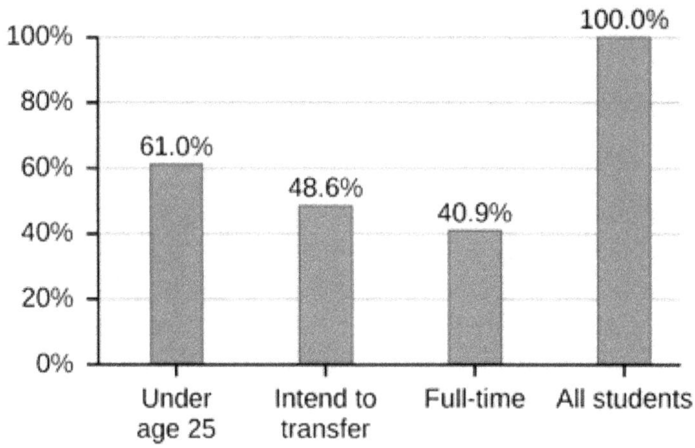

Figure 1.6

Omitting Categories/Missing Data

The table displays Ethnicity of Students but is missing the "Other/Unknown" category. This category contains people who did not feel they fit into any of the ethnicity categories or declined to respond. Notice that the frequencies do not add up to the total number of students. In this situation, create a bar graph and not a pie chart.

Table 1.4 Ethnicity of Students at De Anza College Fall Term 2007 (Census Day)

	Frequency	Percent
Asian	8,794	36.1%
Black	1,412	5.8%
Filipino	1,298	5.3%

	Frequency	Percent
Hispanic	4,180	17.1%
Native American	146	0.6%
Pacific Islander	236	1.0%
White	5,978	24.5%
TOTAL	22,044 out of 24,382	90.4% out of 100%

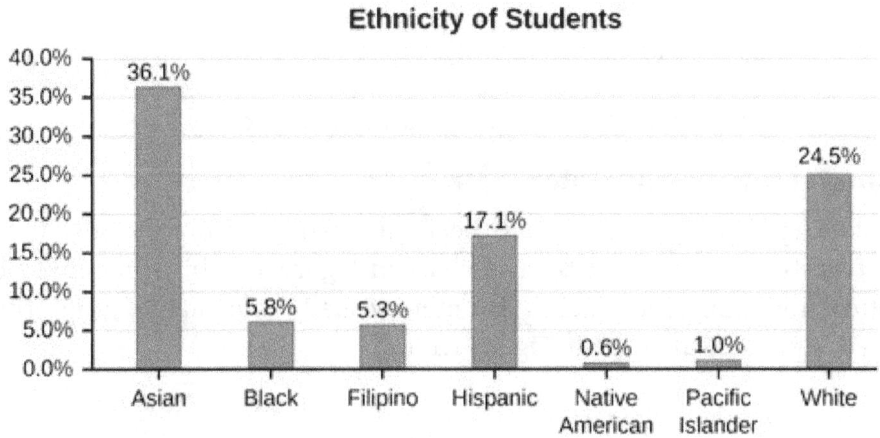

Figure 1.7

The following graph is the same as the previous graph but the "Other/Unknown" percent (9.6%) has been included. The "Other/Unknown" category is large compared to some of the other categories (Native American, 0.6%, Pacific Islander 1.0%). This is important to know when we think about what the data are telling us.

This particular bar graph in Figure 1.8 can be difficult to understand visually. The graph in Figure 1.9 is a Pareto chart. The Pareto chart has the bars sorted from largest to smallest and is easier to read and interpret.

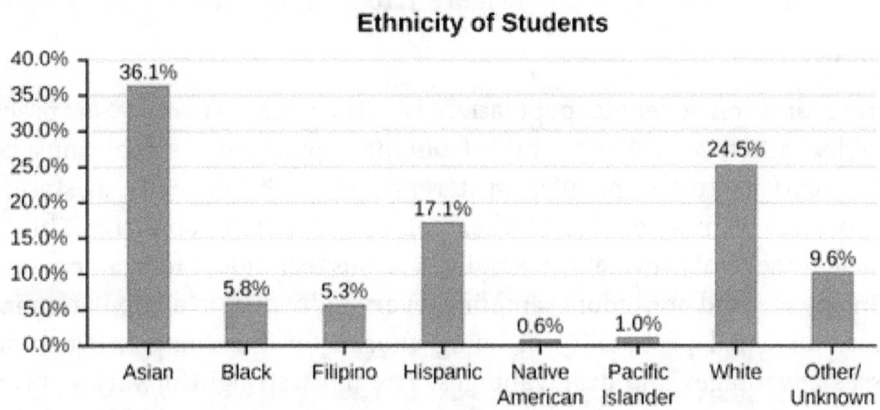

Figure 1.8 Bar Graph with Other/Unknown Category

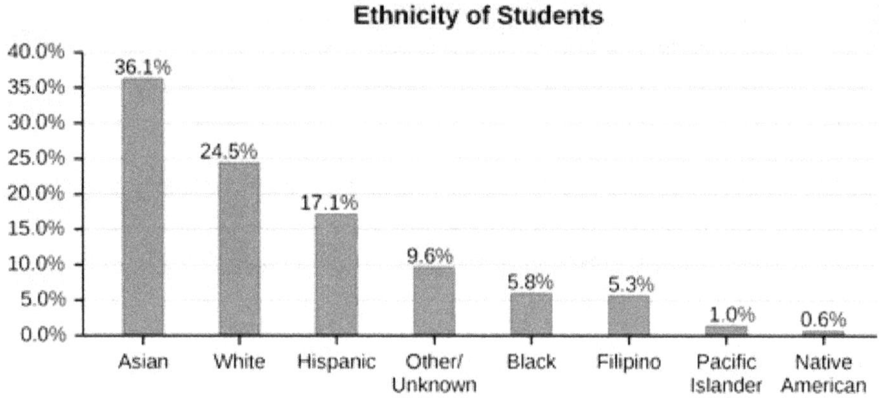

Figure 1.9 Pareto Chart With Bars Sorted by Size

Pie Charts: No Missing Data

The following pie charts have the "Other/Unknown" category included (since the percentages must add to 100%). The chart in Figure 1.10(b) is organized by the size of each wedge, which makes it a more visually informative graph than the unsorted, alphabetical graph in Figure 1.10(a).

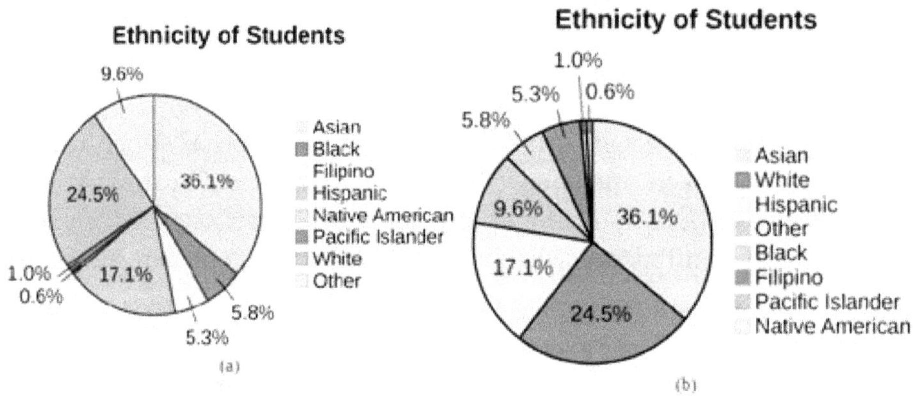

Figure 1.10

Sampling

Collecting data on a whole population is either too expensive or practically unfeasible. Rather, we utilize a sample from the population. A sample must possess identical features to the population it represents. The majority of statisticians employ diverse techniques of random sampling to attain this objective. This section will elucidate several prevalent strategies. Various techniques for random sampling exist. In any method of random sampling, every individual in a population initially possesses an equal probability of being chosen for the sample. Every method possesses advantages and disadvantages. The most straightforward approach to delineate is referred to as a simple random sample. In simple random sampling, any group of n individuals has an equal probability of being selected as any other group of n persons. Consequently, every sample of identical size possesses an equivalent probability of selection.Besides simple random sampling, there are other forms of sampling that involve a chance process for getting the sample.

Other well-known random sampling methods are the stratified sample, the cluster sample, and the systematic sample.

To choose a stratified sample, divide the population into groups called strata and then take a proportionate number from each stratum. For example, you could stratify (group) your college population by department and then choose a proportionate simple random sample from each stratum (each department) to get a stratified random sample. To choose a simple random sample from each department, number each member of the first department, number each member of the second department, and do the same for the remaining departments. Then use simple random sampling to choose proportionate numbers from the first department and do the same for each of the remaining departments. Those numbers picked from the first department, picked from the second department, and so on represent the members who make up the stratified sample.

To choose a cluster sample, divide the population into clusters (groups) and then randomly select some of the clusters. All the members from these clusters are in the cluster sample. For example, if you randomly sample four departments from your college population, the four departments make up the cluster sample. Divide your college faculty by department. The departments are the clusters. Number each department, and then choose four different numbers using simple random sampling. All members of the four departments with those numbers are the cluster sample.

To choose a systematic sample, randomly select a starting point and take every nth piece of data from a listing of the population. For example, suppose you have to do a phone survey. Your phone book contains 20,000 residence listings. You must choose 400 names for the sample. Number the population 1–20,000 and then use a simple random sample to pick a number that represents the first name in the sample. Then choose every fiftieth name thereafter until you have a total of 400 names (you might have to go back to the beginning of your phone list). Systematic sampling is frequently chosen because it is a simple method.

A type of sampling that is non-random is convenience sampling. Convenience sampling involves using results that are readily available. For example, a computer software store conducts a marketing study by interviewing potential customers who happen to be in the store browsing through the available software. The results of convenience sampling may be very good in some cases and highly biased (favor certain outcomes) in others.

Sampling data should be done very carefully. Collecting data carelessly can have devastating results. Surveys mailed to households and then returned may be very biased (they may favor a certain group). It is better for the person conducting the survey to select the sample respondents.

True random sampling is done with replacement. That is, once a member is picked, that member goes back into the population and thus may be chosen more than once. However for practical reasons, in most populations, simple random sampling is done without replacement. Surveys are typically done without replacement. That is, a member of the population may be chosen only once. Most samples are taken from large populations and the sample tends to be small in comparison to the population. Since this is the case, sampling without replacement is approximately the same as sampling with replacement because the chance of picking the same individual more than once with replacement is very low.

Sampling without replacement instead of sampling with replacement becomes a mathematical issue only when the population is small.

When you analyze data, it is important to be aware of sampling errors and nonsampling errors. The actual process of sampling causes sampling errors. For example, the sample may not be large enough. Factors not related to the sampling process cause nonsampling errors. A defective counting device can cause a nonsampling error.

In reality, a sample will never be exactly representative of the population so there will always be some sampling error. As a rule, the larger the sample, the smaller the sampling error.

In statistics, a sampling bias is created when a sample is collected from a population and some members of the population are not as likely to be chosen as others (remember, each member of the population should have an equally likely chance of being chosen). When a sampling bias happens, there can be incorrect conclusions drawn about the population that is being studied.

In a college population of 10,000 people, suppose you want to pick a sample of 1,000 randomly for a survey. For any particular sample of 1,000, if you are sampling with replacement,

- the chance of picking the first person is 1,000 out of 10,000 (0.1000);
- the chance of picking a different second person for this sample is 999 out of 10,000 (0.0999);
- the chance of picking the same person again is 1 out of 10,000 (very low).

If you are sampling without replacement,

- the chance of picking the first person for any particular sample is 1000 out of 10,000 (0.1000);

- the chance of picking a different second person is 999 out of 9,999 (0.0999);
- you do not replace the first person before picking the next person.

Compare the fractions 999/10,000 and 999/9,999. For accuracy, carry the decimal answers to four decimal places. To four decimal places, these numbers are equivalent (0.0999).

Sampling without replacement instead of sampling with replacement becomes a mathematical issue only when the population is small. For example, if the population is 25 people, the sample is ten, and you are sampling with replacement for any particular sample, then the chance of picking the first person is ten out of 25, and the chance of picking a different second person is nine out of 25 (you replace the first person).

If you sample without replacement, then the chance of picking the first person is ten out of 25, and then the chance of picking the second person (who is different) is nine out of 24 (you do not replace the first person).

Compare the fractions 9/25 and 9/24. To four decimal places, 9/25 = 0.3600 and 9/24 = 0.3750. To four decimal places, these numbers are not equivalent.

When you analyze data, it is important to be aware of sampling errors and nonsampling errors. The actual process of sampling causes sampling errors. For example, the sample may not be large enough. Factors not related to the sampling process cause nonsampling errors. A defective counting device can cause a nonsampling error.

In reality, a sample will never be exactly representative of the population so there will always be some sampling error. As a rule, the larger the sample, the smaller the sampling error.

In statistics, a sampling bias is created when a sample is collected from a population and some members of the population are not as likely to be chosen as others (remember, each member of the population should have an equally likely chance of being chosen). When a sampling bias happens, there can be incorrect conclusions drawn about the population that is being studied.

Critical Evaluation

We need to evaluate the statistical studies we read about critically and analyze them before accepting the results of the studies. Common problems to be aware of include

- Problems with samples: A sample must be representative of the population. A sample that is not representative of the population is biased. Biased samples that are not representative of the population give results that are inaccurate and not valid.

- Self-selected samples: Responses only by people who choose to respond, such as call-in surveys, are often unreliable.
- Sample size issues: Samples that are too small may be unreliable. Larger samples are better, if possible. In some situations, having small samples is unavoidable and can still be used to draw conclusions. Examples: crash testing cars or medical testing for rare conditions
- Undue influence: collecting data or asking questions in a way that influences the response
- Non-response or refusal of subject to participate: The collected responses may no longer be representative of the population. Often, people with strong positive or negative opinions may answer surveys, which can affect the results.
- Causality: A relationship between two variables does not mean that one causes the other to occur. They may be related (correlated) because of their relationship through a different variable.
- Self-funded or self-interest studies: A study performed by a person or organization in order to support their claim. Is the study impartial? Read the study carefully to evaluate the work. Do not automatically assume that the study is good, but do not automatically assume the study is bad either. Evaluate it on its merits and the work done.
- Misleading use of data: improperly displayed graphs, incomplete data, or lack of context
- Confounding: When the effects of multiple factors on a response cannot be separated. Confounding makes it difficult or impossible to draw valid conclusions about the effect of each factor.

EXAMPLE 1.11

A study is done to determine the average tuition that San Jose State undergraduate students pay per semester. Each student in the following samples is asked how much tuition he or she paid for the Fall semester. What is the type of sampling in each case?

a. A sample of 100 undergraduate San Jose State students is taken by organizing the students' names by classification (freshman, sophomore, junior, or senior), and then selecting 25 students from each.A random number generator is used to select a student from the alphabetical listing of all undergraduate students in the Fall semester. Starting with that student, every 50th student is chosen until 75 students are included in the sample.

b. A completely random method is used to select 75 students. Each undergraduate student in the fall semester has the same probability of being chosen at any stage of the sampling process.

c. The freshman, sophomore, junior, and senior years are numbered one, two, three, and four, respectively. A random number generator is used to pick two of those years. All students in those two years are in the sample.

d. An administrative assistant is asked to stand in front of the library one Wednesday and to ask the first 100 undergraduate students he encounters what they paid for tuition the Fall semester. Those 100 students are the sample.

Solution 1.11

a. stratified; b. systematic; c. simple random; d. cluster; e. convenience

EXAMPLE 1.12

Determine the type of sampling used (simple random, stratified, systematic, cluster, or convenience).

a. A soccer coach selects six players from a group of boys aged eight to ten, seven players from a group of boys aged 11 to 12, and three players from a group of boys aged 13 to 14 to form a recreational soccer team.

b. A pollster interviews all human resource personnel in five different high tech companies.

c. A high school educational researcher interviews 50 high school female teachers and 50 high school male teachers.

d. A medical researcher interviews every third cancer patient from a list of cancer patients at a local hospital.

e. A high school counselor uses a computer to generate 50 random numbers and then picks students whose names correspond to the numbers.

f. A student interviews classmates in his algebra class to determine how many pairs of jeans a student owns, on the average.

Solution 1.12

a. stratified; b. cluster; c. stratified; d. systematic; e. simple random; f.convenience

If we were to examine two samples representing the same population, even if we used random sampling methods for the samples, they would not be exactly the same. Just as there is variation in data, there is variation in samples. As you become accustomed to sampling, the variability will begin to seem natural.

EXAMPLE 1.13

Suppose ABC College has 10,000 part-time students (the population). We are interested in the average amount of money a part-time student spends on books in the fall term. Asking all 10,000 students is an almost impossible task.

Suppose we take two different samples.

First, we use convenience sampling and survey ten students from a first term organic chemistry class. Many of these students are taking first term calculus in addition to the organic chemistry class. The amount of money they spend on books is as follows:

\$128; \$87; \$173; \$116; \$130; \$204; \$147; \$189; \$93; \$153

The second sample is taken using a list of senior citizens who take P.E. classes and taking every fifth senior citizen on the list, for a total of ten senior citizens. They spend:

\$50; \$40; \$36; \$15; \$50; \$100; \$40; \$53; \$22; \$22

It is unlikely that any student is in both samples.

a. Do you think that either of these samples is representative of (or is characteristic of) the entire 10,000 part-time student population?

Solution 1.13

a. No. The first sample probably consists of science-oriented students. Besides the chemistry course, some of them are also taking first-term calculus. Books for these classes tend to be expensive. Most of these students are, more than likely, paying more than the average part-time student for their books. The second sample is a group of senior citizens who are, more than likely, taking courses for health and interest. The amount of money they spend on books is probably much less than the average parttime student. Both samples are biased. Also, in both cases, not all students have a chance to be in either sample.
b. Since these samples are not representative of the entire population, is it wise to use the results to describe the entire population?

Solution 1.13

b. No. For these samples, each member of the population did not have an equally likely chance of being chosen.

   Now, suppose we take a third sample. We choose ten different part-time students from the disciplines of chemistry, math, English, psychology, sociology, history, nursing, physical education, art, and early childhood development. (We assume that these are the only disciplines in which part-time students at ABC College are enrolled and that an equal number of part-time students are enrolled in each of the disciplines.) Each student is chosen using simple random sampling. Using a calculator, random numbers are generated and a student from a particular discipline is selected if he or she has a corresponding number. The students spend the following amounts: $180; $50; $150; $85; $260; $75; $180; $200; $200; $150

c. Is the sample biased?

Solution 1.13

c. The sample is unbiased, but a larger sample would be recommended to increase the likelihood that the sample will be close to representative of the population. However, for a biased sampling technique, even a large sample runs the risk of not being representative of the population.

   Students often ask if it is "good enough" to take a sample, instead of surveying the entire population. If the survey is done well, the answer is yes.

TRY IT 1.13

A local radio station has a fan base of 20,000 listeners. The station wants to know if its audience would prefer more music or more talk shows. Asking all 20,000 listeners is an almost impossible task.

The station uses convenience sampling and surveys the first 200 people they meet at one of the station's music concert events. 24 people said they'd prefer more talk shows, and 176 people said they'd prefer more music.

Do you think that this sample is representative of (or is characteristic of) the entire 20,000 listener population?

Variation in Data

Variation is present in any set of data. For example, 16-ounce cans of beverage may contain more or less than 16 ounces of liquid. In one study, eight 16 ounce cans were measured and produced the following amount (in ounces) of beverage:

15.8; 16.1; 15.2; 14.8; 15.8; 15.9; 16.0; 15.5

Measurements of the amount of beverage in a 16-ounce can may vary because different people make the measurements or because the exact amount, 16 ounces of liquid, was not put into the cans. Manufacturers regularly run tests to determine if the amount of beverage in a 16-ounce can falls within the desired range.

Be aware that as you take data, your data may vary somewhat from the data someone else is taking for the same purpose. This is completely natural. However, if two or more of you are taking the same data and get very different results, it is time for you and the others to reevaluate your data-taking methods and your accuracy.

Variation in Samples

It was mentioned previously that two or more samples from the same population, taken randomly, and having close to the same characteristics of the population will likely be different from each other. Suppose Doreen and Jung both decide to study the average amount of time students at their college sleep each night. Doreen and Jung each take samples of 500 students. Doreen uses systematic sampling and Jung uses cluster sampling. Doreen's sample will be different from Jung's sample. Even if Doreen and Jung used the same sampling method, in all likelihood their samples would be different. Neither would be wrong, however.

Think about what contributes to making Doreen's and Jung's samples different.

If Doreen and Jung took larger samples (i.e. the number of data values is increased), their sample results (the average amount of time a student sleeps) might be closer to the actual population average. But still, their samples would be, in all likelihood, different from each other. This variability in samples cannot be stressed enough.

Size of a Sample

The size of a sample (often called the number of observations, usually given the symbol n) is important. The examples you have seen in this book so far have been small. Samples of only a few hundred observations, or even smaller, are sufficient for many purposes. In polling, samples that are from 1,200 to 1,500 observations are considered large enough and good enough if the survey is random and is well done. Later we will find that even much smaller sample sizes will give very good results. You will learn why when you study confidence intervals.

Be aware that many large samples are biased. For example, call-in surveys are invariably biased, because people choose to respond or not.

1.3 LEVELS OF MEASUREMENT

Once you have a set of data, you will need to organize it so that you can analyze how frequently each datum occurs in the set. However, when calculating the frequency, you may need to round your answers so that they are as precise as possible.

Levels of Measurement

The way a set of data is measured is called its level of measurement. Correct statistical procedures depend on a researcher being familiar with levels of measurement. Not every statistical operation can be used with every set of data. Data can be classified into four levels of measurement. They are (from lowest to highest level):

- **Nominal scale level**
- **Ordinal scale level**
- **Interval scale level**
- **Ratio scale level**

Data that is measured using a **nominal scale** is **qualitative (categorical).**

Categories, colors, names, labels and favorite foods along with yes or no responses are examples of nominal level data. Nominal scale data are not ordered. For example, trying to classify people according to their favorite food does not make any sense. Putting pizza first and sushi second is not meaningful.

Smartphone companies are another example of nominal scale data. The data are the names of the companies that make smartphones, but there is no agreed upon order of these brands, even though people may have personal preferences. Nominal scale data cannot be used in calculations.

Data that is measured using an ordinal scale is similar to nominal scale data but there is a big difference. The ordinal scale data can be ordered. An example of ordinal scale data is a list of the top five national parks in the United States. The top five national parks in the United States can be ranked from one to five but we cannot measure differences between the data.

Another example of using the ordinal scale is a cruise survey where the responses to questions about the cruise are "excellent," "good," "satisfactory," and "unsatisfactory." These responses are ordered from the most desired response to the least desired. But the differences between two pieces of data cannot be measured. Like the nominal scale data, ordinal scale data cannot be used in calculations.

Data that is measured using the interval scale is similar to ordinal level data because it has a definite ordering but there is a difference between data. The differences between interval scale data can be measured though the data does not have a starting point.

Temperature scales like Celsius (C) and Fahrenheit (F) are measured by using the interval scale. In both temperature measurements, 40° is equal to 100° minus 60°. Differences make sense. But 0 degrees does not because, in both scales, 0 is not the absolute lowest temperature. Temperatures like -10° F and -15° C exist and are colder than 0.

Interval level data can be used in calculations, but one type of comparison cannot be done. 80° C is not four times as hot as 20° C (nor is 80° F four times as hot as 20° F). There is no meaning to the ratio of 80 to 20 (or four to one).

Data that is measured using the ratio scale takes care of the ratio problem and gives you the most information. Ratio scale data is like interval scale data, but it has a 0 point and ratios can be calculated. For example, four multiple choice statistics final exam scores are 80, 68, 20 and 92 (out of a possible 100 points). The exams are machine-graded.

The data can be put in order from lowest to highest: 20, 68, 80, 92.

The differences between the data have meaning. The score 92 is more than the score 68 by 24 points. Ratios can be calculated. The smallest score is 0. So 80 is four times 20. The score of 80 is four times better than the score of 20.

Frequency

Twenty students were asked how many hours they worked per day. Their responses, in hours, are as follows: 5; 6; 3; 3; 2; 4; 7; 5; 2; 3; 5; 6; 5; 4; 4; 3; 5; 2; 5; 3.

Table 1.5 lists the different data values in ascending order and their frequencies.

Table 1.5 Frequency Table of Student Work Hours

Data value	Frequency
2	3
3	5
4	3
5	6
6	2
7	1

A frequency is the number of times a value of the data occurs. According to Table 1.5, there are three students who work two hours, five students who work three hours, and so on. The sum of the values in the frequency column, 20, represents the total number of students included in the sample.

A relative frequency is the ratio (fraction or proportion) of the number of times a value of the data occurs in the set of all outcomes to the total number of outcomes. To find the relative frequencies, divide each frequency by the total number of students in the sample–in this case, 20. Relative frequencies can be written as fractions, percents, or decimals.

Table 1.6 Frequency Table of Student Work Hours with Relative Frequencies

Data value	Frequency	Relative frequency
2	3	$\frac{3}{20}$ or 0.15
3	5	$\frac{5}{20}$ or 0.25
4	3	$\frac{3}{20}$ or 0.15
5	6	$\frac{6}{20}$ or 0.30
6	2	$\frac{2}{20}$ or 0.10
7	1	$\frac{1}{20}$ or 0.05

The sum of the values in the relative frequency column of Table 1.6 is $\frac{20}{20}$, or 1. Cumulative relative frequency is the accumulation of the previous relative frequencies. To find the cumulative relative frequencies, add all the previous relative frequencies to the relative frequency for the current row, as shown in Table 1.7.

Table 1.7 Frequency Table of Student Work Hours with Relative and Cumulative Relative Frequencies

Data value	Frequency	Relative frequency	Cumulative relative frequency
2	3	$\frac{3}{20}$ or 0.15	0.15
3	5	$\frac{5}{20}$ or 0.25	0.15 + 0.25 = 0.40
4	3	$\frac{3}{20}$ or 0.15	0.40 + 0.15 = 0.55
5	6	$\frac{6}{20}$ or 0.30	0.55 + 0.30 = 0.85
6	2	$\frac{2}{20}$ or 0.10	0.85 + 0.10 = 0.95
7	1	$\frac{1}{20}$ or 0.05	0.95 + 0.05 = 1.00

The last entry of the cumulative relative frequency column is one, indicating that one hundred percent of the data has been accumulated.

NOTE

Because of rounding, the relative frequency column may not always sum to one, and the last entry in the cumulative relative frequency column may not be one. However, they each should be close to one.

Table 1.8 represents the heights, in inches, of a sample of 100 male semiprofessional soccer players.

Table 1.8 Frequency Table of Soccer Player Height

Heights (inches)	Frequency	Relative frequency	Cumulative relative frequency
59.95–61.95	5	$\frac{5}{100}$ = 0.05	0.05
61.95–63.95	3	$\frac{3}{100}$ = 0.03	0.05 + 0.03 = 0.08
63.95–65.95	15	$\frac{15}{100}$ = 0.15	0.08 + 0.15 = 0.23
65.95–67.95	40	$\frac{40}{100}$ = 0.40	0.23 + 0.40 = 0.63

Heights (inches)	Frequency	Relative frequency	Cumulative relative frequency
67.95–69.95	17	$\frac{17}{100} = 0.17$	0.63 + 0.17 = 0.80
69.95–71.95	12	$\frac{12}{100} = 0.12$	0.80 + 0.12 = 0.92
71.95–73.95	7	$\frac{7}{100} = 0.07$	0.92 + 0.07 = 0.99
73.95–75.95	1	$\frac{1}{100} = 0.01$	0.99 + 0.01 = 1.00
	Total = 100	**Total = 1.00**	

The data in this table have been grouped into the following intervals:

- 59.95 to 61.95 inches
- 61.95 to 63.95 inches
- 63.95 to 65.95 inches
- 65.95 to 67.95 inches
- 67.95 to 69.95 inches
- 69.95 to 71.95 inches
- 71.95 to 73.95 inches
- 73.95 to 75.95 inches

In this sample, there are five players whose heights fall within the interval 59.95–61.95 inches, three players whose heights fall within the interval 61.95–63.95 inches, 15 players whose heights fall within the interval 63.95–65.95 inches, 40 players whose heights fall within the interval 65.95–67.95 inches, 17 players whose heights fall within the interval 67.95–69.95 inches, 12 players whose heights fall within the interval 69.95–71.95, seven players whose heights fall within the interval 71.95–73.95, and one player whose heights fall within the interval 73.95–75.95. All heights fall between the endpoints of an interval and not at the endpoints.

EXAMPLE 1.14

From Table 1.8, find the percentage of heights that are less than 65.95 inches.

Solution 1.14

If you look at the first, second, and third rows, the heights are all less than 65.95 inches. There are 5 + 3 + 15 = 23 players whose heights are less than 65.95 inches. The percentage of heights less than 65.95 inches is then $\frac{23}{100}$ or 23%. This percentage is the cumulative relative frequency entry in the third row.

TRY IT 1.14

Table 1.9 shows the amount, in inches, of annual rainfall in a sample of towns.

Table 1.9

Rainfall (inches)	Frequency	Relative frequency	Cumulative relative frequency
2.95–4.97	6	$\frac{6}{50} = 0.12$	0.12
4.97–6.99	7	$\frac{7}{50} = 0.14$	0.12 + 0.14 = 0.26
6.99–9.01	15	$\frac{15}{50} = 0.30$	0.26 + 0.30 = 0.56
9.01–11.03	8	$\frac{8}{50} = 0.16$	0.56 + 0.16 = 0.72
11.03–13.05	9	$\frac{9}{50} = 0.18$	0.72 + 0.18 = 0.90
13.05–15.07	5	$\frac{5}{50} = 0.10$	0.90 + 0.10 = 1.00
	Total = 50	**Total = 1.00**	

From Table 1.9, find the percentage of rainfall that is less than 9.01 inches.

EXAMPLE 1.15

From Table 1.8, find the percentage of heights that fall between 61.95 and 65.95 inches.

Solution 1.15

Add the relative frequencies in the second and third rows: 0.03 + 0.15 = 0.18 or 18%.

TRY IT 1.15

From Table 1.9, find the percentage of rainfall that is between 6.99 and 13.05 inches.

EXAMPLE 1.16

Use the heights of the 100 male semiprofessional soccer players in Table 1.8. Fill in the blanks and check your answers.

a. The percentage of heights that are from 67.95 to 71.95 inches is: ____.
b. The percentage of heights that are from 67.95 to 73.95 inches is: ____.
c. The percentage of heights that are more than 65.95 inches is: ____.
d. The number of players in the sample who are between 61.95 and 71.95 inches tall is: ____.
e. What kind of data are the heights?
f. Describe how you could gather this data (the heights) so that the data are characteristic of all male semiprofessional soccer players.

Remember, you count frequencies. To find the relative frequency, divide the frequency by the total number of data values. To find the cumulative relative frequency, add all of the previous relative frequencies to the relative frequency for the current row.

Solution 1.16

a. 29%
b. 36%
c. 77%
d. 87
e. quantitative continuous
f. get rosters from each team and choose a simple random sample from each

EXAMPLE 1.17

Nineteen people were asked how many miles, to the nearest mile, they commute to work each day. The data are as follows: 2; 5; 7; 3; 2; 10; 18; 15; 20; 7; 10; 18; 5; 12; 13; 12; 4; 5; 10. Table 1.10 was produced:

Table 1.10 Frequency of Commuting Distances

Data	Frequency	Relative frequency	Cumulative relative frequency
3	3	$\frac{3}{19}$	0.1579
4	1	$\frac{1}{19}$	0.2105
5	3	$\frac{3}{19}$	0.1579
7	2	$\frac{2}{19}$	0.2632
10	3	$\frac{4}{19}$	0.4737
12	2	$\frac{2}{19}$	0.7895
13	1	$\frac{1}{19}$	0.8421
15	1	$\frac{1}{19}$	0.8948
18	1	$\frac{1}{19}$	0.9474
20	1	$\frac{1}{19}$	1.0000

a. Is the table correct? If it is not correct, what is wrong?

b. True or False: Three percent of the people surveyed commute three miles. If the statement is not correct, what should it be? If the table is incorrect, make the corrections.
c. What fraction of the people surveyed commute five or seven miles?
d. What fraction of the people surveyed commute 12 miles or more? Less than 12 miles? Between five and 13 miles (not including five and 13 miles)?

Solution 1.17

a. No. The frequency column sums to 18, not 19. Not all cumulative relative frequencies are correct.
b. False. The frequency for three miles should be one; for two miles (left out), two. The cumulative relative frequency column should read: 0.1052, 0.1579, 0.2105, 0.3684, 0.4737, 0.6316, 0.7368, 0.7895, 0.8421, 0.9474, 1.0000.
c. $\frac{5}{19}$
d. $\frac{7}{19}, \frac{12}{19}, \frac{7}{19}$

TRY IT 1.17

Table 1.9 represents the amount, in inches, of annual rainfall in a sample of towns. What fraction of towns surveyed get between 11.03 and 13.05 inches of rainfall each year?

EXAMPLE 1.18

Table 1.11 contains the total number of deaths worldwide as a result of earthquakes for the period from 2000 to 2012.

Table 1.11

Year	Total number of deaths
2000	231
2001	21,357
2002	11,685
2003	33,819
2004	228,802
2005	88,003
2006	6,605
2007	712
2008	88,011
2009	1,790
2010	320,120
2011	21,953
2012	768
Total	823,856

Answer the following questions.

a. What is the frequency of deaths measured from 2006 through 2009?
b. What percentage of deaths occurred after 2009?
c. What is the relative frequency of deaths that occurred in 2003 or earlier?
d. What is the percentage of deaths that occurred in 2004?
e. What kind of data are the numbers of deaths?
f. The Richter scale is used to quantify the energy produced by an earthquake. Examples of Richter scale numbers are 2.3, 4.0, 6.1, and 7.0. What kind of data are these numbers?

Solution 1.18

a. 97,118 (11.8%)
b. 41.6%
c. 67,092/823,356 or 0.081 or 8.1 %
d. 27.8%
e. Quantitative discrete
f. Quantitative continuous

TRY IT 1.18

Table 1.12 contains the total number of fatal motor vehicle traffic crashes in the United States for the period from 1994 to 2011.

Table 1.12

Year	Total number of crashes	Year	Total number of crashes
1994	36,254	2004	38,444
1995	37,241	2005	39,252
1996	37,494	2006	38,648
1997	37,324	2007	37,435
1998	37,107	2008	34,172
1999	37,140	2009	30,862
2000	37,526	2010	30,296
2001	37,862	2011	29,757
2002	38,491	Total	653,782
2003	38,477		

Answer the following questions.

a. What is the frequency of deaths measured from 2000 through 2004?
b. What percentage of deaths occurred after 2006?

c. What is the relative frequency of deaths that occurred in 2000 or before?
d. What is the percentage of deaths that occurred in 2011?
e. What is the cumulative relative frequency for 2006? Explain what this number tells you about the data.

1.4 EXPERIMENTAL DESIGN AND ETHICS

Does aspirin reduce the risk of heart attacks? Is one brand of fertilizer more effective at growing roses than another? Is fatigue as dangerous to a driver as the influence of alcohol? Questions like these are answered using randomized experiments. In this module, you will learn important aspects of experimental design. Proper study design ensures the production of reliable, accurate data.

The purpose of an experiment is to investigate the relationship between two variables. When one variable causes change in another, we call the first variable the independent variable or explanatory variable. The affected variable is called the dependent variable or response variable: stimulus, response. In a randomized experiment, the researcher manipulates values of the explanatory variable and measures the resulting changes in the response variable. The different values of the explanatory variable are called treatments. An experimental unit is a single object or individual to be measured.

You want to investigate the effectiveness of vitamin E in preventing disease. You recruit a group of subjects and ask them if they regularly take vitamin E. You notice that the subjects who take vitamin E exhibit better health on average than those who do not. Does this prove that vitamin E is effective in disease prevention? It does not. There are many differences between the two groups compared in addition to vitamin E consumption. People who take vitamin E regularly often take other steps to improve their health: exercise, diet, other vitamin supplements, choosing not to smoke. Any one of these factors could be influencing health. As described, this study does not prove that vitamin E is the key to disease prevention.

Additional variables that can cloud a study are called lurking variables. In order to prove that the explanatory variable is causing a change in the response variable, it is necessary to isolate the explanatory variable. The researcher must design her experiment in such a way that there is only one difference between groups being compared: the planned treatments. This is accomplished by the random assignment of experimental units to treatment groups. When subjects are assigned treatments randomly, all of the potential lurking variables are spread equally among the groups. At this point the only difference between groups is the one imposed by the researcher. Different outcomes measured in the response variable, therefore, must be a direct result of the different treatments. In this way, an experiment can prove a cause-and-effect connection between the explanatory and response variables.

The power of suggestion can have an important influence on the outcome of an experiment. Studies have shown that the expectation of the study participant can be as important as the actual medication. In one study of performance-enhancing drugs, researchers noted:

Results showed that believing one had taken the substance resulted in [performance] times almost as fast as those associated with consuming the drug itself. In contrast, taking the drug without knowledge yielded no significant performance increment.

When participation in a study prompts a physical response from a participant, it is difficult to isolate the effects of the explanatory variable. To counter the power of suggestion, researchers set aside one treatment group as a control group. This group is given a placebo treatment–a treatment that cannot influence the response variable. The control group helps researchers balance the effects of being in an experiment with the effects of the active treatments. Of course, if you are participating in a study and you know that you are receiving a pill which contains no actual medication, then the power of suggestion is no longer a factor. Blinding in a randomized experiment preserves the power of suggestion. When a person involved in a research study is blinded, he does not know who is receiving the active treatment(s) and who is receiving the placebo treatment. A double-blind experiment is one in which both the subjects and the researchers involved with the subjects are blinded.

EXAMPLE 1.19

The Smell & Taste Treatment and Research Foundation conducted a study to investigate whether smell can affect learning. Subjects completed mazes multiple times while wearing masks. They completed the pencil and paper mazes three times wearing floral-scented masks, and three times with unscented masks. Participants were assigned at random to wear the floral mask during the first three trials or during the last three trials. For each trial, researchers recorded the time it took to complete the maze and the subject's impression of the mask's scent: positive, negative, or neutral.

a. Describe the explanatory and response variables in this study.
b. What are the treatments?
c. Identify any lurking variables that could interfere with this study.
d. Is it possible to use blinding in this study?

Solution 1.19

a. The explanatory variable is scent, and the response variable is the time it takes to complete the maze.

b. There are two treatments: a floral-scented mask and an unscented mask.
c. All subjects experienced both treatments. The order of treatments was randomly assigned so there were no differences between the treatment groups. Random assignment eliminates the problem of lurking variables.
d. Subjects will clearly know whether they can smell flowers or not, so subjects cannot be blinded in this study. Researchers timing the mazes can be blinded, though. The researcher who is observing a subject will not know which mask is being worn.

TERMINOLOGY

Average also called mean or arithmetic mean; a number that describes the central tendency of the data

Blinding not telling participants which treatment a subject is receiving

Categorical Variable variables that take on values that are names or labels

Cluster Sampling a method for selecting a random sample and dividing the population into groups (clusters); use simple random sampling to select a set of clusters. Every individual in the chosen clusters is included in the sample.

Continuous Random Variable a random variable (RV) whose outcomes are measured; the height of trees in the forest is a continuous RV.

Control Group a group in a randomized experiment that receives an inactive treatment but is otherwise managed exactly as the other groups

Convenience Sampling a nonrandom method of selecting a sample; this method selects individuals that are easily accessible and may result in biased data.

Cumulative Relative Frequency The term applies to an ordered set of observations from smallest to largest. The cumulative relative frequency is the sum of the relative frequencies for all values that are less than or equal to the given value.

Data a set of observations (a set of possible outcomes); most data can be put into two groups: qualitative (an attribute whose value is indicated by a label) or quantitative (an attribute whose value is indicated by a number). Quantitative data can be separated into two subgroups: discrete and continuous. Data is discrete if it is the result of counting (such as the number of students of a given ethnic group in a class or the number of books on a shelf). Data is continuous if it is the result of measuring (such as distance traveled or weight of luggage)

Discrete Random Variable a random variable (RV) whose outcomes are counted

Double-blinding the act of blinding both the subjects of an experiment and the researchers who work with the subjects

Experimental Unit any individual or object to be measured

Explanatory Variable the independent variable in an experiment; the value controlled by researchers

Frequency the number of times a value of the data occurs

Informed Consent Any human subject in a research study must be cognizant of any risks or costs associated with the study. The subject has the right to know the nature of the treatments included in the study, their potential risks, and their potential benefits. Consent must be given freely by an informed, fit participant.

Institutional Review Board a committee tasked with oversight of research programs that involve human subjects

Lurking Variable a variable that has an effect on a study even though it is neither an explanatory variable nor a response variable

Mathematical Models a description of a phenomenon using mathematical concepts, such as equations, inequalities, distributions, etc.

Nonsampling Error an issue that affects the reliability of sampling data other than natural variation; it includes a variety of human errors including poor study design, biased sampling methods, inaccurate information provided by study participants, data entry errors, and poor analysis.

Numerical Variable variables that take on values that are indicated by numbers

Observational Study a study in which the independent variable is not manipulated by the researcher

Parameter a number that is used to represent a population characteristic and that generally cannot be determined easily

Placebo an inactive treatment that has no real effect on the explanatory variable

Population all individuals, objects, or measurements whose properties are being studied

Probability a number between zero and one, inclusive, that gives the likelihood that a specific event will occur

Proportion the number of successes divided by the total number in the sample

Qualitative Data See Data.

Quantitative Data See Data.

Random Assignment the act of organizing experimental units into treatment groups using random methods

Random Sampling a method of selecting a sample that gives every member of the population an equal chance of being selected.

Relative Frequency the ratio of the number of times a value of the data occurs in the set of all outcomes to the number of all outcomes to the total number of outcomes

Representative Sample a subset of the population that has the same characteristics as the population

Response Variable the dependent variable in an experiment; the value that is measured for change at the end of an experiment

Sample a subset of the population studied

Sampling Bias not all members of the population are equally likely to be selected

Sampling Error the natural variation that results from selecting a sample to represent a larger population; this variation decreases as the sample size increases, so selecting larger samples reduces sampling error.

Sampling with Replacement Once a member of the population is selected for inclusion in a sample, that member is returned to the population for the selection of the next individual.

Sampling without Replacement A member of the population may be chosen for inclusion in a sample only once. If chosen, the member is not returned to the population before the next selection.

Simple Random Sampling a straightforward method for selecting a random sample; give each member of the population a number. Use a random number generator to select a set of labels. These randomly selected labels identify the members of your sample.

Statistic a numerical characteristic of the sample; a statistic estimates the corresponding population parameter.

Statistical Models a description of a phenomenon using probability distributions that describe the expected behavior of the phenomenon and the variability in the expected observations.

Stratified Sampling a method for selecting a random sample used to ensure that subgroups of the population are represented adequately; divide the population into groups (strata). Use simple random sampling to identify a proportionate number of individuals from each stratum.

Survey a study in which data is collected as reported by individuals.

Systematic Sampling a method for selecting a random sample; list the members of the population. Use simple random sampling to select a starting point in the population. Let k = (number of individuals in the population)/(number of individuals needed in the sample). Choose every kth individual in the list starting with the one that was randomly selected. If necessary, return to the beginning of the population list to complete your sample.

Treatments different values or components of the explanatory variable applied in an experiment

Variable a characteristic of interest for each person or object in a population

CHAPTER REVIEW

1.1 Definitions of Statistics, Probability, and Key Terms

The mathematical theory of statistics is easier to learn when you know the language. This module presents important terms that will be used throughout the text.

1.2 Data, Sampling, and Variation in Data and Sampling

Data are individual items of information that come from a population or sample. Data may be classified as qualitative (categorical), quantitative continuous, or quantitative discrete.

Because it is not practical to measure the entire population in a study, researchers use samples to represent the population. A random sample is a representative group from the population chosen by using a method that gives each individual in the population an equal chance of being included in the sample. Random sampling methods include simple random sampling, stratified sampling, cluster sampling, and systematic sampling. Convenience sampling is a nonrandom method of choosing a sample that often produces biased data.

Samples that contain different individuals result in different data. This is true even when the samples are well-chosen and representative of the population. When properly selected, larger samples model the population more closely than smaller samples. There are many different potential problems that can affect the reliability of a sample. Statistical data needs to be critically analyzed, not simply accepted.

1.3 Levels of Measurement

Certain calculations produce figures that are artificially exact. Reporting a figure to eight decimal places is unnecessary when the measurements that produced that value were only precise to the nearest tenth. Round your final answer to one additional decimal place beyond that of the original data. This indicates that if data is measured to the nearest tenth of a unit, the final statistic should be reported to the nearest hundredth.In addition to rounding your answers, you can measure your data using the following four levels of measurement.

- **Nominal scale level:** data that cannot be ordered nor can it be used in calculations

- **Ordinal scale level:** data that can be ordered; the differences cannot be measured
- **Interval scale level:** data with a definite ordering but no starting point; the differences can be measured, but there is no such thing as a ratio.
- **Ratio scale level:** data with a starting point that can be ordered; the differences have meaning and ratios can be calculated.

When organizing data, it is important to know how many times a value appears. How many statistics students study five hours or more for an exam? What percent of families on our block own two pets? Frequency, relative frequency, and cumulative relative frequency are measures that answer questions like these.

1.4 Experimental Design and Ethics

A poorly designed study will not produce reliable data. There are certain key components that must be included in every experiment. To eliminate lurking variables, subjects must be assigned randomly to different treatment groups. One of the groups must act as a control group, demonstrating what happens when the active treatment is not applied. Participants in the control group receive a placebo treatment that looks exactly like the active treatments but cannot influence the response variable. To preserve the integrity of the placebo, both researchers and subjects may be blinded. When a study is designed properly, the only difference between treatment groups is the one imposed by the researcher. Therefore, when groups respond differently to different treatments, the difference must be due to the influence of the explanatory variable.

"An ethics problem arises when you are considering an action that benefits you or some cause you support, hurts or reduces benefits to others, and violates some rule." Ethical violations in statistics are not always easy to spot. Professional associations and federal agencies post guidelines for proper conduct. It is important that you learn basic statistical procedures so that you can recognize proper data analysis.

HOMEWORK

1.1 Definitions of Statistics, Probability, and Key Terms

For each of the following eight exercises, identify: a. the population, b. the sample, c. the parameter, d. the statistic, e. the variable, and f. the data. Give examples where appropriate.

1. A fitness center is interested in the mean amount of time a client exercises in the center each week.
2. Ski resorts are interested in the mean age that children take their first ski and snowboard lessons. They need this information to plan their ski classes optimally.

3. A cardiologist is interested in the mean recovery period of her patients who have had heart attacks.
4. Insurance companies are interested in the mean health costs each year of their clients, so that they can determine the costs of health insurance.
5. A politician is interested in the proportion of voters in his district who think he is doing a good job.
6. A marriage counselor is interested in the proportion of clients she counsels who stay married.
7. Political pollsters may be interested in the proportion of people who will vote for a particular cause.
8. A marketing company is interested in the proportion of people who will buy a particular product.

Use the following information to answer the next three exercises: A Lake Tahoe Community College instructor is interested in the mean number of days Lake Tahoe Community College math students are absent from class during a quarter.

9. What is the population she is interested in?
   a. all Lake Tahoe Community College students
   b. all Lake Tahoe Community College English students
   c. all Lake Tahoe Community College students in her classes
   d. all Lake Tahoe Community College math students
10. Consider the following:

X = number of days a Lake Tahoe Community College math student is absent

In this case, X is an example of a:

   a. variable.
   b. population.
   c. statistic.
   d. data.

11. The instructor's sample produces a mean number of days absent of 3.5 days. This value is an example of a:
   a. parameter.
   b. data.
   c. statistic.
   d. variable.

1.2 Data, Sampling, and Variation in Data and Sampling

For the following exercises, identify the type of data that would be used to describe a response (quantitative discrete, quantitative continuous, or qualitative), and give an example of the data.

12. number of tickets sold to a concert
13. percent of body fat
14. favorite baseball team
15. time in line to buy groceries
16. number of students enrolled at Evergreen Valley College
17. most-watched television show
18. brand of toothpaste
19. distance to the closest movie theatre
20. age of executives in Fortune 500 companies
21. number of competing computer spreadsheet software packages

Use the following information to answer the next two exercises: A study was done to determine the age, number of times per week, and the duration (amount of time) of resident use of a local park in San Jose. The first house in the neighborhood around the park was selected randomly and then every 8th house in the neighborhood around the park was interviewed.

22. "Number of times per week" is what type of data?
    a. qualitative (categorical)
    b. quantitative discrete
    c. quantitative continuous
23. "Duration (amount of time)" is what type of data?
    a. qualitative (categorical)
    b. quantitative discrete
    c. quantitative continuous
24. Airline companies are interested in the consistency of the number of babies on each flight, so that they have adequate safety equipment. Suppose an airline conducts a survey. Over Thanksgiving weekend, it surveys six flights from Boston to Salt Lake City to determine the number of babies on the flights. It determines the amount of safety equipment needed by the result of that study.
    a. Using complete sentences, list three things wrong with the way the survey was conducted.
    b. Using complete sentences, list three ways that you would improve the survey if it were to be repeated.
25. Suppose you want to determine the mean number of students per statistics class in your state. Describe a possible sampling method in three to five complete sentences. Make the description detailed.
26. Suppose you want to determine the mean number of cans of soda drunk each month by students in their twenties at your school. Describe a possible

sampling method in three to five complete sentences. Make the description detailed.

27. List some practical difficulties involved in getting accurate results from a telephone survey.
28. List some practical difficulties involved in getting accurate results from a mailed survey.
29. With your classmates, brainstorm some ways you could overcome these problems if you needed to conduct a phone or mail survey.
30. The instructor takes her sample by gathering data on five randomly selected students from each Lake Tahoe Community College math class. The type of sampling she used is
    a. cluster sampling
    b. stratified sampling
    c. simple random sampling
    d. convenience sampling
31. A study was done to determine the age, number of times per week, and the duration (amount of time) of residents using a local park in San Jose. The first house in the neighborhood around the park was selected randomly and then every eighth house in the neighborhood around the park was interviewed. The sampling method was:
    a. simple random
    b. systematic
    c. stratified
    d. cluster
32. Name the samling method used in each of the following situations:
    a. A woman in the airport is handing out questionnaires to travelers asking them to evaluate the airport's service. She does not ask travelers who are hurrying through the airport with their hands full of luggage, but instead asks all travelers who are sitting near gates and not taking naps while they wait.
    b. A teacher wants to know if her students are doing homework, so she randomly selects rows two and five and then calls on all students in row two and all students in row five to present the solutions to homework problems to the class.
    c. The marketing manager for an electronics chain store wants information about the ages of its customers. Over the next two weeks, at each store location, 100 randomly selected customers are given questionnaires to fill out asking for information about age, as well as about other variables of interest.

d. The librarian at a public library wants to determine what proportion of the library users are children. The librarian has a tally sheet on which she marks whether books are checked out by an adult or a child. She records this data for every fourth patron who checks out books.

e. A political party wants to know the reaction of voters to a debate between the candidates. The day after the debate, the party's polling staff calls 1,200 randomly selected phone numbers. If a registered voter answers the phone or is available to come to the phone, that registered voter is asked whom he or she intends to vote for and whether the debate changed his or her opinion of the candidates.

33. A "random survey" was conducted of 3,274 people of the "microprocessor generation" (people born since 1971, the year the microprocessor was invented). It was reported that 48% of those individuals surveyed stated that if they had $2,000 to spend, they would use it for computer equipment. Also, 66% of those surveyed considered themselves relatively savvy computer users.

a. Do you consider the sample size large enough for a study of this type? Why or why not?

b. Based on your "gut feeling," do you believe the percents accurately reflect the U.S. population for those individuals born since 1971? If not, do you think the percents of the population are actually higher or lower than the sample statistics? Why? Additional information: The survey, reported by Intel Corporation, was filled out by individuals who visited the Los Angeles Convention Center to see the Smithsonian Institute's road show called "America's Smithsonian."

c. With this additional information, do you feel that all demographic and ethnic groups were equally represented at the event? Why or why not?

d. With the additional information, comment on how accurately you think the sample statistics reflect the population parameters.

34. The Well-Being Index is a survey that follows trends of U.S. residents on a regular basis. There are six areas of health and wellness covered in the survey: Life Evaluation, Emotional Health, Physical Health, Healthy Behavior, Work Environment, and Basic Access. Some of the questions used to measure the Index are listed below.

Identify the type of data obtained from each question used in this survey: qualitative(categorical), quantitative discrete, or quantitative continuous.

a. Do you have any health problems that prevent you from doing any of the things people your age can normally do?

b. During the past 30 days, for about how many days did poor health keep you from doing your usual activities?
c. In the last seven days, on how many days did you exercise for 30 minutes or more?
d. Do you have health insurance coverage?

35. In advance of the 1936 Presidential Election, a magazine titled Literary Digest released the results of an opinion poll predicting that the republican candidate Alf Landon would win by a large margin. The magazine sent post cards to approximately 10,000,000 prospective voters. These prospective voters were selected from the subscription list of the magazine, from automobile registration lists, from phone lists, and from club membership lists. Approximately 2,300,000 people returned the postcards.
    a. Think about the state of the United States in 1936. Explain why a sample chosen from magazine subscription lists, automobile registration lists, phone books, and club membership lists was not representative of the population of the United States at that time.
    b. What effect does the low response rate have on the reliability of the sample?
    c. Are these problems examples of sampling error or nonsampling error?
    d. During the same year, George Gallup conducted his own poll of 30,000 prospective voters. These researchers used a method they called "quota sampling" to obtain survey answers from specific subsets of the population. Quota sampling is an example of which sampling method described in this module?
36. Crime-related and demographic statistics for 47 US states in 1960 were collected from government agencies, including the FBI's *Uniform Crime Report*. One analysis of this data found a strong connection between education and crime indicating that higher levels of education in a community correspond to higher crime rates.

Which of the potential problems with samples discussed in 1.2 Data, Sampling, and Variation in Data and Sampling could explain this connection?

37. YouPolls is a website that allows anyone to create and respond to polls. One question posted April 15 asks:

"Do you feel happy paying your taxes when members of the Obama administration are allowed to ignore their tax liabilities?"

As of April 25, 11 people responded to this question. Each participant answered "NO!"

Which of the potential problems with samples discussed in this module could explain this connection?

38. A scholarly article about response rates begins with the following quote: "Declining contact and cooperation rates in random digit dial (RDD) national telephone surveys raise serious concerns about the validity of estimates drawn from such research."

The Pew Research Center for People and the Press admits:

"The percentage of people we interview – out of all we try to interview – has been declining over the past decade or more."

a. What are some reasons for the decline in response rate over the past decade?

b. Explain why researchers are concerned with the impact of the declining response rate on public opinion polls.

1.3 Levels of Measurement

Fifty part-time students were asked how many courses they were taking this term. The (incomplete) results are shown below:

Table 1.13 Part-time Student Course Loads

# of courses	Frequency	Relative frequency	Cumulative relative frequency
1	30	0.6	
2	15		

a. Fill in the blanks in Table 1.13.

b. What percent of students take exactly two courses?

c. What percent of students take one or two courses?

40. Sixty adults with gum disease were asked the number of times per week they used to floss before their diagnosis. The (incomplete) results are shown in Table 1.14.

Table 1.14 Flossing Frequency for Adults with Gum Disease

# flossing per week	Frequency	Relative frequency	Cumulative relative frequency
0	27	0.4500	
1	18		
3			0.9333
6	3	0.0500	
7	1	0.0167	

a. Fill in the blanks in Table 1.14.
b. What percent of adults flossed six times per week?
c. What percent flossed at most three times per week?

41. Nineteen immigrants to the U.S were asked how many years, to the nearest year, they have lived in the U.S. The data are as follows: 2; 5; 7; 2; 2; 10; 20; 15; 0; 7; 0; 20; 5; 12; 15; 12; 4; 5; 10 .

Table 1.15 was produced.

Table 1.15 Frequency of Immigrant Survey Responses

Data	Frequency	Relative frequency	Cumulative relative frequency
0	2	$\frac{2}{19}$	0.1053
2	3	$\frac{3}{19}$	0.2632
4	1	$\frac{1}{19}$	0.3158
5	3	$\frac{3}{19}$	0.4737
7	2	$\frac{2}{19}$	0.5789
10	2	$\frac{2}{19}$	0.6842
12	2	$\frac{2}{19}$	0.7895
15	1	$\frac{1}{19}$	0.8421
20	1	$\frac{1}{19}$	1.0000

a. Fix the errors in Table 1.15. Also, explain how someone might have arrived at the incorrect number(s).
b. Explain what is wrong with this statement: "47 percent of the people surveyed have lived in the U.S. for 5 years."
c. Fix the statement in b to make it correct.
d. What fraction of the people surveyed have lived in the U.S. five or seven years?
e. What fraction of the people surveyed have lived in the U.S. at most 12 years?

f. What fraction of the people surveyed have lived in the U.S. fewer than 12 years?

g. What fraction of the people surveyed have lived in the U.S. from five to 20 years, inclusive?

42. How much time does it take to travel to work? Table 1.16 shows the mean commute time by state for workers at least 16 years old who are not working at home. Find the mean travel time, and round off the answer properly.

Table 1.16

24.0	24.3	25.9	18.9	27.5	17.9	21.8	20.9	16.7	27.3
18.2	24.7	20.0	22.6	23.9	18.0	31.4	22.3	24.0	25.5
24.7	24.6	28.1	24.9	22.6	23.6	23.4	25.7	24.8	25.5
21.2	25.7	23.1	23.0	23.9	26.0	16.3	23.1	21.4	21.5
27.0	27.0	18.6	31.7	23.3	30.1	22.9	23.3	21.7	18.6

43. Forbes magazine published data on the best small firms in 2012. These were firms which had been publicly traded for at least a year, have a stock price of at least $5 per share, and have reported annual revenue between $5 million and $1 billion. Table 1.17 shows the ages of the chief executive officers for the first 60 ranked firms.

Table 1.17

Age	Frequency	Relative frequency	Cumulative relative frequency
40–44	3		
45–49	11		
50–54	13		
55–59	16		
60–64	10		
65–69	6		
70–74	1		

a. What is the frequency for CEO ages between 54 and 65?

b. What percentage of CEOs are 65 years or older?

c. What is the relative frequency of ages under 50?

d. What is the cumulative relative frequency for CEOs younger than 55?

e. Which graph shows the relative frequency and which shows the cumulative relative frequency?

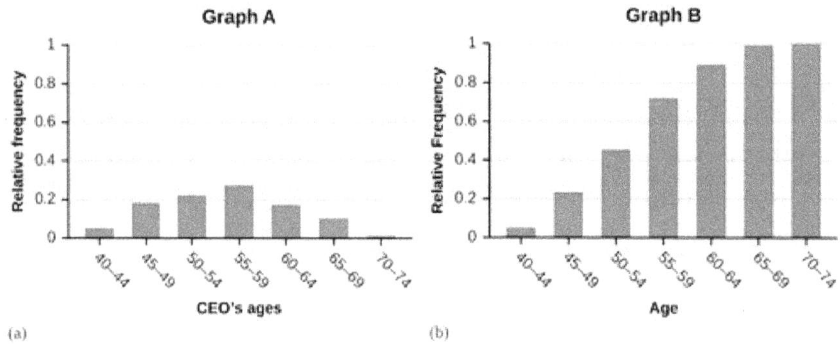

Figure 1.11

Use the following information to answer the next two exercises: Table 1.18 contains data on hurricanes that have made direct hits on the U.S. Between 1851 and 2004. A hurricane is given a strength category rating based on the minimum wind speed generated by the storm.

Table 1.18 Frequency of Hurricane Direct Hits

Category	Number of direct hits	Relative frequency	Cumulative frequency
1	109	0.3993	0.3993
2	72	0.2637	0.6630
3	71	0.2601	
4	18		0.9890
5	3	0.0110	1.0000
	Total = 273		

44. What is the relative frequency of direct hits that were category 4 hurricanes?
    a. 0.0768
    b. 0.0659
    c. 0.2601
    d. Not enough information to calculate

45. What is the relative frequency of direct hits that were AT MOST a category 3 storm?
    a. 0.3480
    b. 0.9231
    c. 0.2601
    d. 0.3370

REFERENCES

1.1 Definitions of Statistics, Probability, and Key Terms

The Data and Story Library, http://lib.stat.cmu.edu/DASL/Stories/CrashTestDummies.html (accessed May 1, 2013).

1.2 Data, Sampling, and Variation in Data and Sampling

Gallup-Healthways Well-Being Index. http://www.well-beingindex.com/default.asp (accessed May 1, 2013).

Gallup-Healthways Well-Being Index. http://www.well-beingindex.com/methodology.asp (accessed May 1, 2013).

Gallup-Healthways Well-Being Index. http://www.gallup.com/poll/146822/gallup-healthways-index-questions.aspx (accessed May 1, 2013).

Data from http://www.bookofodds.com/Relationships-Society/Articles/A0374-How-George-Gallup-Picked-the-President

Dominic Lusinchi, "'President' Landon and the 1936 *Literary Digest* Poll: Were Automobile and Telephone Owners to Blame?" Social Science History 36, no. 1: 23-54 (2012), http://ssh.dukejournals.org/content/36/1/23.abstract (accessed May 1, 2013).

"The Literary Digest Poll," Virtual Laboratories in Probability and Statistics http://www.math.uah.edu/stat/data/LiteraryDigest.html (accessed May 1, 2013).

"Gallup Presidential Election Trial-Heat Trends, 1936–2008," Gallup Politics http://www.gallup.com/poll/110548/gallup-presidential-election-trialheat-trends-19362004.aspx#4 (accessed May 1, 2013).

The Data and Story Library, http://lib.stat.cmu.edu/DASL/Datafiles/USCrime.html (accessed May 1, 2013).

LBCC Distance Learning (DL) program data in 2010-2011, http://de.lbcc.edu/reports/2010-11/future/highlights.html#focus (accessed May 1, 2013).

Data from San Jose Mercury News

1.3 Levels of Measurement

"State & County QuickFacts," U.S. Census Bureau. http://quickfacts.census.gov/qfd/download_data.html (accessed May 1, 2013).

"State & County QuickFacts: Quick, easy access to facts about people, business, and geography," U.S. Census Bureau. http://quickfacts.census.gov/qfd/index.html (accessed May 1, 2013).

"Table 5: Direct hits by mainland United States Hurricanes (1851-2004)," National Hurricane Center, http://www.nhc.noaa.gov/gifs/table5.gif (accessed May 1, 2013).

"Levels of Measurement," http://infinity.cos.edu/faculty/woodbury/stats/tutorial/Data_Levels.htm (accessed May 1, 2013).

Courtney Taylor, "Levels of Measurement," about.com, http://statistics.about.com/od/HelpandTutorials/a/Levels-Of-Measurement.htm (accessed May 1, 2013).

David Lane. "Levels of Measurement," Connexions, http://cnx.org/content/m10809/latest/ (accessed May 1, 2013).

1.4 Experimental Design and Ethics

"Vitamin E and Health," Nutrition Source, Harvard School of Public Health, http://www.hsph.harvard.edu/nutritionsource/vitamin-e/ (accessed May 1, 2013).

Stan Reents. "Don't Underestimate the Power of Suggestion," athleteinme.com, http://www.athleteinme.com/ArticleView.aspx?id=1053 (accessed May 1, 2013).

Ankita Mehta. "Daily Dose of Aspiring Helps Reduce Heart Attacks: Study," International Business Times, July 21, 2011. Also available online at http://www.ibtimes.com/daily-dose-aspirin-helps-reduce-heart-attacks-study-300443 (accessed May 1, 2013).

The Data and Story Library, http://lib.stat.cmu.edu/DASL/Stories/ScentsandLearning.html (accessed May 1, 2013).

M.L. Jackson et al., "Cognitive Components of Simulated Driving Performance: Sleep Loss effect and Predictors," Accident Analysis and Prevention Journal, Jan no. 50 (2013), http://www.ncbi.nlm.nih.gov/pubmed/22721550 (accessed May 1, 2013).

"Earthquake Information by Year," U.S. Geological Survey. http://earthquake.usgs.gov/earthquakes/eqarchives/year/ (accessed May 1, 2013).

"Fatality Analysis Report Systems (FARS) Encyclopedia," National Highway Traffic and Safety Administration. http://www-fars.nhtsa.dot.gov/Main/index.aspx (accessed May 1, 2013).

Data from www.businessweek.com (accessed May 1, 2013).

Data from www.forbes.com (accessed May 1, 2013).

"America's Best Small Companies," http://www.forbes.com/best-small-companies/list/ (accessed May 1, 2013).

U.S. Department of Health and Human Services, Code of Federal Regulations Title 45 Public Welfare Department of Health and Human Services Part 46 Protection of Human Subjects revised January 15, 2009. Section 46.111:Criteria for IRB Approval of Research.

"April 2013 Air Travel Consumer Report," U.S. Department of Transportation, April 11 (2013), http://www.dot.gov/airconsumer/april-2013-air-travel-consumer-report (accessed May 1, 2013).

Lori Alden, "Statistics can be Misleading," econoclass.com, http://www.econoclass.com/misleadingstats.html (accessed May 1, 2013).

Maria de los A. Medina, "Ethics in Statistics," Based on "Building an Ethics Module for Business, Science, and Engineering Students" by Jose A. Cruz-Cruz and William Frey, Connexions, http://cnx.org/content/m15555/latest/ (accessed May 1, 2013).

SOLUTIONS

2. a. all children who take ski or snowboard lessons
b. a group of these children
c. the population mean age of children who take their first snowboard lesson
d. the sample mean age of children who take their first snowboard lesson
e. X = the age of one child who takes his or her first ski or snowboard lesson
f. values for X, such as 3, 7, and so on

4. a. the clients of the insurance companies
b. a group of the clients
c. the mean health costs of the clients
d. the mean health costs of the sample
e. X = the health costs of one client
f. values for X, such as 34, 9, 82, and so on

6. a. all the clients of this counselor
b. a group of clients of this marriage counselor
c. the proportion of all her clients who stay married
d. the proportion of the sample of the counselor's clients who stay married
e. X = the number of couples who stay married
f. yes, no

8. a. all people (maybe in a certain geographic area, such as the United States)
b. a group of the people
c. the proportion of all people who will buy the product
d. the proportion of the sample who will buy the product
e. X = the number of people who will buy it
f. buy, not buy

10. a

12. quantitative discrete, 150

14. qualitative, Oakland A's

16. quantitative discrete, 11,234 students

18. qualitative, Crest

20. quantitative continuous, 47.3 years

22. b

24. a. The survey was conducted using six similar flights. The survey would not be a true representation of the entire population of air travelers. Conducting the survey on a holiday weekend will not produce representative results.

b. Conduct the survey during different times of the year. Conduct the survey using flights to and from various locations. Conduct the survey on different days of the week.

26. Answers will vary. Sample Answer: You could use a systematic sampling method. Stop the tenth person as they leave one of the buildings on campus at 9:50 in the morning. Then stop the tenth person as they leave a different building on campus at 1:50 in the afternoon.

28. Answers will vary. Sample Answer: Many people will not respond to mail surveys. If they do respond to the surveys, you can't be sure who is responding. In addition, mailing lists can be incomplete.

30. b

32. convenience cluster stratified systematic simple random

34. a. qualitative(categorical)

b. quantitative discrete

c. quantitative discrete

d. qualitative(categorical)

36. Causality: The fact that two variables are related does not guarantee that one variable is influencing the other. We cannot assume that crime rate impacts education level or that education level impacts crime rate. Confounding: There are many factors that define a community other than education level and crime rate. Communities with high crime rates and high education levels may have other lurking variables that distinguish them from communities with lower crime rates and lower education levels. Because we cannot isolate these variables of interest, we cannot draw valid conclusions about the connection between education and crime. Possible lurking variables include police expenditures, unemployment levels, region, average age, and size.

38. a. Possible reasons: increased use of caller id, decreased use of landlines, increased use of private numbers, voice mail, privacy managers, hectic nature of personal schedules, decreased willingness to be interviewed

b. When a large number of people refuse to participate, then the sample may not have the same characteristics of the population. Perhaps the majority of people willing to participate are doing so because they feel strongly about the subject of the survey.

40. a.

Table 1.19

# flossing per week	Frequency	Relative frequency	Cumulative relative frequency
0	27	0.4500	0.4500
1	18	0.3000	0.7500
3	11	0.1833	0.9333
6	3	0.0500	0.9833
7	1	0.0167	1

a. 5.00%

b. 93.33%

42. The sum of the travel times is 1,173.1. Divide the sum by 50 to calculate the mean value: 23.462. Because each state's travel time was measured to the nearest tenth, round this calculation to the nearest hundredth: 23.46.

44. b

Chapter 2 DESCRIPTIVE STATISTICS METHODS

Figure 2.1 When you have large amounts of data, you will need to organize it in a way that makes sense. These ballots from an election are rolled together with similar ballots to keep them organized. (credit: William Greeson)

INTRODUCTION: AN OVERVIEW

What actions will you take upon collecting data? Data can be characterized and exhibited in several formats. For instance, supposing you wish to purchase a residence in a specific locality. If you are unfamiliar with house pricing, you may request your real estate agent to provide a sample dataset of prices. Examining all the prices in the sample can often be daunting. An improved approach may involve examining the median price and the price variability. The median and variance are two methods for describing data. Your agent may also furnish you with a data graph. This chapter will cover numerical and graphical methods for describing and presenting data. This domain of statistics is referred to as "Descriptive Statistics." You will acquire the skills to compute and, more crucially, to comprehend these metrics and graphs. A statistical graph is an instrument that facilitates the understanding of the form or distribution of a sample or population. A graph can more effectively communicate data than a collection of numbers, as it visually illustrates data clusters and sparse values. Newspapers and the Internet utilize graphs to illustrate trends and facilitate rapid comparisons of facts and numbers for readers. Statisticians frequently visualize data initially to obtain an overview of the information. Subsequently, more formal instruments may be utilized. Various forms of graphs utilized to summarize and arrange data include the dot plot, bar graph, histogram, stem-and-leaf plot, frequency polygon (a variant of broken line graph), pie chart, and box plot. This chapter will succinctly examine stem-and-leaf plots, line graphs, bar graphs, frequency polygons, and time series graphs. We will focus on histograms and box graphs.

2.1 DISPLAY DATA

Stem-and-Leaf Graphs (Stemplots), Line Graphs, and Bar Graphs

One simple graph, the **stem-and-leaf graph** or **stemplot**, comes from the field of exploratory data analysis. It is a good choice when the data sets are small. To create the plot, divide each observation of data into a stem and a leaf. The leaf consists of a **final significant digit**. For example, 23 has stem two and leaf three. The number 432 has stem 43 and leaf two. Likewise, the number 5,432 has stem 543 and leaf two. The decimal 9.3 has stem nine and leaf three. Write the stems in a vertical line from smallest to largest. Draw a vertical line to the right of the stems. Then write the leaves in increasing order next to their corresponding stem.

EXAMPLE 2.1

For Susan Dean's spring pre-calculus class, scores for the first exam were as follows (smallest to largest): 33; 42; 49; 49; 53; 55; 55; 61; 63; 67; 68; 68; 69; 69; 72; 73; 74; 78; 80; 83; 88; 88; 88; 90; 92; 94; 94; 94; 94; 96; 100

Table 2.1 Stem-and-Leaf Graph

Stem	Leaf
3	3
4	2 9 9
5	3 5 5
6	1 3 7 8 8 9 9
7	2 3 4 8
8	0 3 8 8 8
9	0 2 4 4 4 4 6
10	0

The stemplot shows that most scores fell in the 60s, 70s, 80s, and 90s. Eight out of the 31 scores or approximately 26% $\left(\frac{8}{31}\right)$ were in the 90s or 100, a fairly high number of As.

TRY IT 2.1

For the Park City basketball team, scores for the last 30 games were as follows (smallest to largest):32; 32; 33; 34; 38; 40; 42; 42; 43; 44; 46; 47; 47; 48; 48; 48; 49; 50; 50; 51; 52; 52; 52; 53; 54; 56; 57; 57; 60; 61 Construct a stem plot for the data.

The stemplot is a quick way to graph data and gives an exact picture of the data. You want to look for an overall pattern and any outliers. An outlier is an observation of data that does not fit the rest of the data. It is sometimes called an extreme value. When you graph an outlier, it will appear not to fit the pattern of the graph. Some outliers are due to mistakes (for example, writing down 50 instead of 500) while others may indicate that something unusual is happening. It takes some background information to explain outliers, so we will cover them in more detail later.

EXAMPLE 2.2

The data are the distances (in kilometers) from a home to local supermarkets. Create a stemplot using the data: 1.1; 1.5; 2.3; 2.5; 2.7; 3.2; 3.3; 3.3; 3.5; 3.8; 4.0; 4.2; 4.5; 4.5; 4.7; 4.8; 5.5; 5.6; 6.5; 6.7; 12.3

Do the data seem to have any concentration of values?

NOTE

The leaves are to the right of the decimal.

Solution 2.2

The value 12.3 may be an outlier. Values appear to concentrate at three and four kilometers.

Table 2.2

Stem	Leaf
1	15
2	357
3	23358
4	025578
5	56
6	57
7	
8	
9	
10	
11	
12	3

TRY IT 2.2

The following data show the distances (in miles) from the homes of off-campus statistics students to the college. Create a stem plot using the data and identify any outliers:

0.5; 0.7; 1.1; 1.2; 1.2; 1.3; 1.3; 1.5; 1.5; 1.7; 1.7; 1.8; 1.9; 2.0; 2.2; 2.5; 2.6; 2.8; 2.8; 2.8; 3.5; 3.8; 4.4; 4.8; 4.9; 5.2; 5.5; 5.7; 5.8; 8.0

EXAMPLE 2.3

A side-by-side stem-and-leaf plot allows a comparison of the two data sets in two columns. In a side-by-side stem-and-leaf plot, two sets of leaves share the same stem. The leaves are to the left and the right of the stems. Table 2.4 and Table 2.5 show the ages of presidents at their inauguration and at their death. Construct a side-by-side stem-and-leaf plot using this data.

Solution 2.3

Table 2.3

Ages at Inauguration		Ages Death
99877632	4	69
877776665555444442211110	5	366778
9854421110	6	003344567778
	7	0011147889
	8	01358
	9	0033

Table 2.4 Presidential Ages at Inauguration

President	Age	President	Age	President	Age
Washington	57	Lincoln	52	Hoover	54
J. Adams	61	A. Johnson	56	F. Roosevelt	51
Jefferson	57	Grant	46	Truman	60
Madison	57	Hayes	54	Eisenhower	62
Monroe	58	Garfield	49	Kennedy	43
J. Q. Adams	57	Arthur	51	L. Johnson	55
Jackson	61	Cleveland	47	Nixon	56
Van Buren	54	B. Harrison	55	Ford	61
W. H. Harrison	68	Cleveland	55	Carter	52
Tyler	51	McKinley	54	Reagan	69
Polk	49	T. Roosevelt	42	G.H.W. Bush	64
Taylor	64	Taft	51	Clinton	47
Fillmore	50	Wilson	56	G. W. Bush	54
Pierce	48	Harding	55	Obama	47
Buchanan	65	Coolidge	51		

Table 2.5 Presidential Age at Death

President	Age	President	Age	President	Age
Washington	67	Lincoln	56	Hoover	90
J. Adams	90	A. Johnson	66	F. Roosevelt	63
Jefferson	83	Grant	63	Truman	88
Madison	85	Hayes	70	Eisenhower	78
Monroe	73	Garfield	49	Kennedy	46
J. Q. Adams	80	Arthur	56	L. Johnson	64
Jackson	78	Cleveland	71	Nixon	81
Van Buren	79	B. Harrison	67	Ford	93
W. H. Harrison	68	Cleveland	71	Reagan	93
Tyler	71	McKinley	58		
Polk	53	T. Roosevelt	60		
Taylor	65	Taft	72		
Fillmore	74	Wilson	67		
Pierce	64	Harding	57		
Buchanan	77	Coolidge	60		

Another type of graph that is useful for specific data values is a line graph. In the particular line graph shown in Example 2.4, the x-axis (horizontal axis) consists of data values and the y-axis (vertical axis) consists of frequency points. The frequency points are connected using line segments.

EXAMPLE 2.4

In a survey, 40 mothers were asked how many times per week a teenager must be reminded to do his or her chores. The results are shown in Table 2.6 and in Figure 2.2.

Table 2.6

Number of times teenager is reminded	Frequency
0	2
1	5
2	8
3	14
4	7
5	4

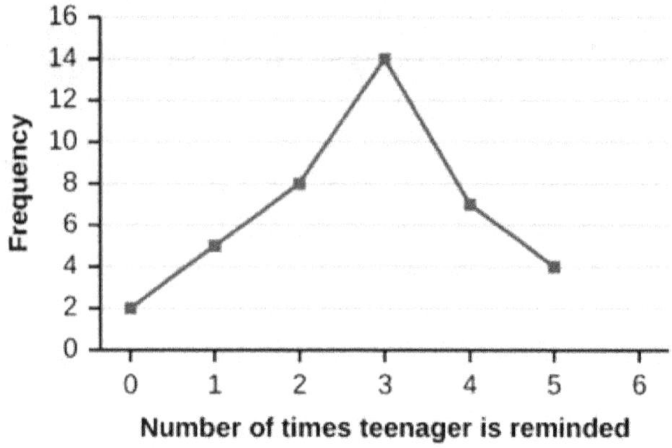

Figure 2.2

TRY IT 2.4

In a survey, 40 people were asked how many times per year they had their car in the shop for repairs. The results are shown in Table 2.7. Construct a line graph.

Table 2.7

Number of times in shop	Frequency
0	7
1	10
2	14
3	9

Bar graphs consist of bars that are separated from each other. The bars can be rectangles or they can be rectangular boxes (used in three-dimensional plots), and they can be vertical or horizontal. The bar graph shown in Example 2.5 has age groups represented on the x-axis and proportions on the y-axis.

EXAMPLE 2.5

By the end of 2011, Facebook had over 146 million users in the United States. Table 2.8 shows three age groups, the number of users in each age group, and the proportion (%) of users in each age group. Construct a bar graph using this data.

Table 2.8

Age groups	Number of Facebook users	Proportion (%) of Facebook users
13–25	65,082,280	45%
26–44	53,300,200	36%
45–64	27,885,100	19%

Solution 2.5

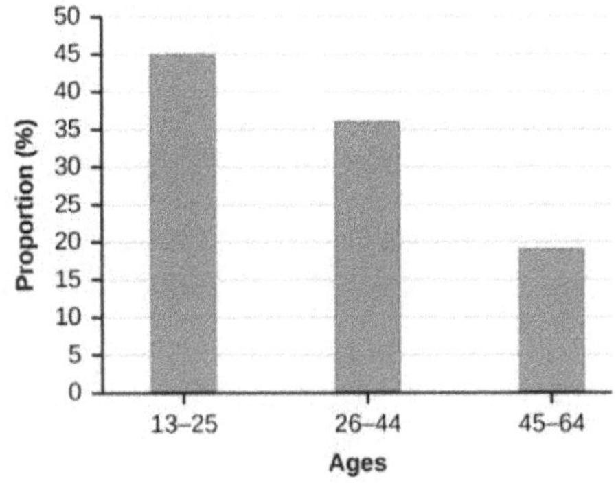

Figure 2.3

TRY IT 2.5

The population in Park City is made up of children, working-age adults, and retirees. Table 2.9 shows the three age groups, the number of people in the town from each age group, and the proportion (%) of people in each age group. Construct a bar graph showing the proportions.

Table 2.9

Age groups	Number of people	Proportion of population
Children	67,059	19%
Working-age adults	152,198	43%
Retirees	131,662	38%

EXAMPLE 2.6

The columns in Table 2.10 contain: the race or ethnicity of students in U.S. Public Schools for the class of 2011, percentages for the Advanced Placement examine population for that class, and percentages for the overall student population. Create a bar graph with the student race or ethnicity (qualitative data) on the *x*-axis, and the Advanced Placement examinee population percentages on the *y*-axis.

Table 2.10

Race/ethnicity	AP examinee population	Overall student population
1 = Asian, Asian American or Pacific Islander	10.3%	5.7%
2 = Black or African American	9.0%	14.7%
3 = Hispanic or Latino	17.0%	17.6%
4 = American Indian or Alaska Native	0.6%	1.1%
5 = White	57.1%	59.2%
6 = Not reported/other	6.0%	1.7%

Solution 2.6

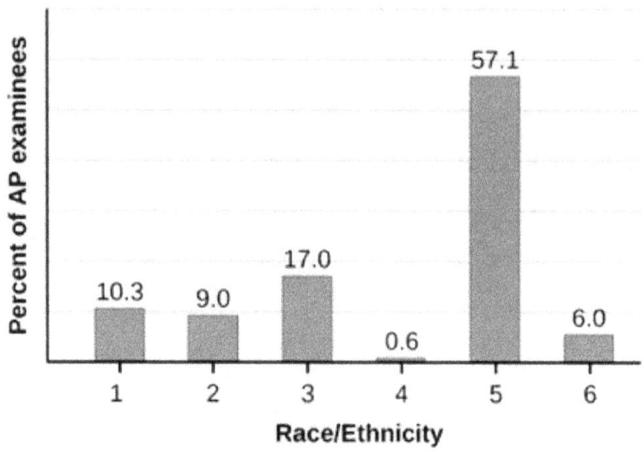

Figure 2.4

TRY IT 2.6

Park city is broken down into six voting districts. The table shows the percent of the total registered voter population that lives in each district as well as the percent total of the entire population that lives in each district. Construct a bar graph that shows the registered voter population by district.

Table 2.11

District	Registered voter population	Overall city population
1	15.5%	19.4%
2	12.2%	15.6%
3	9.8%	9.0%
4	17.4%	18.5%
5	22.8%	20.7%
6	22.3%	16.8%

EXAMPLE 2.7

Below is a two-way table showing the types of pets owned by men and women:

Table 2.12

	Dogs	Cats	Fish	Total
Men	4	2	2	8
Women	4	6	2	12
Total	8	8	4	20

Given these data, calculate the conditional distributions for the subpopulation of men who own each pet type.

Solution 2.7

Men who own dogs = 4/8 = 0.5

Men who own cats = 2/8 = 0.25

Men who own fish = 2/8 = 0.25

Note: The sum of all of the conditional distributions must equal one. In this case, 0.5 + 0.25 + 0.25 = 1; therefore, the solution "checks".

Histograms, Frequency Polygons, and Time Series Graphs

For most of the work you do in this book, you will use a histogram to display the data. One advantage of a histogram is that it can readily display large data sets. A rule of thumb is to use a histogram when the data set consists of 100 values or more.

A histogram consists of contiguous (adjoining) boxes. It has both a horizontal axis and a vertical axis. The horizontal axis is labeled with what the data represents (for instance, distance from your home to school). The vertical axis is labeled either frequency or relative frequency (or percent frequency or probability). The graph will have the same shape with either label. The histogram (like the stemplot) can give you the shape of the data, the center, and the spread of the data.

The relative frequency is equal to the frequency for an observed value of the data divided by the total number of data values in the sample.(Remember, frequency is defined as the number of times an answer occurs.) If:

- f = frequency
- n = total number of data values (or the sum of the individual frequencies), and
- RF = relative frequency,

then:

$$RF = \frac{f}{n}$$

For example, if three students in Mr. Ahab's English class of 40 students received from 90% to 100%, then, f = 3, n = 40, and $RF = \frac{f}{n} = \frac{3}{40} = 0.075$. 7.5% of the students received 90–100%. 90–100% are quantitative measures.

To construct a histogram, first decide how many bars or intervals, also called classes, represent the data. Many histograms consist of five to 15 bars or classes for clarity. The number of bars needs to be chosen. Choose a starting point for the first interval to be less than the smallest data value. A convenient starting point is a lower value carried out to one more decimal place than the value with the most decimal places. For example, if the value with the most decimal places is 6.1

and this is the smallest value, a convenient starting point is 6.05 (6.1 – 0.05 = 6.05). We say that 6.05 has more precision. If the value with the most decimal places is 2.23 and the lowest value is 1.5, a convenient starting point is 1.495 (1.5 – 0.005 = 1.495). If the value with the most decimal places is 3.234 and the lowest value is 1.0, a convenient starting point is 0.9995 (1.0 – 0.0005 = 0.9995). If all the data happen to be integers and the smallest value is two, then a convenient starting point is 1.5 (2 – 0.5 = 1.5). Also, when the starting point and other boundaries are carried to one additional decimal place, no data value will fall on a boundary. The next two examples go into detail about how to construct a histogram using continuous data and how to create a histogram using discrete data.

EXAMPLE 2.8

The following data are the heights (in inches to the nearest half inch) of 100 male semiprofessional soccer players. The heights are continuous data, since height is measured. 60; 60.5; 61; 61; 61.5 63.5; 63.5; 63.5 64; 64; 64; 64; 64; 64; 64; 64.5; 64.5; 64.5; 64.5; 64.5; 64.5; 64.5; 64.5 66; 66; 66; 66; 66; 66; 66; 66; 66; 66; 66.5; 66.5; 66.5; 66.5; 66.5; 66.5; 66.5; 66.5; 66.5; 66.5; 66.5; 67; 67; 67; 67; 67; 67; 67; 67; 67; 67; 67; 67; 67.5; 67.5; 67.5; 67.5; 67.5; 67.5; 67.5 68; 68; 69; 69; 69; 69; 69; 69; 69; 69; 69; 69; 69.5; 69.5; 69.5; 69.5; 69.5 70; 70; 70; 70; 70; 70; 70.5; 70.5; 70.5; 71; 71; 71 72; 72; 72; 72.5; 72.5; 73; 73.5 74

The smallest data value is 60. Since the data with the most decimal places has one decimal (for instance, 61.5), we want our starting point to have two decimal places. Since the numbers 0.5, 0.05, 0.005, etc. are convenient numbers, use 0.05 and subtract it from 60, the smallest value, for the convenient starting point.

60 – 0.05 = 59.95 which is more precise than, say, 61.5 by one decimal place. The starting point is, then, 59.95.

The largest value is 74, so 74 + 0.05 = 74.05 is the ending value.

Next, calculate the width of each bar or class interval. To calculate this width, subtract the starting point from the ending value and divide by the number of bars (you must choose the number of bars you desire). Suppose you choose eight bars.

$$\frac{74.05 - 59.95}{8} = 1.76$$

NOTE

We will round up to two and make each bar or class interval two units wide. Rounding up to two is one way to prevent a value from falling on a boundary.

Rounding to the next number is often necessary even if it goes against the standard rules of rounding. For this example, using 1.76 as the width would also work. A guideline that is followed by some for the width of a bar or class interval is to take the square root of the number of data values and then round to the nearest whole number, if necessary. For example, if there are 150 values of data, take the square root of 150 and round to 12 bars or intervals.

The boundaries are:

- 59.95
- 59.95 + 2 = 61.95
- 61.95 + 2 = 63.95
- 63.95 + 2 = 65.95
- 65.95 + 2 = 67.95
- 67.95 + 2 = 69.95
- 69.95 + 2 = 71.95
- 71.95 + 2 = 73.95
- 73.95 + 2 = 75.95

The heights 60 through 61.5 inches are in the interval 59.95–61.95. The heights that are 63.5 are in the interval 61.95–63.95. The heights that are 64 through 64.5 are in the interval 63.95–65.95. The heights 66 through 67.5 are in the interval 65.95–67.95. The heights 68 through 69.5 are in the interval 67.95–69.95. The heights 70 through 71 are in the interval 69.95–71.95. The heights 72 through 73.5 are in the interval 71.95–73.95. The height 74 is in the interval 73.95–75.95.

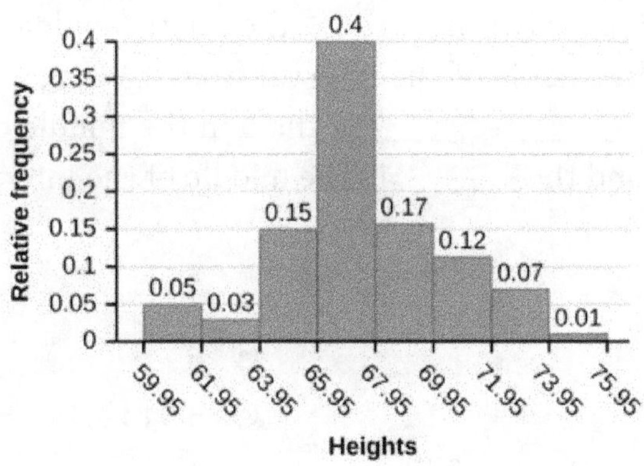

Figure 2.5

The following histogram displays the heights on the *x*-axis and relative frequency on the *y*-axis.

TRY IT 2.8

The following data are the shoe sizes of 50 male students. The sizes are continuous data since shoe size is measured. Construct a histogram and calculate the width of each bar or class interval. Suppose you choose six bars. 9; 9; 9.5; 9.5; 10; 10; 10; 10; 10; 10; 10.5; 10.5; 10.5; 10.5; 10.5; 10.5; 10.5; 10.5

11; 11; 11; 11; 11; 11; 11; 11; 11; 11; 11; 11; 11; 11.5; 11.5; 11.5; 11.5; 11.5; 11.5; 11.5 12; 12; 12; 12; 12; 12; 12; 12.5; 12.5; 12.5; 12.5; 14

EXAMPLE 2.9

Create a histogram for the following data: the number of books bought by 50 part-time college students at ABC College. The number of books is discrete data, since books are counted. 1; 1; 1; 1; 1; 1; 1; 1; 1; 1; 1 2; 2; 2; 2; 2; 2; 2; 2; 2; 2 3; 3; 3; 3; 3; 3; 3; 3; 3; 3; 3; 3; 3; 3; 3; 3 4; 4; 4; 4; 4; 4 5; 5; 5; 5; 5 6; 6

Eleven students buy one book. Ten students buy two books. Sixteen students buy three books. Six students buy four books. Five students buy five books. Two students buy six books.

Because the data are integers, subtract 0.5 from 1, the smallest data value and add 0.5 to 6, the largest data value. Then the starting point is 0.5 and the ending value is 6.5.

Next, calculate the width of each bar or class interval. If the data are discrete and there are not too many different values, a width that places the data values in the middle of the bar or class interval is the most convenient. Since the data consist of the numbers 1, 2, 3, 4, 5, 6, and the starting point is 0.5, a width of one places the 1 in the middle of the interval from 0.5 to 1.5, the 2 in the middle of the interval from 1.5 to 2.5, the 3 in the middle of the interval from 2.5 to 3.5, the 4 in the middle of the interval from _______ to _______, the 5 in the middle of the interval from _______ to _______, and the _______ in the middle of the interval from _______ to _______ .

Solution 2.9

- 3.5 to 4.5
- 4.5 to 5.5
- 6
- 5.5 to 6.5

Calculate the number of bars as follows:

$$\frac{6.5 - 0.5}{\text{number of bars}} = 1$$

where 1 is the width of a bar. Therefore, bars = 6.

The following histogram displays the number of books on the *x*-axis and the frequency on the *y*-axis.

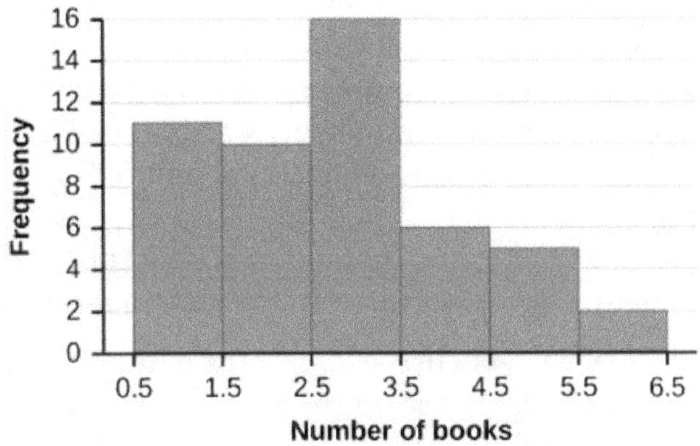

Figure 2.6

EXAMPLE 2.10

Using this data set, construct a histogram.

Table 2.13

Number of hours my classmates spent playing video games on weekends				
9.95	10	2.25	16.75	0
19.5	22.5	7.5	15	12.75
5.5	11	10	20.75	17.5
23	21.9	24	23.75	18
20	15	22.9	18.8	20.5

Solution 2.10

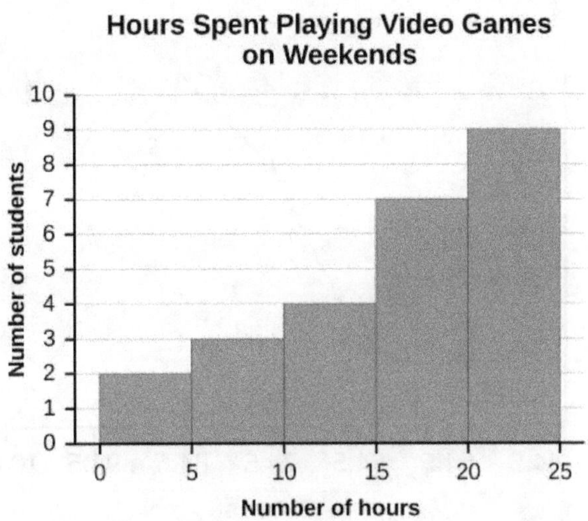

Figure 2.7

Some values in this data set fall on boundaries for the class intervals. A value is counted in a class interval if it falls on the left boundary, but not if it falls on the right boundary. Different researchers may set up histograms for the same data in different ways. There is more than one correct way to set up a histogram.

Frequency Polygons

Frequency polygons are analogous to line graphs, and just as line graphs make continuous data visually easy to interpret, so too do frequency polygons.

To construct a frequency polygon, first examine the data and decide on the number of intervals, or class intervals, to use on the *x*-axis and *y*-axis. After choosing the appropriate ranges, begin plotting the data points. After all the points are plotted, draw line segments to connect them.

EXAMPLE 2.11

A frequency polygon was constructed from the frequency table below.

Table 2.14

Frequency distribution for calculus final test scores			
Lower bound	Upper bound	Frequency	Cumulative frequency
49.5	59.5	5	5
59.5	69.5	10	15
69.5	79.5	30	45
79.5	89.5	40	85
89.5	99.5	15	100

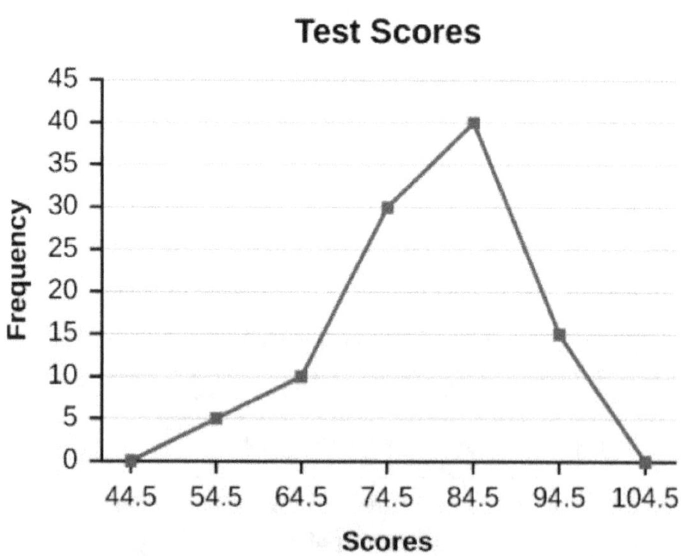

Figure 2.8

The first label on the *x*-axis is 44.5. This represents an interval extending from 39.5 to 49.5. Since the lowest test score is 54.5, this interval is used only to allow the graph to touch the *x*-axis. The point labeled 54.5 represents the next interval, or the first "real" interval from the table, and contains five scores. This reasoning is followed for each of the remaining intervals with the point 104.5 representing the interval from 99.5 to 109.5. Again, this interval contains no data and is only used so that the graph will touch the *x*-axis. Looking at the graph, we say that this distribution is skewed because one side of the graph does not mirror the other side.

TRY IT 2.11

Construct a frequency polygon of U.S. Presidents' ages at inauguration shown in Table 2.15.

Table 2.15

Age at inauguration	Frequency
41.5–46.5	4
46.5–51.5	11
51.5–56.5	14
56.5–61.5	9
61.5–66.5	4
66.5–71.5	2

Frequency polygons are useful for comparing distributions. This is achieved by overlaying the frequency polygons drawn for different data sets.

EXAMPLE 2.12

We will construct an overlay frequency polygon comparing the scores from Example 2.11 with the students' final numeric grade.

Table 2.16

Frequency distribution for calculus final test scores			
Lower bound	Upper bound	Frequency	Cumulative frequency
49.5	59.5	5	5
59.5	69.5	10	15
69.5	79.5	30	45
79.5	89.5	40	85
89.5	99.5	15	100

Table 2.17

Frequency distribution for calculus final grades			
Lower bound	Upper bound	Frequency	Cumulative frequency
49.5	59.5	10	10
59.5	69.5	10	20
69.5	79.5	30	50
79.5	89.5	45	95
89.5	99.5	5	100

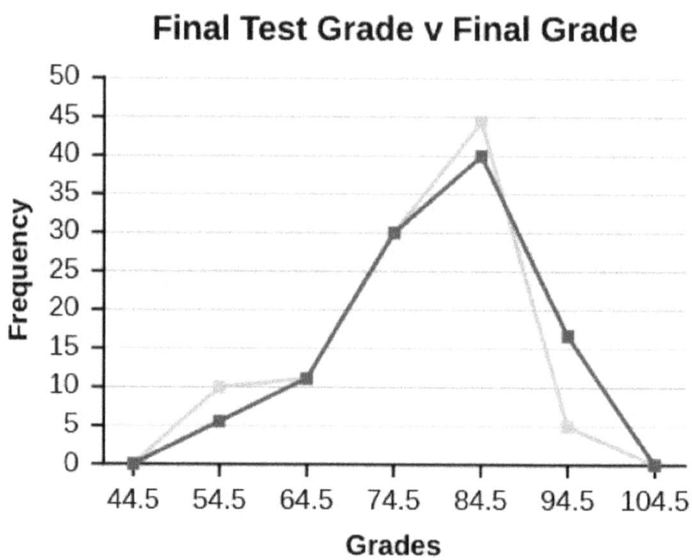

Figure 2.9

Constructing a Time Series Graph

Suppose that we want to study the temperature range of a region for an entire month. Every day at noon we note the temperature and write this down in a log. A variety of statistical studies could be done with these data. We could find the mean or the median temperature for the month. We could construct a histogram displaying the number of days that temperatures reach a certain range of values. However, all of these methods ignore a portion of the data that we have collected.

One feature of the data that we may want to consider is that of time. Since each date is paired with the temperature reading for the day, we don't have to think of the data as being random. We can instead use the times given to impose a chronological order on the data. A graph that recognizes this ordering and displays the changing temperature as the month progresses is called a time series graph.

To construct a time series graph, we must look at both pieces of our paired data **set.** We start with a standard Cartesian coordinate system. The horizontal axis is used to plot the date or time increments, and the vertical axis is used to plot the values of the variable that we are measuring. By doing this, we make each point on the graph correspond to a date and a measured quantity. The points on the graph are typically connected by straight lines in the order in which they occur.

EXAMPLE 2.13

Problem

The following data shows the Annual Consumer Price Index, each month, for ten years. Construct a time series graph for the Annual Consumer Price Index data only.

Table 2.18

Year	Jan	Feb	Mar	Apr	May	Jun	Jul
2003	181.7	183.1	184.2	183.8	183.5	183.7	183.9
2004	185.2	186.2	187.4	188.0	189.1	189.7	189.4
2005	190.7	191.8	193.3	194.6	194.4	194.5	195.4
2006	198.3	198.7	199.8	201.5	202.5	202.9	203.5
2007	202.416	203.499	205.352	206.686	207.949	208.352	208.299
2008	211.080	211.693	213.528	214.823	216.632	218.815	219.964
2009	211.143	212.193	212.709	213.240	213.856	215.693	215.351
2010	216.687	216.741	217.631	218.009	218.178	217.965	218.011
2011	220.223	221.309	223.467	224.906	225.964	225.722	225.922
2012	226.665	227.663	229.392	230.085	229.815	229.478	229.104

Table 2.19

Year	Aug	Sep	Oct	Nov	Dec	Annual
2003	184.6	185.2	185.0	184.5	184.3	184.0
2004	189.5	189.9	190.9	191.0	190.3	188.9
2005	196.4	198.8	199.2	197.6	196.8	195.3
2006	203.9	202.9	201.8	201.5	201.8	201.6
2007	207.917	208.490	208.936	210.177	210.036	207.342
2008	219.086	218.783	216.573	212.425	210.228	215.303
2009	215.834	215.969	216.177	216.330	215.949	214.537
2010	218.312	218.439	218.711	218.803	219.179	218.056
2011	226.545	226.889	226.421	226.230	225.672	224.939
2012	230.379	231.407	231.317	230.221	229.601	229.594

Solution 2.3

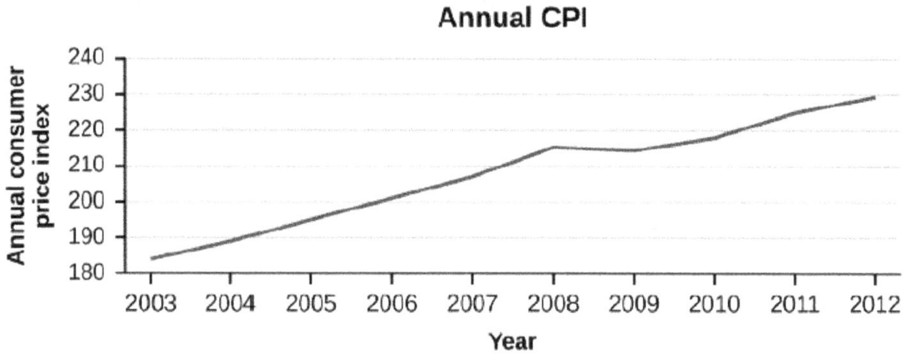

Figure 2.10

TRY IT 2.13

The following table is a portion of a data set from www.worldbank.org. Use the table to construct a time series graph for CO_2 emissions for the United States.

Table 2.20

Year	Ukraine	CO_2 emissions United Kingdom	United States
2003	352,259	540,640	5,681,664
2004	343,121	540,409	5,790,761
2005	339,029	541,990	5,826,394
2006	327,797	542,045	5,737,615
2007	328,357	528,631	5,828,697
2008	323,657	522,247	5,656,839
2009	272,176	474,579	5,299,563

Uses of a Time Series Graph

Time series graphs are important tools in various applications of statistics. When recording values of the same variable over an extended period of time, sometimes it is difficult to discern any trend or pattern. However, once the same data points are displayed graphically, some features jump out. Time series graphs make trends easy to spot.

How NOT to Lie with Statistics

It is important to remember that the very reason we develop a variety of methods to present data is to develop insights into the subject of what the observations represent. We want to get a "sense" of the data. Are the observations all very much alike or are they spread across a wide range of values, are they bunched at one end of the spectrum or are they distributed evenly and so on. We are trying to get a visual picture of the numerical data. Shortly we will develop formal mathematical measures of the data, but our visual graphical presentation can say much. It can, unfortunately, also say much that is distracting, confusing and simply wrong in terms of the impression the visual leaves. Many years ago Darrell Huff wrote the book *How to Lie with Statistics.* It has been through 25 plus printings and sold more than one and one-half million copies. His perspective was a harsh one and used many actual examples that were designed to mislead. He wanted to make people aware of such deception, but perhaps more importantly to educate so that others do not make the same errors inadvertently.

Again, the goal is to enlighten with visuals that tell the story of the data. Pie charts have a number of common problems when used to convey the message of the data. Too many pieces of the pie overwhelm the reader. More than perhaps five or six categories ought to give an idea of the relative importance of each piece. This is after all the goal of a pie chart, what subset matters most relative to the others. If there are more components than this then perhaps an alternative approach would be better or perhaps some can be consolidated into an "other" category. Pie charts cannot show changes over time, although we see this attempted all too often. In federal, state, and city finance documents pie charts are often presented to show the components of revenue available to the governing body for appropriation: income tax, sales tax motor vehicle taxes and so on. In and of itself this is interesting information and can be nicely done with a pie chart. The error occurs when two years are set side-by-side. Because the total revenues change year to year, but the size of the pie is fixed, no real information is provided and the relative size of each piece of the pie cannot be meaningfully compared.

Histograms can be very helpful in understanding the data. Properly presented, they can be a quick visual way to present probabilities of different categories by the simple visual of comparing relative areas in each category. Here the error, purposeful or not, is to vary the width of the categories. This of course makes comparison to the other categories impossible. It does embellish the importance of the category with the expanded width because it has a greater area, inappropriately, and thus visually "says" that that category has a higher probability of occurrence.

Time series graphs perhaps are the most abused. A plot of some variable across time should never be presented on axes that change part way across the page either in the vertical or horizontal dimension. Perhaps the time frame is changed from years to months. Perhaps this is to save space or because monthly data was not available for early years. In either case this confounds the presentation and destroys any value of the graph. If this is not done to purposefully confuse the reader, then it certainly is either lazy or sloppy work.

Changing the units of measurement of the axis can smooth out a drop or accentuate one. If you want to show large changes, then measure the variable in small units, penny rather than thousands of dollars. And of course to continue the fraud, be sure that the axis does not begin at zero, zero. If it begins at zero, zero, then it becomes apparent that the axis has been manipulated.

Perhaps you have a client that is concerned with the volatility of the portfolio you manage. An easy way to present the data is to use long time periods on the time series graph. Use months or better, quarters rather than daily or weekly data. If that doesn't get the volatility down then spread the time axis relative to the rate of return or portfolio valuation axis. If you want to show "quick" dramatic growth, then shrink the time axis. Any positive growth will show visually "high" growth rates. Do note that if the growth is negative then this trick will show the portfolio is collapsing at a dramatic rate.

Again, the goal of descriptive statistics is to convey meaningful visuals that tell the story of the data. Purposeful manipulation is fraud and unethical at the worst, but even at its best, making these type of errors will lead to confusion on the part of the analysis.

2.2 MEASURES OF THE LOCATION OF THE DATA

The common measures of location are quartiles and percentiles

Quartiles are special percentiles. The first quartile, Q_1, is the same as the 25th percentile, and the third quartile, Q_3, is the same as the 75th percentile. The median, *M*, is called both the second quartile and the 50th percentile.

To calculate quartiles and percentiles, the data must be ordered from smallest to largest. Quartiles divide ordered data into quarters. Percentiles divide ordered data into hundredths. To score in the 90th percentile of an exam does not mean, necessarily, that you received 90% on a test. It means that 90% of test scores are the same or less than your score and 10% of the test scores are the same or greater than your test score.

Percentiles are useful for comparing values. For this reason, universities and colleges use percentiles extensively. One instance in which colleges and universities use percentiles is when SAT results are used to determine a minimum testing score that will be used as an acceptance factor. For example, suppose Duke accepts SAT scores at or above the 75th percentile. That translates into a score of at least 1220.

Percentiles are mostly used with very large populations. Therefore, if you were to say that 90% of the test scores are less (and not the same or less) than your score, it would be acceptable because removing one particular data value is not significant.

The median is a number that measures the "center" of the data. You can think of the median as the "middle value," but it does not actually have to be one of the observed values. It is a number that separates ordered data into halves. Half the values are the same number or smaller than the median, and half the values are the same number or larger. For example, consider the following data.

1; 11.5; 6; 7.2; 4; 8; 9; 10; 6.8; 8.3; 2; 2; 10; 1
Ordered from smallest to largest:
1; 1; 2; 2; 4; 6; 6.8; 7.2; 8; 8.3; 9; 10; 10; 11.5

Since there are 14 observations, the median is between the seventh value, 6.8, and the eighth value, 7.2. To find the median, add the two values together and divide by two.

$$\frac{6.8 + 7.2}{2} = 7$$

The median is seven. Half of the values are smaller than seven and half of the values are larger than seven.

Quartiles are numbers that separate the data into quarters. Quartiles may or may not be part of the data. To find the quartiles, first find the median or second quartile. The first quartile, Q_1, is the middle value of the lower half of the data, and the third quartile, Q_3, is the middle value, or median, of the upper half of the data. To get the idea, consider the same data set: 1; 1; 2; 2; 4; 6; 6.8; 7.2; 8; 8.3; 9; 10; 10; 11.5

The median or second quartile is seven. The lower half of the data are 1, 1, 2, 2, 4, 6, 6.8. The middle value of the lower half is two. 1; 1; 2; 2; 4; 6; 6.8

The number two, which is part of the data, is the first quartile. One-fourth of the entire sets of values are the same as or less than two and three-fourths of the values are more than two.

The upper half of the data is 7.2, 8, 8.3, 9, 10, 10, 11.5. The middle value of the upper half is nine.

The third quartile, Q3, is nine. Three-fourths (75%) of the ordered data set are less than nine. One-fourth (25%) of the ordered data set are greater than nine. The third quartile is part of the data set in this example.

The interquartile range is a number that indicates the spread of the middle half or the middle 50% of the data. It is the difference between the third quartile (Q_3) and the first quartile (Q_1).

$IQR = Q_3 - Q_1$

The *IQR* can help to determine potential outliers. A value is suspected to be a potential outlier if it is less than (1.5)(IQR) below the first quartile or more than (1.5)(IQR) above the third quartile. Potential outliers always require further investigation.

NOTE

A potential outlier is a data point that is significantly different from the other data points. These special data points may be errors or some kind of abnormality or they may be a key to understanding the data.

EXAMPLE 2.14

For the following 13 real estate prices, calculate the *IQR* and determine if any prices are potential outliers. Prices are in dollars. 389,950; 230,500; 158,000; 479,000; 639,000; 114,950; 5,500,000; 387,000; 659,000; 529,000; 575,000; 488,800; 1,095,000

Solution 2.14

Order the data from smallest to largest.

114,950; 158,000; 230,500; 387,000; 389,950; 479,000; 488,800; 529,000; 575,000; 639,000; 659,000;

1,095,000; 5,500,000

$M = 488{,}800$

$$Q_1 = \frac{230{,}500 + 387{,}000}{2} = 308{,}750$$

$$Q_3 = \frac{639{,}000 + 659{,}000}{2} = 649{,}000$$

$IQR = 649{,}000 - 308{,}750 = 340{,}250$

$(1.5)(IQR) = (1.5)(340{,}250) = 510{,}375$

$Q_1 - (1.5)(IQR) = 308{,}750 - 510{,}375 = -201{,}625$

$Q_3 + (1.5)(IQR) = 649{,}000 + 510{,}375 = 1{,}159{,}375$

No house price is less than –201,625. However, 5,500,000 is more than 1,159,375. Therefore, 5,500,000 is a potential outlier.

EXAMPLE 2.15

For the two data sets in the test scores example, find the following:

a. The interquartile range. Compare the two interquartile ranges.
b. Any outliers in either set.

Solution 2.15

The five number summary for the day and night classes is

Table 2.21

	Minimum	Q1	Median	Q3	Maximum
Day	32	56	74.5	82.5	99
Night	25.5	78	81	89	98

a. The IQR for the day group is $Q_3 - Q_1 = 82.5 - 56 = 26.5$
   The IQR for the night group is $Q_3 - Q_1$ - $89 - 78 = 11$

The interquartile range (the spread or variability) for the day class is larger than the night class *IQR*. This suggests more variation will be found in the day class's test scores.

b. Day class outliers are found using the IQR times 1.5 rule. So,
   $Q_1 - IQR\ (1.5) = 56 - 26.5\ (1.5) = 16.25$
   $Q_3 + IQR\ (1.5) = 82.5 + 26.5\ (1.5) = 122.25$

Since the minimum and maximum values for the day class are greater than 16.25 and less than 122.25, there are no outliers.

Night class outliers are calculated as:

$Q_1 - IQR\ (1.5) = 78 - 11(1.5) = 61.5$
$Q_3 + \text{IQR}\ (1.5) = 89 + 11(1.5) = 105.5$

For this class, any test score less than 61.5 is an outlier. Therefore, the scores of 45 and 25.5 are outliers. Since no test score is greater than 105.5, there is no upper end outlier.

EXAMPLE 2.16

Fifty statistics students were asked how much sleep they get per school night (rounded to the nearest hour). The results were:

Table 2.22

Amount of sleep per school night (hours)	Frequency	Relative frequency	Cumulative relative frequency
4	2	0.04	0.04
5	5	0.10	0.14
6	7	0.14	0.28
7	12	0.24	0.52
8	14	0.28	0.80
9	7	0.14	0.94
10	3	0.06	1.00

Find the 28th percentile. Notice the 0.28 in the "cumulative relative frequency" column. Twenty-eight percent of 50 data values is 14 values. There are 14 values less than the 28th percentile. They include the two 4s, the five 5s, and the seven 6s. The 28th percentile is between the last six and the first seven. The 28th percentile is 6.5.

Find the median. Look again at the "cumulative relative frequency" column and find 0.52. The median is the 50th percentile or the second quartile. 50% of 50 is 25. There are 25 values less than the median. They include the two 4s, the five 5s, the seven 6s, and eleven of the 7s. The median or 50th percentile is between the 25th, or seven, and 26th, or seven, values. The median is seven.

Find the third quartile. The third quartile is the same as the 75th percentile. You can "eyeball" this answer. If you look at the "cumulative relative frequency" column, you find 0.52 and 0.80. When you have all the fours, fives, sixes and sevens, you have 52% of the data. When you include all the 8s, you have 80% of the data. The 75th percentile, then, must be an eight. Another way to look at the problem is to find 75% of 50, which is 37.5, and round up to 38. The third quartile, Q3, is the 38th value, which is an eight. You can check this answer by counting the values. (There are 37 values below the third quartile and 12 values above.)

TRY IT 2.16

Forty bus drivers were asked how many hours they spend each day running their routes (rounded to the nearest hour). Find the 65th percentile.

Table 2.23

Amount of time spent on route (hours)	Frequency	Relative frequency	Cumulative relative frequency
2	12	0.30	0.30
3	14	0.35	0.65
4	10	0.25	0.90
5	4	0.10	1.00

EXAMPLE 2.17

Using Table 2.22:

a. Find the 80th percentile.
b. Find the 90th percentile.
c. Find the first quartile. What is another name for the first quartile?

Solution

Using the data from the frequency table, we have.

a. The 80th percentile is between the last eight and the first nine in the table (between the 40th and 411st values. Therefore, we need to take the mean of the 40th and 411st values. The 80th percentile $= \frac{8+9}{2} = 8.5$
b. The 90th percentile will be the 45th data value (location is 0.90(50) = 45) and the 45th data value is nine.
c. Q1 is also the 25th percentile. The 25th percentile location calculation: *P*25 = 0.25(50) = 12.5 ≈ 13 the 13th data value. Thus, the 25th percentile is six.

A Formula for Finding the kth Percentile

If you were to do a little research, you would find several formulas for calculating the kth percentile. Here is one of them.

k = the kth percentile. It may or may not be part of the data.

i = the index (ranking or position of a data value)

n = the total number of data points, or observations

Order the data from smallest to largest.

- Calculate $i = \frac{k}{100}(n + 1)$
- If i is an integer, then the kth percentile is the data value in the ith position in the ordered set of data.
- If i is not an integer, then round i up and round i down to the nearest integers. Average the two data values in these two positions in the ordered data set. This is easier to understand in an example.

EXAMPLE 2.18

Listed are 29 ages for Academy Award winning best actors *in order from smallest to largest.* 18; 21; 22; 25; 26; 27; 29; 30; 31; 33; 36; 37; 41; 42; 47; 52; 55; 57; 58; 62; 64; 67; 69; 71; 72; 73; 74; 76; 77

a. Find the 70th percentile.
b. Find the 83rd percentile.

Solution 2.18

a. k = 70

   i = the index

   n = 29

$i = \frac{k}{100}(n+1) = (\frac{70}{100})(20+1) = 21$ Twenty-one is an integer, and the data value in the 21st position in the ordered data set is 64. The 70th percentile is 64 years.

k = 83rd percentile

i = the index

n = 29

$i = \frac{k}{100}(n+1) = (\frac{83}{100})(29+1) = 24.9$ which is NOT an integer. Round it down to 24 and up to 25. The age in the 24th position is 71 and the age in the 25th position is 72. Average 71 and 72. The 83rd percentile is 71.5 years.

TRY IT 2.18

Listed are 29 ages for Academy Award winning best actors *in order from smallest to largest.*

18; 21; 22; 25; 26; 27; 29; 30; 31; 33; 36; 37; 41; 42; 47; 52; 55; 57; 58; 62; 64; 67; 69; 71; 72; 73; 74; 76; 77 Calculate the 20th percentile and the 55th percentile.

A Formula for Finding the Percentile of a Value in a Data Set

- Order the data from smallest to largest.
- x = the number of data values counting from the bottom of the data list up to but not including the data value for which you want to find the percentile.
- y = the number of data values equal to the data value for which you want to find the percentile.
- n = the total number of data.
- Calculate $\frac{x+0.5y}{n}(100)$. Then round to the nearest integer.

EXAMPLE 2.19

Listed are 29 ages for Academy Award winning best actors *in order from smallest to largest.* 18; 21; 22; 25; 26; 27; 29; 30; 31; 33; 36; 37; 41; 42; 47; 52; 55; 57; 58; 62; 64; 67; 69; 71; 72; 73; 74; 76; 77

a. Find the percentile for 58.
b. Find the percentile for 25.

Solution 2.19

a. Counting from the bottom of the list, there are 18 data values less than 58. There is one value of 58.

   $x = 18$ and $y = 1. \frac{x+0.5y}{n}(100) = \frac{18+0.5(1)}{29}(100) = 63.80.58$ is the 64th percentile.
b. Counting from the bottom of the list, there are three data values less than 25. There is one value of 25.

$x = 3$ and $y = 1. \frac{x+0.5y}{n}(100) = \frac{3+0.5(1)}{29}(100) = 12.07$ Twenty-five is the 12th percentile.

Interpreting Percentiles, Quartiles, and Median

A percentile indicates the relative standing of a data value when data are sorted into numerical order from smallest to largest. Percentages of data values are less than or equal to the pth percentile. For example, 15% of data values are less than or equal to the 15th percentile.

- Low percentiles always correspond to lower data values.
- High percentiles always correspond to higher data values.

A percentile may or may not correspond to a value judgment about whether it is "good" or "bad." The interpretation of whether a certain percentile is "good" or "bad" depends on the context of the situation to which the data applies. In some situations, a low percentile would be considered "good;" in other contexts a high percentile might be considered "good". In many situations, there is no value judgment that applies.

Understanding how to interpret percentiles properly is important not only when describing data, but also when calculating probabilities in later chapters of this text.

NOTE

When writing the interpretation of a percentile in the context of the given data, the sentence should contain the following information.

- information about the context of the situation being considered
- the data value (value of the variable) that represents the percentile
- the percent of individuals or items with data values below the percentile
- the percent of individuals or items with data values above the percentile.

Example 2.20

On a timed math test, the first quartile for time it took to finish the exam was 35 minutes. Interpret the first quartile in the context of this situation.

Solution 2.20

- Twenty-five percent of students finished the exam in 35 minutes or less.
- Seventy-five percent of students finished the exam in 35 minutes or more.
- A low percentile could be considered good, as finishing more quickly on a timed exam is desirable. (If you take too long, you might not be able to finish.)

Example 2.21

Problem

On a 20 question math test, the 70th percentile for number of correct answers was 16. Interpret the 70th percentile in the context of this situation.

Solution 2.21

- Seventy percent of students answered 16 or fewer questions correctly.
- Thirty percent of students answered 16 or more questions correctly.
- A higher percentile could be considered good, as answering more questions correctly is desirable.

TRY IT 2.21

On a 60 point written assignment, the 80th percentile for the number of points earned was 49. Interpret the 80th percentile in the context of this situation.

Example 2.22

At a community college, it was found that the 30th percentile of credit units that students are enrolled for is seven units. Interpret the 30th percentile in the context of this situation.

Solution 2.22

- Thirty percent of students are enrolled in seven or fewer credit units.
- Seventy percent of students are enrolled in seven or more credit units.
- In this example, there is no "good" or "bad" value judgment associated with a higher or lower percentile. Students attend community college for varied reasons and needs, and their course load varies according to their needs.

EXAMPLE 2.23

Sharpe Middle School is applying for a grant that will be used to add fitness equipment to the gym. The principal surveyed 15 anonymous students to determine how many minutes a day the students spend exercising. The results from the 15 anonymous students are shown.

0 minutes; 40 minutes; 60 minutes; 30 minutes; 60 minutes
10 minutes; 45 minutes; 30 minutes; 300 minutes; 90 minutes;
30 minutes; 120 minutes; 60 minutes; 0 minutes; 20 minutes
Determine the following five values.

Min = 0

Q_1 = 20

Med = 40

$Q_3 = 60$

Max = 300

If you were the principal, would you be justified in purchasing new fitness equipment? Since 75% of the students exercise for 60 minutes or less daily, and since the *IQR* is 40 minutes (60 – 20 = 40), we know that half of the students surveyed exercise between 20 minutes and 60 minutes daily. This seems a reasonable amount of time spent exercising, so the principal would be justified in purchasing the new equipment.

However, the principal needs to be careful. The value 300 appears to be a potential outlier.

$Q_3 + 1.5(IQR) = 60 + (1.5)(40) = 120.$

The value 300 is greater than 120 so it is a potential outlier. If we delete it and calculate the five values, we get the following values:

Min = 0

$Q_1 = 20$

$Q_3 = 60$

Max = 120

We still have 75% of the students exercising for 60 minutes or less daily and half of the students exercising between 20 and 60 minutes a day. However, 15 students is a small sample and the principal should survey more students to be sure of his survey results.

2.3 MEASURES OF THE CENTER OF THE DATA

The "center" of a data set is also a way of describing location. The two most widely used measures of the "center" of the data are the mean (average) and the median. To calculate the mean weight of 50 people, add the 50 weights together and divide by 50. Technically this is the arithmetic mean. We will discuss the geometric mean later. To find the median weight of the 50 people, order the data and find the number that splits the data into two equal parts meaning an equal number of observations on each side. The weight of 25 people are below this weight and 25 people are heavier than this weight. The median is generally a better measure of the center when there are extreme values or outliers because it is not affected by the precise numerical values of the outliers. The mean is the most common measure of the center.

NOTE

The words "mean" and "average" are often used interchangeably. The substitution of one word for the other is common practice. The technical term is "arithmetic mean" and "average" is technically a center location. Formally, the arithmetic mean is called the first moment of the distribution by mathematicians. However, in practice among non-statisticians, "average" is commonly accepted for "arithmetic mean."

When each value in the data set is not unique, the mean can be calculated by multiplying each distinct value by its frequency and then dividing the sum by the total number of data values. The letter used to represent the sample mean is an x with a bar over it (pronounced "x bar"): $\bar{x}$.

The Greek letter μ (pronounced "mew") represents the population mean. One of the requirements for the sample mean to be a good estimate of the population mean is for the sample taken to be truly random.

To see that both ways of calculating the mean are the same, consider the sample: 1; 1; 1; 2; 2; 3; 4; 4; 4; 4; 4

$$\bar{x} = \frac{1+1+1+2+2+3+4+4+4+4+4}{11} = 2.7$$

$$\bar{x} = \frac{3(1)+2(2)+1(3)+5(4)}{11} = 2.7$$

In the second calculation, the frequencies are 3, 2, 1, and 5.

You can quickly find the location of the median by using the expression $\frac{n+1}{2}$. The letter n is the total number of data values in the sample. If n is an odd number, the median is the middle value of the ordered data (ordered smallest to largest). If n is an even number, the median is equal to the two middle values added together and divided by two after the data has been ordered. For example, if the total number of data values is 97, then $\frac{n+1}{2} = \frac{97+1}{2} = 49$. The median is the 49th value in the ordered data. If the total number of data values is 100, then $\frac{n+1}{2} = \frac{100+1}{2} = 50.5$. The median occurs midway between the 50th and 51st values. The location of the median and the value of the median are not the same. The upper case letter M is often used to represent the median. The next example illustrates the location of the median and the value of the median.

Example 2.24

Problem

AIDS data indicating the number of months a patient with AIDS lives after taking a new antibody drug are as follows (smallest to largest): 3; 4; 8; 8; 10; 11; 12; 13; 14; 15; 15; 16; 16; 17; 17; 18; 21; 22; 22; 24; 24; 25; 26; 26; 27; 27; 29; 29; 31; 32; 33; 33; 34; 34; 35; 37; 40; 44; 44; 47; Calculate the mean and the median.

Solution 2.24

The calculation for the mean is:

$$\bar{x} = \frac{[3 + 4 + (8)(2) + 10 + 11 + 12 + 13 + 14 + (15)(2) + (16)(2) + ... + 35 + 37 + 40 + (44)(2) + 47]}{40} = 23.6$$

To find the median, M, first use the formula for the location. The location is:

$$\frac{n+1}{2} = \frac{40+1}{2} = 20.5$$

Starting at the smallest value, the median is located between the 20th and 21st values (the two 24s):

3; 4; 8; 8; 10; 11; 12; 13; 14; 15; 15; 16; 16; 17; 17; 18; 21; 22; 22; 24; 24; 25; 26; 26; 27; 27; 29; 29; 31; 32; 33; 33; 34; 34; 35; 37; 40; 44; 44; 47;

$$M = \frac{24 + 24}{2} = 24$$

Example 2.25

Problem

Suppose that in a small town of 50 people, one person earns \$5,000,000 per year and the other 49 each earn \$30,000. Which is the better measure of the "center": the mean or the median?

Solution 2.25

$$\bar{x} = \frac{5,000,000 + 49(30,000)}{50} = 129,400$$

M = 30,000

(There are 49 people who earn \$30,000 and one person who earns \$5,000,000.)

The median is a better measure of the "center" than the mean because 49 of the values are 30,000 and one is 5,000,000. The 5,000,000 is an outlier. The 30,000 gives us a better sense of the middle of the data.

Another measure of the center is the mode. The mode is the most frequent value. There can be more than one mode in a data set as long as those values have the same frequency and that frequency is the highest. A data set with two modes is called bimodal.

EXAMPLE 2.26

Statistics exam scores for 20 students are as follows:

50; 53; 59; 59; 63; 63; 72; 72; 72; 72; 72; 76; 78; 81; 83; 84; 84; 84; 90; 93

Find the mode.

Solution 2.26

The most frequent score is 72, which occurs five times. Mode = 72.

EXAMPLE 2.27

Five real estate exam scores are 430, 430, 480, 480, 495. The data set is bimodal because the scores 430 and 480 each occur twice.

When is the mode the best measure of the "center"? Consider a weight loss program that advertises a mean weight loss of six pounds the first week of the program. The mode might indicate that most people lose two pounds the first week, making the program less appealing.

NOTE

The mode can be calculated for qualitative data as well as for quantitative data. For example, if the data set is: red, red, red, green, green, yellow, purple, black, blue, the mode is red.

Calculating the Arithmetic Mean of Grouped Frequency Tables

When only grouped data is available, you do not know the individual data values (we only know intervals and interval frequencies); therefore, you cannot compute an exact mean for the data set. What we must do is estimate the actual mean by calculating the mean of a frequency table. A frequency table is a data representation in which grouped data is displayed along with the corresponding frequencies. To calculate the mean from a grouped frequency table we can apply the basic definition of $mean = \frac{data\ sum}{number\ of\ data\ values}$ We simply need to modify the definition to fit within the restrictions of a frequency table.

Since we do not know the individual data values we can instead find the midpoint of each interval. The midpoint is $\frac{lower\ boundary + upper\ boundary}{2}$. We can now modify the mean definition to be Mean of Frequency $Table = \frac{\sum fm}{\sum f}$ where f = the frequency of the interval and m = the midpoint of the interval.

EXAMPLE 2.28

A frequency table displaying professor Blount's last statistic test is shown. Find the best estimate of the class mean.

Table 2.24

Grade interval	Number of students
50–56.5	1
56.5–62.5	0
62.5–68.5	4
68.5–74.5	4

74.5–80.5	2
80.5–86.5	3
86.5–92.5	4

Solution 2.28

Find the midpoints for all intervals

Table 2.25

Grade interval	Midpoint
50–56.5	53.25
56.5–62.5	59.5
62.5–68.5	65.5
68.5–74.5	71.5
74.5–80.5	77.5
80.5–86.5	83.5
86.5–92.5	89.5
92.5–98.5	95.5

- Calculate the sum of the product of each interval frequency and midpoint. $\sum fm$ 53.25(1)+59.5(0)+65.5(4)+71.5(4)+77.5(2)+83.5(3)+89.5(4)+95.5(1)=1460
- $\mu = \frac{\sum fm}{\sum f} = \frac{1460.25}{19} = 76.86$

TRY IT 2.28

Maris conducted a study on the effect that playing video games has on memory recall. As part of her study, she compiled the following data:

Table 2.26

Hours teenagers spend on video games	Number of teenagers
0–3.5	3
3.5–7.5	7
7.5–11.5	12
11.5–15.5	7
15.5–19.5	9

What is the best estimate for the mean number of hours spent playing video games?

2.4 SIGMA NOTATION AND CALCULATING THE ARITHMETIC MEAN

Formula for Population Mean

$$\mu = \frac{1}{N}\sum_{i=1}^{N} x_i$$

Formula for Sample Mean

$$\bar{x} = \frac{1}{n}\sum_{i=1}^{n} x_i$$

This unit is here to remind you of material that you once studied and said at the time "I am sure that I will never need this!"

Here are the formulas for a population mean and the sample mean. The Greek letter μ is the symbol for the population mean and $\bar{x}$ is the symbol for the sample mean. Both formulas have a mathematical symbol that tells us how to make the calculations. It is called Sigma notation because the symbol is the Greek capital letter sigma: Σ. Like all mathematical symbols it tells us what to do: just as the plus sign tells us to add and the x tells us to multiply. These are called mathematical operators. The Σ symbol tells us to add a specific list of numbers.

Let's say we have a sample of animals from the local animal shelter and we are interested in their average age. If we list each value, or observation, in a column, you can give each one an index number. The first number will be number 1 and the second number 2 and so on.

Table 2.27

Animal	Age
1	9
2	1
3	8.5
4	10.5
5	10
6	8.5
7	12
8	8
9	1
10	9.5

Each observation represents a particular animal in the sample. Purr is animal number one and is a 9 year old cat, Toto is animal number 2 and is a 1 year old puppy and so on.

To calculate the mean we are told by the formula to add up all these numbers, ages in this case, and then divide the sum by 10, the total number of animals in the sample.

Animal number one, the cat Purr, is designated as X_1, animal number 2, Toto, is designated as X_2 and so on through Dundee who is animal number 10 and is designated as X_{10}.

The i in the formula tells us which of the observations to add together. In this case it is X_1 through X_{10} which is all of them. We know which ones to add by the indexing notation, the i = 1 and the n or capital N for the population. For this example the indexing notation would be i = 1 and because it is a sample we use a small n on the top of the Σ which would be 10.

The standard deviation requires the same mathematical operator and so it would be helpful to recall this knowledge from your past.

The sum of the ages is found to be 78 and dividing by 10 gives us the sample mean age as 7.8 years.

2.5 GEOMETRIC MEAN

The mean (Arithmetic), median and mode are all measures of the "center" of the data, the "average". They are all in their own way trying to measure the "common" point within the data, that which is "normal". In the case of the arithmetic mean this is solved by finding the value from which all points are equal linear distances. We can imagine that all the data values are combined through addition and then distributed back to each data point in equal amounts. The sum of all the values is what is redistributed in equal amounts such that the total sum remains the same.

The geometric mean redistributes not the sum of the values but the product of multiplying all the individual values and then redistributing them in equal portions such that the total product remains the same. This can be seen from the formula for the geometric mean, $\tilde{x}$: *(Pronounced x-tilde)*

$$\tilde{x} = \left(\prod_{i=1}^{n} x_i\right)^{\frac{1}{n}} = \sqrt[n]{x_1 * x_2 \cdots x_n} = (x_1 * x_2 \cdots x_n)^{\frac{1}{n}}$$

where $\pi\pi$ is another mathematical operator, that tells us to multiply all the $xixi$ numbers in the same way capital Greek sigma tells us to add all the $xixi$ numbers. Remember that a fractional exponent is calling for the nth root of the number thus an exponent of 1/3 is the cube root of the number.

The geometric mean answers the question, "if all the quantities had the same value, what would that value have to be in order to achieve the same product?" The geometric mean gets its name from the fact that when redistributed in this way the sides form a geometric shape for which all sides have the same length. To see this, take the example of the numbers 10, 51.2 and 8. The geometric mean is the product of multiplying these three numbers together (4,096) and taking the cube root because there are three numbers among which this product is to be distributed. Thus the geometric mean of these three numbers is 16. This describes a cube 16x16x16 and has a volume of 4,096 units.

The geometric mean is relevant in Economics and Finance for dealing with growth: growth of markets, in investment, population and other variables the growth in which there is an interest. Imagine that our box of 4,096 units (perhaps dollars) is the value of an investment after three years and that the investment returns in percents were the three numbers in our example. The geometric mean will provide us with the answer to the question, what is the average rate of return: 16 percent. The arithmetic mean of these three numbers is 23.6 percent. The reason for this difference, 16 versus 23.6, is that the arithmetic mean is additive and thus does not account for the interest on the interest, compound interest, embedded in the investment growth process. The same issue arises when asking for the average rate of growth of a population or sales or market penetration, etc., knowing the annual rates of growth. The formula for the geometric mean rate of return, or any other growth rate, is:

$$r_s = (x_1 * x_2 \cdots x_n)^{\frac{1}{n}} - 1$$

Manipulating the formula for the geometric mean can also provide a calculation of the average rate of growth between two periods knowing only the initial value a_0 and the ending value an*an* and the number of periods, n*n*. The following formula provides this information:

$$\left(\frac{a_n}{a_0}\right)^{\frac{1}{n}} = \tilde{x}$$

Finally, we note that the formula for the geometric mean requires that all numbers be positive, greater than zero. The reason of course is that the root of a negative number is undefined for use outside of mathematical theory. There are ways to avoid this problem however. In the case of rates of return and other simple growth problems we can convert the negative values to meaningful positive equivalent values. Imagine that the annual returns for the past three years are +12%, -8%, and +2%. Using the decimal multiplier equivalents of 1.12, 0.92, and 1.02, allows us to compute a geometric mean of 1.0167. Subtracting 1 from this value gives the geometric mean of +1.67% as a net rate of population growth (or financial return). From this example we can see that the geometric mean provides us with this formula for calculating the geometric (mean) rate of return for a series of annual rates of return:

$$r_s = \tilde{x} - 1$$

where r_s is average rate of return and $\tilde{x}$ is the geometric mean of the returns during some number of time periods. Note that the length of each time period must be the same.

As a general rule one should convert the percent values to its decimal equivalent multiplier. It is important to recognize that when dealing with percents, the geometric mean of percent values does not equal the geometric mean of the decimal multiplier equivalents and it is the decimal multiplier equivalent geometric mean that is relevant.

2.6 SKEWNESS AND THE MEAN, MEDIAN, AND MODE

Consider the following data set. 4; 5; 6; 6; 6; 7; 7; 7; 7; 7; 7; 8; 8; 8; 9; 10

This data set can be represented by following histogram. Each interval has width one, and each value is located in the middle of an interval.

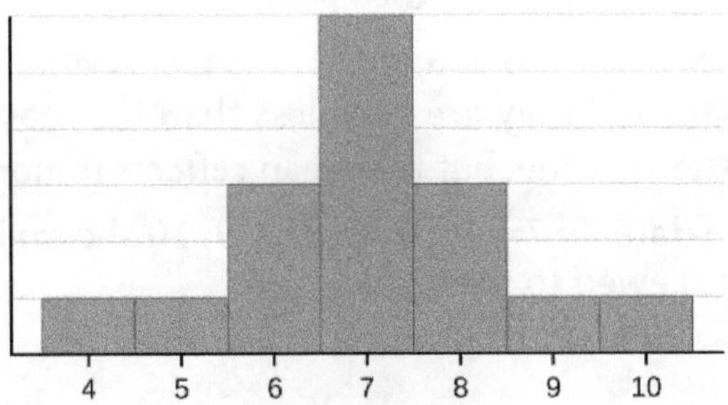

Figure 2.11

The histogram displays a symmetrical distribution of data. A distribution is symmetrical if a vertical line can be drawn at some point in the histogram such that the shape to the left and the right of the vertical line are mirror images of each other. The mean, the median, and the mode are each seven for these data. In a perfectly symmetrical distribution, the mean and the median are the same. This example has one mode (unimodal), and the mode is the same as the mean and median. In a symmetrical distribution that has two modes (bimodal), the two modes would be different from the mean and median.

The histogram for the data: 4; 5; 6; 6; 6; 7; 7; 7; 7; 8 shown in Figure 2.11 is not symmetrical. The right-hand side seems "chopped off" compared to the left side. A distribution of this type is called skewed to the left because it is pulled out to the left. We can formally measure the skewness of a distribution just as we can mathematically measure the center weight of the data or its general "speadness". The mathematical formula for skewness is: $a_3 = \sum \frac{(x_i - \bar{x})^3}{ns^3}$. The greater the deviation from zero indicates a greater degree of skewness. If the skewness is negative then the distribution is skewed left as in Figure 2.12. A positive measure of skewness indicates right skewness such as Figure 2.13.

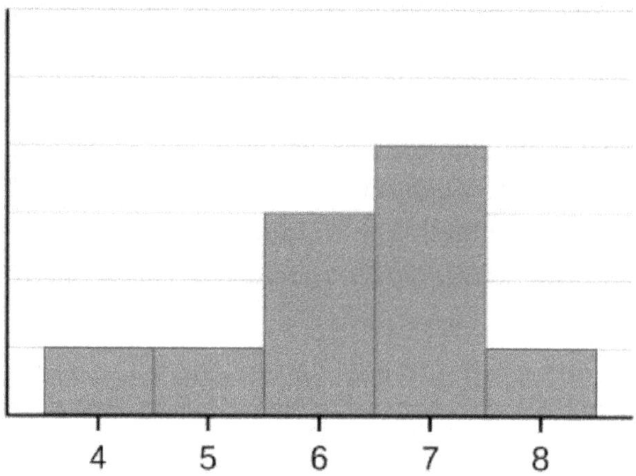

Figure 2.12

The mean is 6.3, the median is 6.5, and the mode is seven. Notice that the mean is less than the median, and they are both less than the mode. The mean and the median both reflect the skewing, but the mean reflects it more so.

The histogram for the data: 6; 7; 7; 7; 7; 8; 8; 8; 9; 10 shown in Figure 2.13, is also not symmetrical. It is skewed to the right.

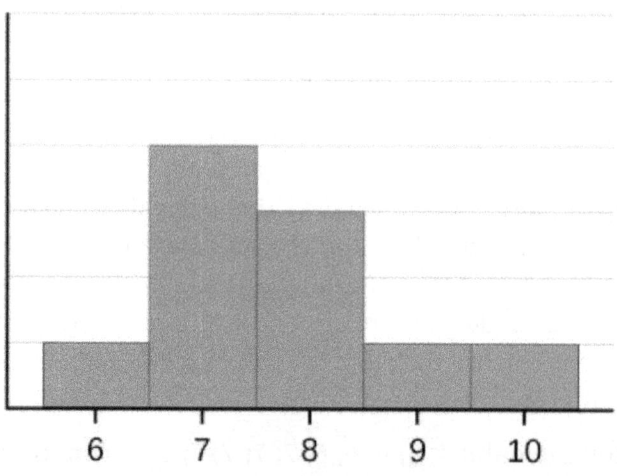

Figure 2.13

The mean is 7.7, the median is 7.5, and the mode is seven. Of the three statistics, the mean is the largest, while the mode is the smallest. Again, the mean reflects the skewing the most.

The mean is affected by outliers that do not influence the mean. Therefore, when the distribution of data is skewed to the left, the mean is often less than the median. When the distribution is skewed to the right, the mean is often greater than the median. In symmetric distributions, we expect the mean and median to be approximately equal in value. This is an important connection between the shape of the distribution and the relationship of the mean and median. It is not, however, true for every data set. The most common exceptions occur in sets of discrete data.

As with the mean, median and mode, and as we will see shortly, the variance, there are mathematical formulas that give us precise measures of these characteristics of the distribution of the data. Again looking at the formula for skewness we see that this is a relationship between the mean of the data and the individual observations cubed.

$$a_3 = \sum \frac{\left(x_i - \bar{x}\right)^3}{ns^3}$$

where ss is the sample standard deviation of the data, X_i, and $\bar{x}$ is the arithmetic mean and nn is the sample size.

Formally the arithmetic mean is known as the first moment of the distribution. The second moment we will see is the variance, and skewness is the third moment. The variance measures the squared differences of the data from the mean and skewness measures the cubed differences of the data from the mean. While a variance can never be a negative number, the measure of skewness can and this is how we determine if the data are skewed right of left. The skewness for a normal distribution is zero, and any symmetric data should have skewness near zero. Negative values for the skewness indicate data that are skewed left and positive values for the skewness indicate data that are skewed right. By skewed left, we mean that the left tail is long relative to the right tail. Similarly, skewed right means that the right tail is long relative to the left tail. The skewness characterizes the degree of asymmetry of a distribution around its mean. While the mean and standard deviation are *dimensional* quantities (this is why we will take the square root of the variance) that is, have the same units as the measured quantities X_i, the skewness is conventionally defined in such a way as to make it *nondimensional*. It is a pure number that characterizes only the shape of the distribution. A positive value of skewness signifies a distribution with an asymmetric tail extending out towards more positive X and a negative value signifies a distribution whose tail extends out towards more negative X. A zero measure of skewness will indicate a symmetrical distribution.

Skewness and symmetry become important when we discuss probability distributions in later chapters.

2.7 MEASURES OF THE SPREAD OF THE DATA

An important characteristic of any set of data is the variation in the data. In some data sets, the data values are concentrated closely near the mean; in other data sets, the data values are more widely spread out from the mean. The most common measure of variation, or spread, is the standard deviation. The standard deviation is a number that measures how far data values are from their mean.

The standard deviation

- provides a numerical measure of the overall amount of variation in a data set, and
- can be used to determine whether a particular data value is close to or far from the mean.

The standard deviation provides a measure of the overall variation in a data set

The standard deviation is always positive or zero. The standard deviation is small when the data are all concentrated close to the mean, exhibiting little variation or spread. The standard deviation is larger when the data values are more spread out from the mean, exhibiting more variation.

Suppose that we are studying the amount of time customers wait in line at the checkout at supermarket *A* and supermarket *B*. The average wait time at both supermarkets is five minutes. At supermarket *A*, the standard deviation for the wait time is two minutes; at supermarket *B*. The standard deviation for the wait time is four minutes.

Because supermarket *B* has a higher standard deviation, we know that there is more variation in the wait times at supermarket *B*. Overall, wait times at supermarket *B* are more spread out from the average; wait times at supermarket *A* are more concentrated near the average.

Calculating the Standard Deviation

If *x* is a number, then the difference "*x* minus the mean" is called its deviation. In a data set, there are as many deviations as there are items in the data set. The deviations are used to calculate the standard deviation. If the numbers belong to a population, in symbols a deviation is $x - \mu$. For sample data, in symbols a deviation is $x - \bar{x}$.

The procedure to calculate the standard deviation depends on whether the numbers are the entire population or are data from a sample. The calculations are similar, but not identical. Therefore the symbol used to represent the standard deviation depends on whether it is calculated from a population or a sample. The lower case letter s represents the sample standard deviation and the Greek letter σ (sigma, lower case) represents the population standard deviation. If the sample has the same characteristics as the population, then s should be a good estimate of σ.

To calculate the standard deviation, we need to calculate the variance first. The variance is the average of the squares of the deviations (the $x - \bar{x}$ values for a sample, or the $x - \mu$ values for a population). The symbol σ^2 represents the population variance; the population standard deviation σ is the square root of the population variance. The symbol s^2 represents the sample variance; the sample standard deviation s is the square root of the sample variance. You can think of the standard deviation as a special average of the deviations. Formally, the variance is the second moment of the distribution or the first moment around the mean. Remember that the mean is the first moment of the distribution.

If the numbers come from a census of the entire population and not a sample, when we calculate the average of the squared deviations to find the variance, we divide by N, the number of items in the population. If the data are from a sample rather than a population, when we calculate the average of the squared deviations, we divide by n – 1, one less than the number of items in the sample.

Formulas for the Sample Standard Deviation

- $s = \sqrt{\dfrac{\Sigma(x - \bar{x})^2}{n - 1}}$ or $s = \sqrt{\dfrac{\Sigma f(x - \bar{x})^2}{n - 1}}$ or $s = \sqrt{\dfrac{\left(\sum_{i=1}^{n} x^2\right) - n\bar{x}^2}{n - 1}}$
- For the sample standard deviation, the denominator is n - 1, that is the sample size minus 1.

Formulas for the Population Standard Deviation

- $\sigma = \sqrt{\dfrac{\Sigma(x - \mu)^2}{N}}$ or $\sigma = \sqrt{\dfrac{\Sigma f(x - \mu)^2}{N}}$ or $\sigma = \sqrt{\dfrac{\sum_{i=1}^{N} x_i^2}{N} - \mu^2}$
- For the population standard deviation, the denominator is N, the number of items in the population.

In these formulas, f represents the frequency with which a value appears. For example, if a value appears once, f is one. If a value appears three times in the data set or population, f is three. Two important observations concerning the variance and standard deviation: the deviations are measured from the mean and the deviations are squared. In principle, the deviations could be measured from any point, however, our interest is measurement from the center weight of the data, what is the "normal" or most usual value of the observation. Later we will be trying to measure the "unusualness" of an observation or a sample mean and thus we need a measure from the mean. The second observation is that the deviations are squared. This does two things, first it makes the deviations all positive and second it changes the units of measurement from that of the mean and the original observations. If the data are weights then the mean is measured in pounds, but the variance is measured in pounds-squared. One reason to use the standard deviation is to return to the original units of measurement by taking the square root of the variance. Further, when the deviations are squared it explodes their value. For example, a deviation of 10 from the mean when squared is 100, but a deviation of 100 from the mean is 10,000. What this does is place great weight on outliers when calculating the variance.

Types of Variability in Samples

When trying to study a population, a sample is often used, either for convenience or because it is not possible to access the entire population. Variability is the term used to describe the differences that may occur in these outcomes. Common types of variability include the following:

- Observational or measurement variability
- Natural variability
- Induced variability
- Sample variability

Here are some examples to describe each type of variability.

Example 1: Measurement variability: Measurement variability occurs when there are differences in the instruments used to measure or in the people using those instruments. If we are gathering data on how long it takes for a ball to drop from a height by having students measure the time of the drop with a stopwatch, we may experience measurement variability if the two stopwatches used were made by different manufacturers: For example, one stopwatch measures to the nearest second, whereas the other one measures to the nearest tenth of a second. We also may experience measurement variability because two different people are gathering the data. Their reaction times in pressing the button on the stopwatch may differ; thus, the outcomes will vary accordingly. The differences in outcomes may be affected by measurement variability.

Example 2: Natural variability: Natural variability arises from the differences that naturally occur because members of a population differ from each other. For example, if we have two identical corn plants and we expose both plants to the same amount of water and sunlight, they may still grow at different rates simply because they are two different corn plants. The difference in outcomes may be explained by natural variability.

Example 3: Induced variability: Induced variability is the counterpart to natural variability; this occurs because we have artificially induced an element of variation (that, by definition, was not present naturally): For example, we assign people to two different groups to study memory, and we induce a variable in one group by limiting the amount of sleep they get. The difference in outcomes may be affected by induced variability.

Example 4: Sample variability Sample variability occurs when multiple random samples are taken from the same population. For example, if I conduct four surveys of 50 people randomly selected from a given population, the differences in outcomes may be affected by sample variability.

EXAMPLE 2.29

In a fifth grade class, the teacher was interested in the average age and the sample standard deviation of the ages of her students. The following data are the ages for a SAMPLE of $n = 20$ fifth grade students. The ages are rounded to the nearest half year:

9; 9.5; 9.5; 10; 10; 10; 10; 10.5; 10.5; 10.5; 10.5; 11; 11; 11; 11; 11; 11; 11.5; 11.5; 11.5;

$$\bar{x} = \frac{9 + 9.5(2) + 10(4) + 10.5(4) + 11(6) + 11.5(3)}{20} = 10.525$$

The average age is 10.53 years, rounded to two places.

The variance may be calculated by using a table. Then the standard deviation is calculated by taking the square root of the variance. We will explain the parts of the table after calculating *s*.

Table 2.28

Data	Freq.	Deviations	Deviations²	(Freq.)(Deviations²)
x	f	$(x - \bar{x})$	$(x - \bar{x})^2$	$(f)(x - \bar{x})^2$
9	1	9 – 10.525 = –1.525	$(-1.525)^2 = 2.325625$	1 × 2.325625 = 2.325625
9.5	2	9.5 – 10.525 = –1.025	$(-1.025)^2 = 1.050625$	2 × 1.050625 = 2.101250
10	4	10 – 10.525 = –0.525	$(-0.525)^2 = 0.275625$	4 × 0.275625 = 1.1025
10.5	4	10.5 – 10.525 = –0.025	$(-0.025)^2 = 0.000625$	4 × 0.000625 = 0.0025
11	6	11 – 10.525 = 0.475	$(0.475)^2 = 0.225625$	6 × 0.225625 = 1.35375
11.5	3	11.5 – 10.525 = 0.975	$(0.975)^2 = 0.950625$	3 × 0.950625 = 2.851875
				The total is 9.7375

The sample variance, s^2, is equal to the sum of the last column (9.7375) divided by the total number of data values minus one (20 – 1):

$$s^2 = \frac{9.7375}{20 - 1} = 0.5125$$

The sample standard deviation s is equal to the square root of the sample variance: $s = \sqrt{0.5125} = 0.715891$, which is rounded to two decimal places, $s = 0.72$.

Explanation of the standard deviation calculation shown in the table

The deviations show how spread out the data are about the mean. The data value 11.5 is farther from the mean than is the data value 11 which is indicated by the deviations 0.97 and 0.47. A positive deviation occurs when the data value is greater than the mean, whereas a negative deviation occurs when the data value is less than the mean. The deviation is –1.525 for the data value nine. If you add the deviations, the sum is always zero. (For Example 2.29, there are $n = 20$ deviations.) So you cannot simply add the deviations to get the spread of the data. By squaring the deviations, you make them positive numbers, and the sum will also be positive. The variance, then, is the average squared deviation. By squaring the deviations we are placing an extreme penalty on observations that are far from the mean; these observations get greater weight in the calculations of the variance. We will see later on that the variance (standard deviation) plays the critical role in determining our conclusions in inferential statistics. We can begin now by using the standard deviation as a measure of "unusualness." "How did you do on the test?" "Terrific!

Two standard deviations above the mean." This, we will see, is an unusually good exam grade.

The variance is a squared measure and does not have the same units as the data. Taking the square root solves the problem. The standard deviation measures the spread in the same units as the data.

Notice that instead of dividing by $n = 20$, the calculation divided by $n - 1 = 20 - 1 = 19$ because the data is a sample. For the sample variance, we divide by the sample size minus one ($n - 1$). Why not divide by n? The answer has to do with the population variance. The sample variance is an estimate of the population variance. This estimate requires us to use an estimate of the population mean rather than the actual population mean. Based on the theoretical mathematics that lies behind these calculations, dividing by ($n - 1$) gives a better estimate of the population variance.

The standard deviation, s or σ, is either zero or larger than zero. Describing the data with reference to the spread is called "variability". The variability in data depends upon the method by which the outcomes are obtained; for example, by measuring or by random sampling. When the standard deviation is zero, there is no spread; that is, the all the data values are equal to each other. The standard deviation is small when the data are all concentrated close to the mean, and is larger when the data values show more variation from the mean. When the standard deviation is a lot larger than zero, the data values are very spread out about the mean; outliers can make s or σ very large.

EXAMPLE 2.30

Use the following data (first exam scores) from Susan Dean's spring pre-calculus class:

33; 42; 49; 49; 53; 55; 55; 61; 63; 67; 68; 68; 69; 69; 72; 73; 74; 78; 80; 83; 88; 88; 88; 90; 92; 94; 94; 94; 94; 96; 100

a. Create a chart containing the data, frequencies, relative frequencies, and cumulative relative frequencies to three decimal places.
b. Calculate the following to one decimal place:
    i. The sample mean
    ii. The sample standard deviation
    iii. The median
    iv. The first quartile
    v. The third quartile
    vi. IQR

Solution 2.30

a. See **Table 2.29**

b. i. The sample mean = 73.5

ii. The sample standard deviation = 17.9

iii. The median = 73

iv. The first quartile = 61

v. The third quartile = 90

vi. IQR = 90 – 61 = 29

Table 2.29

Data	Frequency	Relative frequency	Cumulative relative frequency
33	1	0.032	0.032
42	1	0.032	0.064
49	2	0.065	0.129
53	1	0.032	0.161
55	2	0.065	0.226
61	1	0.032	0.258
63	1	0.032	0.29
67	1	0.032	0.322
68	2	0.065	0.387
69	2	0.065	0.452
72	1	0.032	0.484
73	1	0.032	0.516
74	1	0.032	0.548
78	1	0.032	0.580
80	1	0.032	0.612
83	1	0.032	0.644
88	3	0.097	0.741
90	1	0.032	0.773
92	1	0.032	0.805
94	4	0.129	0.934
96	1	0.032	0.966
100	1	0.032	0.998 (Why isn't this value 1? ANSWER: Rounding)

Standard deviation of Grouped Frequency Tables

Recall that for grouped data we do not know individual data values, so we cannot describe the typical value of the data with precision. In other words, we cannot find the exact mean, median, or mode. We can, however, determine the best estimate of the measures of center by finding the mean of the grouped data with the formula:

Mean of Frequency $Table = \frac{\sum fm}{\sum f}$ where f = interval frequencies and m = interval midpoints.

Just as we could not find the exact mean, neither can we find the exact standard deviation. Remember that standard deviation describes numerically the expected deviation a data value has from the mean. In simple English, the standard deviation allows us to compare how "unusual" individual data is compared to the mean.

Example 2.31

Find the standard deviation for the data in Table 2.30.

Table 2.30

Class	Frequency, f	Midpoint, m	f*m	$f(m-\bar{x})^2$
0–2	1	1	1*1=1	1(1-7.58)2=43.26
3–5	6	4	6*4=24	6(4-7.58)2=76.77
6-8	10	7	10*7=70	10(7-7.58)2=3.33
9-11	7	10	7*10=70	7(10-7.58)2=41.10
12-14	0	13	0*13=0	0(13-7.58)2=0
	26=n		$\bar{x} = \frac{197}{26} = 7.58$	$s^2 = \frac{306.35}{26-1} = 12.25$

For this data set, we have the mean $\bar{x} = 7.58$ and the standard deviation, sx = 3.5. This means that a randomly selected data value would be expected to be 3.5 units from the mean. If we look at the first class, we see that the class midpoint is equal to one. This is almost two full standard deviations from the mean since 7.58–3.5–3.5=0.58. While the formula for calculating the standard deviation is not complicated, $s_x = \sqrt{\frac{\Sigma(m-\bar{x})^2 f}{n-1}}$ where sx = sample standard deviation, $\bar{x}$ = sample mean, the calculations are tedious. It is usually best to use technology when performing the calculations.

Comparing Values from Different Data Sets

The standard deviation is useful when comparing data values that come from different data sets. If the data sets have different means and standard deviations, then comparing the data values directly can be misleading.

- For each data value x, calculate how many standard deviations away from its mean the value is.
- Use the formula: x = mean + (#ofSTDEVs)(standard deviation); solve for #ofSTDEVs.

- $\# ofSTDEVs = \frac{\text{x - mean}}{\text{standard deviation}}$
- Compare the results of this calculation.

#ofSTDEVs is often called a "z-score"; we can use the symbol *z*. In symbols, the formulas become:

Table 2.31

Sample	$x = \bar{x} + zs$	$z = \frac{x - \bar{x}}{s}$
Population	$x = \mu + z\sigma$	$z = \frac{x - \mu}{\sigma}$

EXAMPLE 2.32

Problem

Two students, John and Ali, from different high schools, wanted to find out who had the highest GPA when compared to his school. Which student had the highest GPA when compared to his school?

Table 2.32

Student	GPA	School mean GPA	School standard deviation
John	2.85	3.0	0.7
Ali	77	80	10

Solution 2.32

For each student, determine how many standard deviations (#ofSTDEVs) his GPA is away from the average, for his school. Pay careful attention to signs when comparing and interpreting the answer.

$$z = \text{\# of STDEVs} = \frac{\text{value - mean}}{\text{standard deviation}} = \frac{x - \mu}{\sigma}$$

$$\text{For John, } z = \# ofSTDEVs = \frac{2.85 - 3.0}{0.7} = -0.21$$

$$\text{For Ali, } z = \# ofSTDEVs = \frac{77 - 80}{10} = -0.3$$

John has the better GPA when compared to his school because his GPA is 0.21 standard deviations below his school's mean while Ali's GPA is 0.3 standard deviations below his school's mean.

John's z-score of –0.21 is higher than Ali's z-score of –0.3. For GPA, higher values are better, so we conclude that

John has the better GPA when compared to his school.

TRY IT 2.32

Two swimmers, Angie and Beth, from different teams, wanted to find out who had the fastest time for the 50 meter freestyle when compared to her team. Which swimmer had the fastest time when compared to her team?

Table 2.33

Swimmer	Time (seconds)	Team mean time	Team standard deviation
Angie	26.2	27.2	0.8
Beth	27.3	30.1	1.4

The following lists give a few facts that provide a little more insight into what the standard deviation tells us about the distribution of the data.

For ANY data set, no matter what the distribution of the data is:

- At least 75% of the data is within two standard deviations of the mean.
- At least 89% of the data is within three standard deviations of the mean.
- At least 95% of the data is within 4.5 standard deviations of the mean.
- This is known as Chebyshev's Rule.

For data having a Normal Distribution, which we will examine in great detail later:

- Approximately 68% of the data is within one standard deviation of the mean.
- Approximately 95% of the data is within two standard deviations of the mean.
- More than 99% of the data is within three standard deviations of the mean.
- This is known as the Empirical Rule.
- It is important to note that this rule only applies when the shape of the distribution of the data is bell-shaped and symmetric. We will learn more about this when studying the "Normal" or "Gaussian" probability distribution in later chapters.

Coefficient of Variation

Another useful way to compare distributions besides simple comparisons of means or standard deviations is to adjust for differences in the scale of the data being measured. Quite simply, a large variation in data with a large mean is different than the same variation in data with a small mean. To adjust for the scale of the underlying data the Coefficient of Variation (CV) has been developed. Mathematically:

$CV = \frac{s}{\bar{x}} * 100$ conditioned upon $\bar{x} \neq 0$, where s is the standard deviation of the data and $\bar{x}$ is the mean.

We can see that this measures the variability of the underlying data as a percentage of the mean value; the center weight of the data set. This measure is useful in comparing risk where an adjustment is warranted because of differences in scale of two data sets. In effect, the scale is changed to common scale, percentage differences, and allows direct comparison of the two or more magnitudes of variation of different data sets.

Terminology

Frequency the number of times a value of the data occurs

Frequency Table a data representation in which grouped data is displayed along with the corresponding frequencies

Histogram a graphical representation in x-y form of the distribution of data in a data set; x represents the data and y represents the frequency, or relative frequency. The graph consists of contiguous rectangles.

Interquartile Range or IQR, is the range of the middle 50 percent of the data values; the IQR is found by subtracting the first quartile from the third quartile.

Mean (arithmetic) a number that measures the central tendency of the data; a common name for mean is 'average.' The term 'mean' is a shortened form of 'arithmetic mean.' By definition, the mean for a sample (denoted by $\bar{x}$) is $\bar{x} = \frac{\text{Sum of all values in the sample}}{\text{Number of values in the sample}}$, and the mean for a population (denoted by μ) is $\mu = \frac{\text{Sum of all values in the population}}{\text{Number of values in the population}}$.

Mean (geometric) a measure of central tendency that provides a measure of average geometric growth over multiple time periods.

Median a number that separates ordered data into halves; half the values are the same number or smaller than the median and half the values are the same number or larger than the median. The median may or may not be part of the data.

Midpoint the mean of an interval in a frequency table

Mode the value that appears most frequently in a set of data

Outlier an observation that does not fit the rest of the data

Percentile a number that divides ordered data into hundredths; percentiles may or may not be part of the data. The median of the data is the second quartile and the 50th percentile. The first and third quartiles are the 25th and the 75th percentiles, respectively.

Quartiles the numbers that separate the data into quarters; quartiles may or may not be part of the data. The second quartile is the median of the data.

Relative Frequency the ratio of the number of times a value of the data occurs in the set of all outcomes to the number of all outcomes

Standard Deviation a number that is equal to the square root of the variance and measures how far data values are from their mean; notation: s for sample standard deviation and σ for population standard deviation.

Variance mean of the squared deviations from the mean, or the square of the standard deviation; for a set of data, a deviation can be represented as $x - \bar{x}$ where x is a value of the data and $\bar{x}$ is the sample mean. The sample variance is equal to the sum of the squares of the deviations divided by the difference of the sample size and one.

CHAPTER REVIEW

2.1 Display Data

A stem-and-leaf plot is a way to plot data and look at the distribution. In a stem-and-leaf plot, all data values within a class are visible. The advantage in a stem-and-leaf plot is that all values are listed, unlike a histogram, which gives classes of data values. A line graph is often used to represent a set of data values in which a quantity varies with time. These graphs are useful for finding trends. That is, finding a general pattern in data sets including temperature, sales, employment, company profit or cost over a period of time. A bar graph is a chart that uses either horizontal or vertical bars to show comparisons among categories. One axis of the chart shows the specific categories being compared, and the other axis represents a discrete value. Some bar graphs present bars clustered in groups of more than one (grouped bar graphs), and others show the bars divided into subparts to show cumulative effect (stacked bar graphs). Bar graphs are especially useful when categorical data is being used.

A histogram is a graphic version of a frequency distribution. The graph consists of bars of equal width drawn adjacent to each other. The horizontal scale represents classes of quantitative data values and the vertical scale represents frequencies. The heights of the bars correspond to frequency values. Histograms are typically used for large, continuous, quantitative data sets. A frequency polygon can also be used when graphing large data sets with data points that repeat. The data usually goes on y-axis with the frequency being graphed on the x-axis. Time series graphs can be helpful when looking at large amounts of data for one variable over a period of time.

2.2 Measures of the Location of the Data

The values that divide a rank-ordered set of data into 100 equal parts are called percentiles. Percentiles are used to compare and interpret data. For example, an observation at the 50th percentile would be greater than 50 percent of the other observations in the set. Quartiles divide data into quarters. The first quartile (Q_1) is the 25th percentile,the second quartile (Q_2 or median) is 50th percentile, and the third quartile (Q_3) is the 75th percentile. The interquartile range, or *IQR*, is the range of the middle 50 percent of the data values. The *IQR* is found by subtracting Q_1 from Q_3, and can help determine outliers by using the following two expressions.

- $Q_3 + IQR(1.5)$
- $Q_1 - IQR(1.5)$

2.3 Measures of the Center of the Data

The mean and the median can be calculated to help you find the "center" of a data set. The mean is the best estimate for the actual data set, but the median is the best measurement when a data set contains several outliers or extreme values. The mode will tell you the most frequently occurring datum (or data) in your data set. The mean, median, and mode are extremely helpful when you need to analyze your data, but if your data set consists of ranges which lack specific values, the mean may seem impossible to calculate. However, the mean can be approximated if you add the lower boundary with the upper boundary and divide by two to find the midpoint of each interval. Multiply each midpoint by the number of values found in the corresponding range. Divide the sum of these values by the total number of data values in the set.

2.6 Skewness and the Mean, Median, and Mode

Looking at the distribution of data can reveal a lot about the relationship between the mean, the median, and the mode. There are three types of distributions. A left (or negative) skewed distribution has a shape like Figure 2.12. A right (or positive) skewed distribution has a shape like Figure 2.13. A symmetrical distribution looks like Figure 2.11.

2.7 Measures of the Spread of the Data

The standard deviation can help you calculate the spread of data. There are different equations to use if are calculating the standard deviation of a sample or of a population.

- The Standard Deviation allows us to compare individual data or classes to the data set mean numerically.
- $s = \sqrt{\frac{\sum (x - \bar{x})^2}{n - 1}}$ or $s = \sqrt{\frac{\sum f(x - \bar{x})^2}{n - 1}}$ is the formula for calculating the

standard deviation of a sample. To calculate the standard deviation of a population, we would use the population mean, μ, and the formula $\sigma = \sqrt{\frac{\sum (x-\mu)^2}{N}}$ or $\sigma = \sqrt{\frac{\sum f(x-\mu)^2}{N}}$.

FORMULA REVIEW

2.2 Measures of the Location of the Data

$$i = \left(\frac{k}{100}\right)(n+1)$$

where i = the ranking or position of a data value,

k = the kth percentile,

n = total number of data.

Expression for finding the percentile of a data value: $\left(\frac{x + 0.5y}{n}\right)(100)$

where x = the number of values counting from the bottom of the data list up to but not including the data value for which you want to find the percentile,

y = the number of data values equal to the data value for which you want to find the percentile,

n = total number of data

2.3 Measures of the Center of the Data

$\mu = \frac{\sum fm}{\sum f}$ Where f = interval frequencies and m = interval midpoints.

The arithmetic mean for a sample (denoted by $\bar{x}$) $\bar{x} = \frac{\text{Sum of all values in the sample}}{\text{Number of values in the sample}}$

The arithmetic mean for a population (denoted by μ) $\mu = \frac{\text{Sum of all values in the population}}{\text{Number of values in the population}}$

2.5 Geometric Mean

The Geometric Mean: $\tilde{x} = \left(\prod_{i=1}^{n} x_i\right)^{\frac{1}{n}} = \sqrt[n]{x_1 * x_2 \cdots x_n} = (x_1 * x_2 \cdots x_n)^{\frac{1}{n}}$

2.6 Skewness and the Mean, Median, and Mode

Formula for skewness: $a_3 = \sum \frac{(x_i - \bar{x})^3}{ns^3}$

Formula for Coefficient of Variation: $CV = \frac{s}{\bar{x}} * 100$ conditioned upon $\bar{x} \neq 0$

2.7 Measures of the Spread of the Data

$s_x = \sqrt{\frac{\sum fm^2}{n} - \bar{x}^2}$ where s_x = sample standard deviation $\bar{x}$ = sample means

Formulas for Sample Standard Deviation $s = \sqrt{\frac{\Sigma(x - \bar{x})^2}{n - 1}}$ or $s = \sqrt{\frac{\Sigma f(x - \bar{x})^2}{n - 1}}$

$s = \sqrt{\frac{\left(\sum_{i=1}^{n} x^2\right) - n\bar{x}^2}{n - 1}}$ For the sample standard deviation, the denominator is *n* - **1**, that is the sample size - 1.

Formulas for Population Standard $\sigma = \sqrt{\frac{\Sigma(x - \mu)^2}{N}}$ or $\sigma = \sqrt{\frac{\Sigma f(x - \mu)^2}{N}}$ or $\sigma = \sqrt{\frac{\sum_{i=1}^{N} x_i^2}{N} - \mu^2}$

For the population standard deviation, the denominator is *N*, the number of items in the population.

PRACTICE

2.1 Display Data

For the next three exercises, use the data to construct a line graph.

1. In a survey, 40 people were asked how many times they visited a store before making a major purchase. The results are shown in Table 2.34.

Table 2.34

Number of times in store	Frequency
1	4
2	10
3	16
4	6
5	4

2. In a survey, several people were asked how many years it has been since they purchased a mattress. The results are shown in **Table 2.35.**

Table 2.35

Years since last purchase	Frequency
0	2
1	8
2	13
3	22
4	16
5	9

3. Several children were asked how many TV shows they watch each day. The results of the survey are shown in Table 2.36.

Table 2.36

Number of TV shows	Frequency
0	12
1	18

Number of TV shows	Frequency
2	36
3	7
4	2

4. The students in Ms. Ramirez's math class have birthdays in each of the four seasons. Table 2.37 shows the four seasons, the number of students who have birthdays in each season, and the percentage (%) of students in each group. Construct a bar graph showing the number of students.

Table 2.37

Seasons	Number of students	Proportion of population
Spring	8	24%
Summer	9	26%
Autumn	11	32%
Winter	6	18%

5. Using the data from Mrs. Ramirez's math class supplied in Exercise 2.4, construct a bar graph showing the percentages.

6. David County has six high schools. Each school sent students to participate in a county-wide science competition. Table 2.38 shows the percentage breakdown of competitors from each school, and the percentage of the entire student population of the county that goes to each school. Construct a bar graph that shows the population percentage of competitors from each school.

Table 2.38

High school	Science competition population	Overall student population
Alabaster	28.9%	8.6%
Concordia	7.6%	23.2%
Genoa	12.1%	15.0%
Mocksville	18.5%	14.3%
Tynneson	24.2%	10.1%
West End	8.7%	28.8%

7. Use the data from the David County science competition supplied in **Exercise 2.6.** Construct a bar graph that shows the county-wide population percentage of students at each school.

8. Sixty-five randomly selected car salespersons were asked the number of cars they generally sell in one week. Fourteen people answered that they generally sell three cars; nineteen generally sell four cars; twelve generally sell five cars; nine generally sell six cars; eleven generally sell seven cars. Complete the table.

Table 2.39

Data Value (# cars)	Frequency	Relative Frequency	Cumulative Relative Frequency

9. What does the frequency column in Table 2.39 sum to? Why?
10. What does the relative frequency column in Table 2.39 sum to? Why?
11. What is the difference between relative frequency and frequency for each data value in Table 2.39?
12. What is the difference between cumulative relative frequency and relative frequency for each data value?
13. To construct the histogram for the data in Table 2.39, determine appropriate minimum and maximum *x* and *y* values and the scaling. Sketch the histogram. Label the horizontal and vertical axes with words. Include numerical scaling.

Figure 2.14

14. Construct a frequency polygon for the following:

    a.

Table 2.40

Pulse rates for women	Frequency
60–69	12
70–79	14
80–89	11
90–99	1
100–109	1
110–119	0
120–129	1

b.

Table 2.41

Actual speed in a 30 MPH zone	Frequency
42–45	25
46–49	14
50–53	7
54–57	3
58–61	1

c.

Table 2.42

Tar (mg) in nonfiltered cigarettes	Frequency
10–13	1
14–17	0
18–21	15
22–25	7
26–29	2

15. Construct a frequency polygon from the frequency distribution for the 50 highest ranked countries for depth of hunger.

Table 2.43

Depth of hunger	Frequency
230–259	21
260–289	13
290–319	5
320–349	7
350–379	1
380–409	1
410–439	1

16. Use the two frequency tables to compare the life expectancy of men and women from 20 randomly selected countries. Include an overlayed frequency polygon and discuss the shapes of the distributions, the center, the spread, and any outliers. What can we conclude about the life expectancy of women compared to men?

Table 2.44

Life expectancy at birth – women	Frequency
49–55	3
56–62	3
63–69	1

Life expectancy at birth - women	Frequency
70–76	3
77–83	8
84–90	2

Table 2.45

Life expectancy at birth - men	Frequency
49–55	3
56–62	3
63–69	1
70–76	1
77–83	7
84–90	5

17. Construct a times series graph for (a) the number of male births, (b) the number of female births, and (c) the total number of births.

Table 2.46

Sex/ Year	1855	1856	1857	1858	1859	1860	1861
Female	45,545	49,582	50,257	50,324	51,915	51,220	52,403
Male	47,804	52,239	53,158	53,694	54,628	54,409	54,606
Total	93,349	101,821	103,415	104,018	106,543	105,629	107,009

Table 2.47

Sex/ Year	1862	1863	1864	1865	1866	1867	1868	1869
Female	51,812	53,115	54,959	54,850	55,307	55,527	56,292	55,033
Male	55,257	56,226	57,374	58,220	58,360	58,517	59,222	58,321
Total	107,069	109,341	112,333	113,070	113,667	114,044	115,514	113,354

Table 2.48

Sex/Year	1870	1871	1872	1873	1874	1875
Female	56,431	56,099	57,472	58,233	60,109	60,146
Male	58,959	60,029	61,293	61,467	63,602	63,432
Total	115,390	116,128	118,765	119,700	123,711	123,578

18. The following data sets list full time police per 100,000 citizens along with homicides per 100,000 citizens for the city of Detroit, Michigan during the period from 1961 to 1973.

Table 2.49

Year	1961	1962	1963	1964	1965	1966	1967
Police	260.35	269.8	272.04	272.96	272.51	261.34	268.89
Homicides	8.6	8.9	8.52	8.89	13.07	14.57	21.36

Table 2.50

Year	1968	1969	1970	1971	1972	1973
Police	295.99	319.87	341.43	356.59	376.69	390.19
Homicides	28.03	31.49	37.39	46.26	47.24	52.33

a. Construct a double time series graph using a common *x*-axis for both sets of data.

b. Which variable increased the fastest? Explain.

c. Did Detroit's increase in police officers have an impact on the murder rate? Explain.

2.2 Measures of the Location of the Data

19. Listed are 29 ages for Academy Award winning best actors *in order from smallest to largest.*

18; 21; 22; 25; 26; 27; 29; 30; 31; 33; 36; 37; 41; 42; 47; 52; 55; 57; 58; 62; 64; 67; 69; 71; 72; 73; 74; 76; 77

a. Find the 40th percentile.

b. Find the 78th percentile.

20. Listed are 32 ages for Academy Award winning best actors *in order from smallest to largest.*

18; 18; 21; 22; 25; 26; 27; 29; 30; 31; 31; 33; 36; 37; 37; 41; 42; 47; 52; 55; 57; 58; 62; 64; 67; 69; 71; 72; 73; 74; 76; 77

a. Find the percentile of 37.

b. Find the percentile of 72.

21. Jesse was ranked 37th in his graduating class of 180 students. At what percentile is Jesse's ranking?

22. a. For runners in a race, a low time means a faster run. The winners in a race have the shortest running times. Is it more desirable to have a finish time with a high or a low percentile when running a race?

b. The 20th percentile of run times in a particular race is 5.2 minutes. Write a sentence interpreting the 20th percentile in the context of the situation.

c. A bicyclist in the 90th percentile of a bicycle race completed the race in 1 hour and 12 minutes. Is he among the fastest or slowest cyclists in the race? Write a sentence interpreting the 90th percentile in the context of the situation.

23. a. For runners in a race, a higher speed means a faster run. Is it more desirable to have a speed with a high or a low percentile when running a race?

b. The 40th percentile of speeds in a particular race is 7.5 miles per hour. Write a sentence interpreting the 40th percentile in the context of the situation.

24. On an exam, would it be more desirable to earn a grade with a high or low percentile? Explain.

25. Mina is waiting in line at the Department of Motor Vehicles (DMV). Her wait time of 32 minutes is the 85th percentile of wait times. Is that good or bad? Write a sentence interpreting the 85th percentile in the context of this situation.

26. In a survey collecting data about the salaries earned by recent college graduates, Li found that her salary was in the 78th percentile. Should Li be pleased or upset by this result? Explain.

27. In a study collecting data about the repair costs of damage to automobiles in a certain type of crash tests, a certain model of car had $1,700 in damage and was in the 90th percentile. Should the manufacturer and the consumer be pleased or upset by this result? Explain and write a sentence that interprets the 90th percentile in the context of this problem.

28. The University of California has two criteria used to set admission standards for freshman to be admitted to a college in the UC system:

a. Students' GPAs and scores on standardized tests (SATs and ACTs) are entered into a formula that calculates an "admissions index" score. The admissions index score is used to set eligibility standards intended to meet the goal of admitting the top 12% of high school students in the state. In this context, what percentile does the top 12% represent?

b. Students whose GPAs are at or above the 96th percentile of all students at their high school are eligible (called eligible in the local context), even if they are not in the top 12% of all students in the state. What percentage of students from each high school are "eligible in the local context"?

29. Suppose that you are buying a house. You and your realtor have determined that the most expensive house you can afford is the 34th percentile. The 34th percentile of housing prices is $240,000 in the town you want to move to. In this town, can you afford 34% of the houses or 66% of the houses?

Use the following information to answer the next six exercises. Sixty-five randomly selected car salespersons were asked the number of cars they generally sell in one week. Fourteen people answered that they generally sell three cars; nineteen generally sell four cars; twelve generally sell five cars; nine generally sell six cars; eleven generally sell seven cars.

30. First quartile = _______

31. Second quartile = median = 50th percentile = _______

32. Third quartile = _______

33. Interquartile range (*IQR*) = _____ – _____ = _____

34. 10th percentile = _______

35. 70th percentile = _______

2.3 Measures of the Center of the Data

36. Find the mean for the following frequency tables.

Table 2.51

a.

Grade	Frequency
49.5–59.5	2
59.5–69.5	3
69.5–79.5	8
79.5–89.5	12
89.5–99.5	5

b.

Table 2.52

Daily low temperature	Frequency
49.5–59.5	53
59.5–69.5	32
69.5–79.5	15
79.5–89.5	1
89.5–99.5	0

c.

Table 2.53

Points per game	Frequency
49.5–59.5	14
59.5–69.5	32
69.5–79.5	15
79.5–89.5	23
89.5–99.5	2

Use the following information to answer the next three exercises: The following data show the lengths of boats moored in a marina. The data are ordered from smallest to largest: 16; 17; 19; 20; 20; 21; 23; 24; 25; 25; 25; 26; 26; 27; 27; 27; 28; 29; 30; 32; 33; 33; 34; 35; 37; 39; 40

37. Calculate the mean.

38. Identify the median.

39. Identify the mode.

Use the following information to answer the next three exercises: Sixty-five randomly selected car salespersons were asked the number of cars they generally sell in one week. Fourteen people answered that they generally sell three cars; nineteen generally sell four cars; twelve generally sell five cars; nine generally sell six cars; eleven generally sell seven cars. Calculate the following:

40. sample mean = $\bar{x}$ = _______

41. median = _______

42. mode = _______

2.4 Sigma Notation and Calculating the Arithmetic Mean

43. A group of 10 children are on a scavenger hunt to find different color rocks. The results are shown in the Table 2.54 below. The column on the right shows the number of colors of rocks each child has. What is the mean number of rocks?

Table 2.54

Child	Rock colors
1	5
2	5
3	6
4	2
5	4
6	3
7	7
8	2
9	1
10	10

44. A group of children are measured to determine the average height of the group. The results are in Table 2.55 below. What is the mean height of the group to the nearest hundredth of an inch?

Table 2.55

Child	Height in inches
Adam	45.21
Betty	39.45
Charlie	43.78
Donna	48.76
Earl	37.39
Fran	39.90
George	45.56
Heather	46.24

45. A person compares prices for five automobiles. The results are in Table 2.56. What is the mean price of the cars the person has considered?

Table 2.56

Price
$20,987
$22,008
$19,998
$23,433
$21,444

46. A customer protection service has obtained 8 bags of candy that are supposed to contain 16 ounces of candy each. The candy is weighed to determine if the average weight is at least the claimed 16 ounces. The results are in given in Table 2.57. What is the mean weight of a bag of candy in the sample?

Table 2.57

Weight in ounces
15.65
16.09
16.01
15.99
16.02
16.00
15.98
16.08

47. A teacher records grades for a class of 70, 72, 79, 81, 82, 82, 83, 90, and 95. What is the mean of these grades?

48. A family is polled to see the mean of the number of hours per day the television set is on. The results, starting with Sunday, are 6, 3, 2, 3, 1, 3, and 7 hours. What is the average number of hours the family had the television set on to the nearest whole number?

49. A city received the following rainfall for a recent year. What is the mean number of inches of rainfall the city received monthly, to the nearest hundredth of an inch? Use Table 2.58.

Table 2.58

Month	Rainfall in inches
January	2.21
February	3.12
March	4.11
April	2.09
May	0.99
June	1.08
July	2.99
August	0.08
September	0.52
October	1.89
November	2.00
December	3.06

50. A football team scored the following points in its first 8 games of the new season. Starting at game 1 and in order the scores are 14, 14, 24, 21, 7, 0, 38, and 28. What is the mean number of points the team scored in these eight games?

2.5 Geometric Mean

51. What is the geometric mean of the data set given? 5, 10, 20

52. What is the geometric mean of the data set given? 9.000, 15.00, 21.00

53. What is the geometric mean of the data set given? 7.0, 10.0, 39.2

54. What is the geometric mean of the data set given? 17.00, 10.00, 19.00

55. What is the average rate of return for the values that follow? 1.0, 2.0, 1.5

56. What is the average rate of return for the values that follow? 0.80, 2.0, 5.0

57. What is the average rate of return for the values that follow? 0.90, 1.1, 1.2

58. What is the average rate of return for the values that follow? 4.2, 4.3, 4.5

2.6 Skewness and the Mean, Median, and Mode

Use the following information to answer the next three exercises: State whether the data are symmetrical, skewed to the left, or skewed to the right.

59. 1; 1; 1; 2; 2; 2; 2; 3; 3; 3; 3; 3; 3; 3; 3; 4; 4; 4; 5; 5

60. 16; 17; 19; 22; 22; 22; 22; 22; 23

61. 87; 87; 87; 87; 87; 88; 89; 89; 90; 91

62. When the data are skewed left, what is the typical relationship between the mean and median?

63. When the data are symmetrical, what is the typical relationship between the mean and median?

64. What word describes a distribution that has two modes?

65. Describe the shape of this distribution.

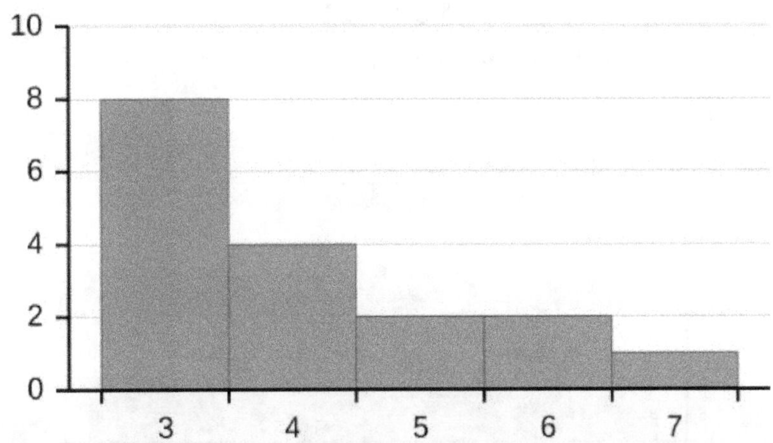

Figure 2.15

66. Describe the relationship between the mode and the median of this distribution.

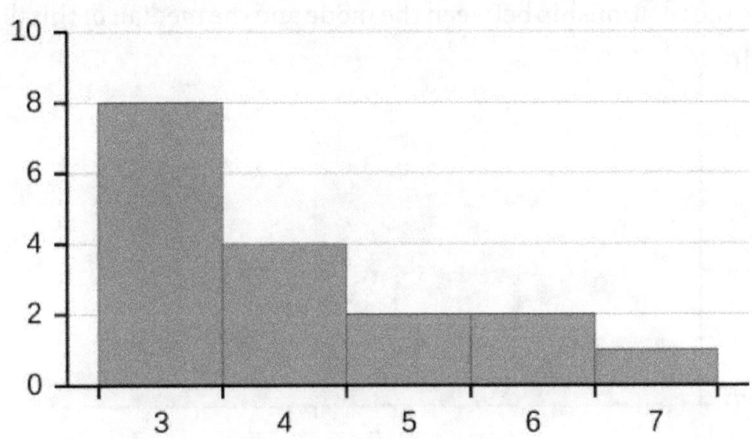

Figure 2.16

67. Describe the relationship between the mean and the median of this distribution.

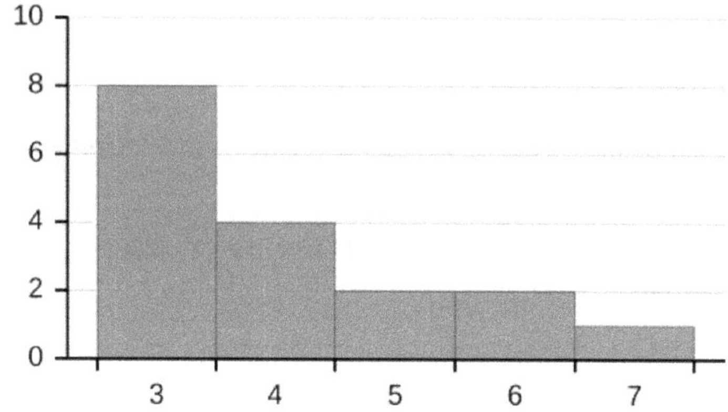

Figure 2.17

68. Describe the shape of this distribution.

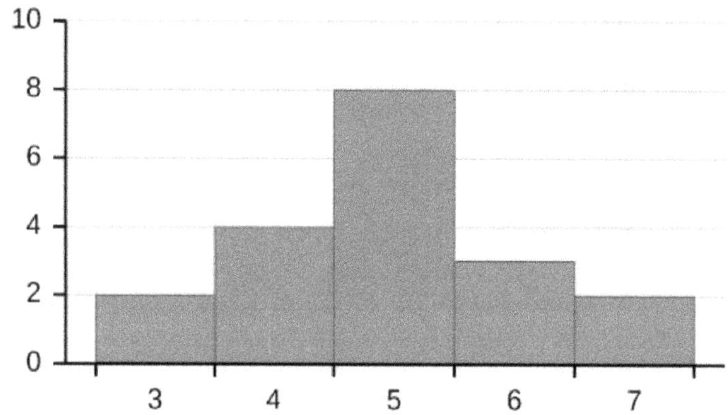

Figure 2.18

69. Describe the relationship between the mode and the median of this distribution.

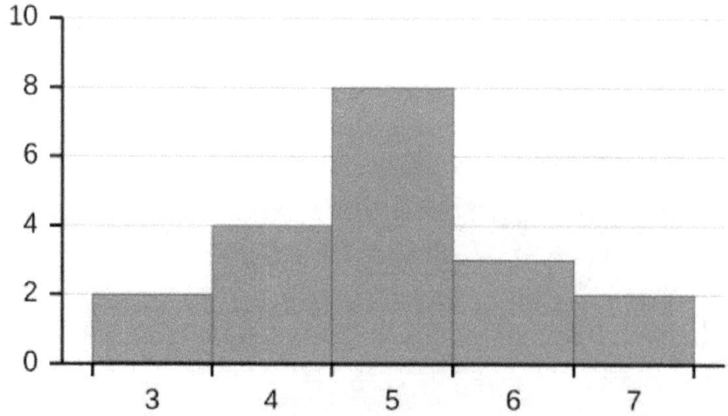

Figure 2.19

70. Are the mean and the median the exact same in this distribution? Why or why not?

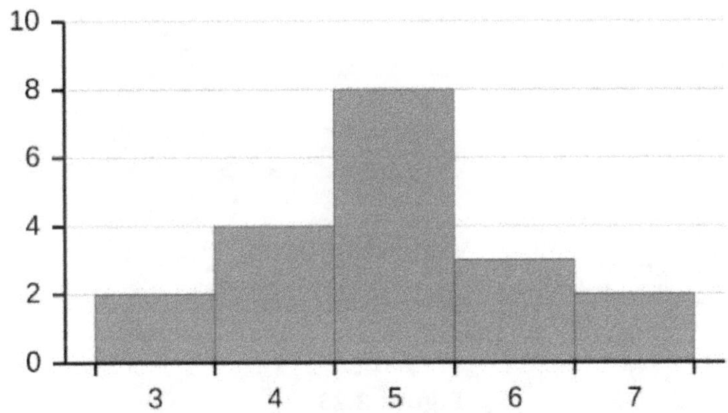

Figure 2.20

71. Describe the shape of this distribution.

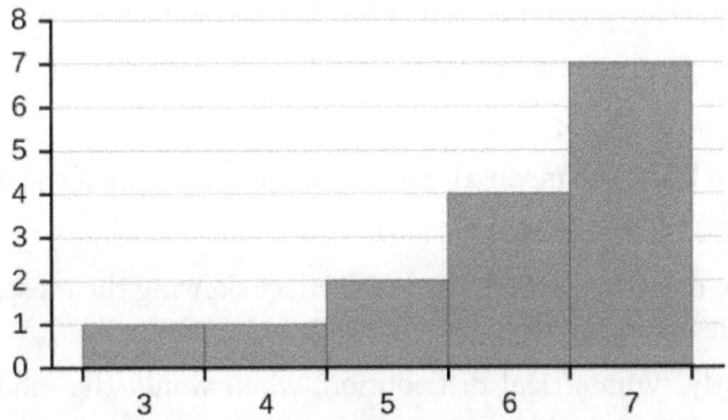

Figure 2.21

72. Describe the relationship between the mode and the median of this distribution.

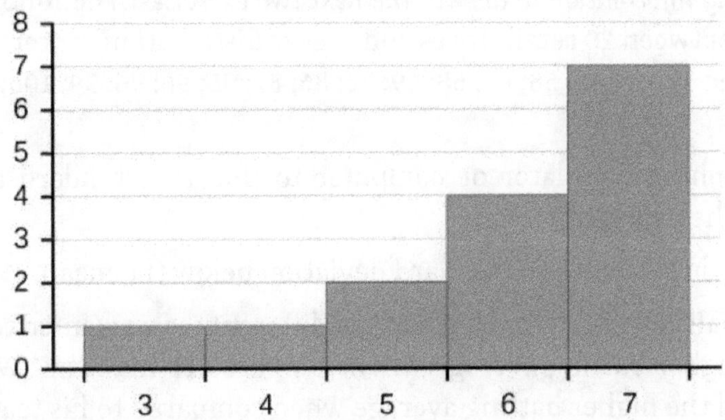

Figure 2.22

73. Describe the relationship between the mean and the median of this distribution.

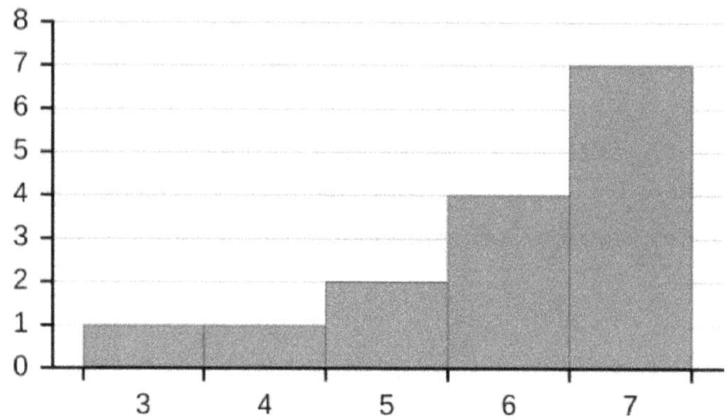

Figure 2.23

74. The mean and median for the data are the same.

3; 4; 5; 5; 6; 6; 6; 6; 7; 7; 7; 7; 7; 7; 7

Is the data perfectly symmetrical? Why or why not?

75. Which is the greatest, the mean, the mode, or the median of the data set?

11; 11; 12; 12; 12; 12; 13; 15; 17; 22; 22; 22

76. Which is the least, the mean, the mode, and the median of the data set?

56; 56; 56; 58; 59; 60; 62; 64; 64; 65; 67

77. Of the three measures, which tends to reflect skewing the most, the mean, the mode, or the median? Why?

78. In a perfectly symmetrical distribution, when would the mode be different from the mean and median?

2.7 Measures of the Spread of the Data

Use the following information to answer the next two exercises: The following data are the distances between 20 retail stores and a large distribution center. The distances are in miles. 29; 37; 38; 40; 58; 67; 68; 69; 76; 86; 87; 95; 96; 96; 99; 106; 112; 127; 145; 150

79. Use a graphing calculator or computer to find the standard deviation and round to the nearest tenth.

80. Find the value that is one standard deviation below the mean.

81. Two baseball players, Fredo and Karl, on different teams wanted to find out who had the higher batting average when compared to his team. Which baseball player had the higher batting average when compared to his team?

Table 2.59

Baseball player	Batting average	Team batting average	Team standard deviation
Fredo	0.158	0.166	0.012
Karl	0.177	0.189	0.015

82. Use **Table 2.59** to find the value that is three standard deviations:

a. above the mean b. below the mean

Find the standard deviation for the following frequency tables using the formula. Check the calculations with the TI 83/84.

83. Find the standard deviation for the following frequency tables using the formula. Check the calculations with the TI 83/84.

a.

Table 2.60

Grade	Frequency
49.5–59.5	2
59.5–69.5	3
69.5–79.5	8
79.5–89.5	12
89.5–99.5	5

b.

Table 2.61

Daily low temperature	Frequency
49.5–59.5	53
59.5–69.5	32
69.5–79.5	15
79.5–89.5	1
89.5–99.5	0

c.

Table 2.62

Points per game	Frequency
49.5–59.5	14
59.5–69.5	32
69.5–79.5	15
79.5–89.5	23
89.5–99.5	2

Homework

2.1 Display Data

84. **Table 2.63** contains the 2010 obesity rates in U.S. states and Washington, DC.

Table 2.63

State	Percent (%)	State	Percent (%)	State	Percent (%)
Alabama	32.2	Kentucky	31.3	North Dakota	27.2
Alaska	24.5	Louisiana	31.0	Ohio	29.2
Arizona	24.3	Maine	26.8	Oklahoma	30.4
Arkansas	30.1	Maryland	27.1	Oregon	26.8
California	24.0	Massachusetts	23.0	Pennsylvania	28.6
Colorado	21.0	Michigan	30.9	Rhode Island	25.5
Connecticut	22.5	Minnesota	24.8	South Carolina	31.5
Delaware	28.0	Mississippi	34.0	South Dakota	27.3
Washington, DC	22.2	Missouri	30.5	Tennessee	30.8
Florida	26.6	Montana	23.0	Texas	31.0
Georgia	29.6	Nebraska	26.9	Utah	22.5
Hawaii	22.7	Nevada	22.4	Vermont	23.2
Idaho	26.5	New Hampshire	25.0	Virginia	26.0
Illinois	28.2	New Jersey	23.8	Washington	25.5
Indiana	29.6	New Mexico	25.1	West Virginia	32.5
Iowa	28.4	New York	23.9	Wisconsin	26.3
Kansas	29.4	North Carolina	27.8	Wyoming	25.1

a. Use a random number generator to randomly pick eight states. Construct a bar graph of the obesity rates of those eight states.

b. Construct a bar graph for all the states beginning with the letter "A."

c. Construct a bar graph for all the states beginning with the letter "M."

85. Suppose that three book publishers were interested in the number of fiction paperbacks adult consumers purchase per month. Each publisher conducted a survey. In the survey, adult consumers were asked the number of fiction paperbacks they had purchased the previous month. The results are as follows:

Table 2.64 Publisher A

# of books	Freq.	Rel. freq.
0	10	
1	12	
2	16	
3	12	
4	8	
5	6	
6	2	
8	2	

Table 2.65 Publisher B

# of books	Freq.	Rel. freq.
0	18	
1	24	
2	24	
3	22	
4	15	
5	10	
7	5	
9	1	

Table 2.66 Publisher C

# of books	Freq.	Rel. freq.
0–1	20	
2–3	35	
4–5	12	
6–7	2	
8–9	1	

a. Find the relative frequencies for each survey. Write them in the charts.
b. Use the frequency column to construct a histogram for each publisher's survey. For Publishers A and B, make bar widths of one. For Publisher C, make bar widths of two.
c. In complete sentences, give two reasons why the graphs for Publishers A and B are not identical.
d. Would you have expected the graph for Publisher C to look like the other two graphs? Why or why not?

e. Make new histograms for Publisher A and Publisher B. This time, make bar widths of two.

f. Now, compare the graph for Publisher C to the new graphs for Publishers A and B. Are the graphs more similar or more different? Explain your answer.

86. Often, cruise ships conduct all on-board transactions, with the exception of gambling, on a cashless basis. At the end of the cruise, guests pay one bill that covers all onboard transactions. Suppose that 60 single travelers and 70 couples were surveyed as to their on-board bills for a seven-day cruise from Los Angeles to the Mexican Riviera. Following is a summary of the bills for each group.

Table 2.67 Singles

Amount($)	Frequency	Rel. frequency
51–100	5	
101–150	10	
151–200	15	
201–250	15	
251–300	10	
301–350	5	

Table 2.68 Couples

Amount($)	Frequency	Rel. frequency
100–150	5	
201–250	5	
251–300	5	
301–350	5	
351–400	10	
401–450	10	
451–500	10	
501–550	10	
551–600	5	
601–650	5	

a. Fill in the relative frequency for each group.
b. Construct a histogram for the singles group. Scale the *x*-axis by $50 widths. Use relative frequency on the *y*-axis.
c. Construct a histogram for the couples group. Scale the *x*-axis by $50 widths. Use relative frequency on the *y*-axis.
d. Compare the two graphs:
   i. List two similarities between the graphs.
   ii. List two differences between the graphs.
   iii. Overall, are the graphs more similar or different?
e. Construct a new graph for the couples by hand. Since each couple is paying for two individuals, instead of scaling the *x*-axis by $50, scale it by $100. Use relative frequency on the *y*-axis.
f. Compare the graph for the singles with the new graph for the couples:
   i. List two similarities between the graphs.
   ii. Overall, are the graphs more similar or different?
g. How did scaling the couples graph differently change the way you compared it to the singles graph?
h. Based on the graphs, do you think that individuals spend the same amount, more or less, as singles as they do person by person as a couple? Explain why in one or two complete sentences.

87. Twenty-five randomly selected students were asked the number of movies they watched the previous week. The results are as follows.

Table 2.69

# of movies	Frequency	Relative frequency	Cumulative relative frequency
0	5		
1	9		
2	6		
3	4		
4	1		

a. Construct a histogram of the data.
b. Complete the columns of the chart.

Use the following information to answer the next two exercises: Suppose one hundred eleven people who shopped in a special t-shirt store were asked the number of t-shirts they own costing more than $19 each.

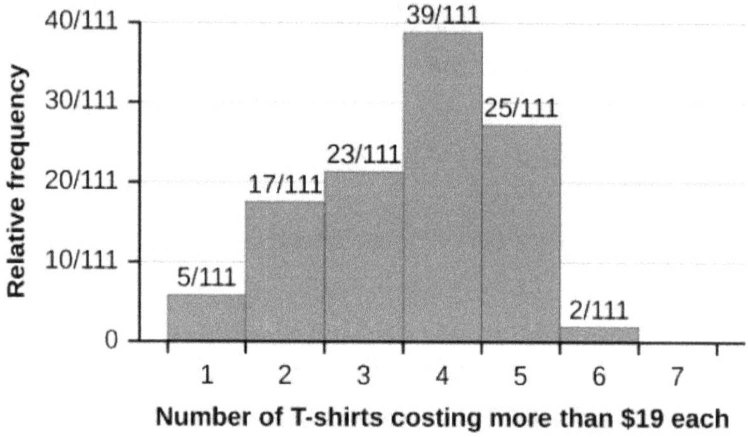

88. The percentage of people who own at most three t-shirts costing more than $19 each is approximately:

a. 21
b. 59
c. 41
d. Cannot be determined

89. If the data were collected by asking the first 111 people who entered the store, then the type of sampling is:

a. cluster
b. simple random
c. stratified
d. convenience

90. Following are the 2010 obesity rates by U.S. states and Washington, DC.

Table 2.70

State	Percent (%)	State	Percent (%)	State	Percent (%)
Alabama	32.2	Kentucky	31.3	North Dakota	27.2
Alaska	24.5	Louisiana	31.0	Ohio	29.2
Arizona	24.3	Maine	26.8	Oklahoma	30.4
Arkansas	30.1	Maryland	27.1	Oregon	26.8
California	24.0	Massachusetts	23.0	Pennsylvania	28.6
Colorado	21.0	Michigan	30.9	Rhode Island	25.5
Connecticut	22.5	Minnesota	24.8	South Carolina	31.5
Delaware	28.0	Mississippi	34.0	South Dakota	27.3

State	Percent (%)	State	Percent (%)	State	Percent (%)
Washington, DC	22.2	Missouri	30.5	Tennessee	30.8
Florida	26.6	Montana	23.0	Texas	31.0
Georgia	29.6	Nebraska	26.9	Utah	22.5
Hawaii	22.7	Nevada	22.4	Vermont	23.2
Idaho	26.5	New Hampshire	25.0	Virginia	26.0
Illinois	28.2	New Jersey	23.8	Washington	25.5
Indiana	29.6	New Mexico	25.1	West Virginia	32.5
Iowa	28.4	New York	23.9	Wisconsin	26.3
Kansas	29.4	North Carolina	27.8	Wyoming	25.1

Construct a bar graph of obesity rates of your state and the four states closest to your state. Hint: Label the *x*-axis with the states.

2.2 Measures of the Location of the Data

91. The median age for U.S. Black people currently is 30.9 years; for U.S. White people it is 42.3 years.

a. Based upon this information, give two reasons why the Black median age could be lower than the White median age.
b. Does the lower median age for Black people necessarily mean that Black people die younger than White people? Why or why not?
c. How might it be possible for Black people and White people to die at approximately the same age, but for the median age for White people to be higher?

92. Six hundred adult Americans were asked by telephone poll, "What do you think constitutes a middle-class income?" The results are in **Table 2.71**. Also, include left endpoint, but not the right endpoint.

Table 2.71

Salary ($)	Relative frequency
< 20,000	0.02
20,000–25,000	0.09
25,000–30,000	0.19
30,000–40,000	0.26
40,000–50,000	0.18
50,000–75,000	0.17
75,000–99,999	0.02
100,000+	0.01

a. What percentage of the survey answered "not sure"?
b. What percentage think that middle-class is from $25,000 to $50,000?
c. Construct a histogram of the data.
    i. Should all bars have the same width, based on the data? Why or why not?
    ii. How should the <20,000 and the 100,000+ intervals be handled? Why?
d. Find the 40th and 80th percentiles
e. Construct a bar graph of the data

2.3 Measures of the Center of the Data

93. The most obese countries in the world have obesity rates that range from 11.4% to 74.6%. This data is summarized in the following table.

Table 2.72

Percent of population obese	Number of countries
11.4–20.45	29
20.45–29.45	13
29.45–38.45	4
38.45–47.45	0
47.45–56.45	2
56.45–65.45	1
65.45–74.45	0
74.45–83.45	1

a. What is the best estimate of the average obesity percentage for these countries?
b. The United States has an average obesity rate of 33.9%. Is this rate above average or below?
c. How does the United States compare to other countries?

94. Table 2.73 gives the percent of children under five considered to be underweight. What is the best estimate for the mean percentage of underweight children?

Table 2.73

Percent of underweight children	Number of countries
16–21.45	23
21.45–26.9	4
26.9–32.35	9
32.35–37.8	7
37.8–43.25	6
43.25–48.7	1

2.4 Sigma Notation and Calculating the Arithmetic Mean

95. A sample of 10 prices is chosen from a population of 100 similar items. The values obtained from the sample, and the values for the population, are given in Table 2.74 and Table 2.75 respectively.

a. Is the mean of the sample within $1 of the population mean?

b. What is the difference in the sample and population means?

Table 2.74

Prices of the sample
$21
$23
$21
$24
$22
$22
$25
$21
$20
$24

Table 2.75

Prices of the population	Frequency
$20	20
$21	35
$22	15
$23	10
$24	18
$25	2

96. A standardized test is given to ten people at the beginning of the school year with the results given in Table 2.76 below. At the end of the year the same people were again tested.

a. What is the average improvement?

b. Does it matter if the means are subtracted, or if the individual values are subtr

acted?

Table 2.76

Student	Beginning score	Ending score
1	1100	1120
2	980	1030
3	1200	1208
4	998	1000

Student	Beginning score	Ending score
5	893	948
6	1015	1030
7	1217	1224
8	1232	1245
9	967	988
10	988	997

97. A small class of 7 students has a mean grade of 82 on a test. If six of the grades are 80, 82,86, 90, 90, and 95, what is the other grade?

98. A class of 20 students has a mean grade of 80 on a test. Nineteen of the students has a mean grade between 79 and 82, inclusive.

a. What is the lowest possible grade of the other student?
b. What is the highest possible grade of the other student?

99. If the mean of 20 prices is $10.39, and 5 of the items with a mean of $10.99 are sampled, what is the mean of the other 15 prices?

2.5 Geometric Mean

100. An investment grows from $10,000 to $22,000 in five years. What is the average rate of return?

101. An initial investment of $20,000 grows at a rate of 9% for five years. What is its final value?

102. A culture contains 1,300 bacteria. The bacteria grow to 2,000 in 10 hours. What is the rate at which the bacteria grow per hour to the nearest tenth of a percent?

103. An investment of $3,000 grows at a rate of 5% for one year, then at a rate of 8% for three years. What is the average rate of return to the nearest hundredth of a percent?

104. An investment of $10,000 goes down to $9,500 in four years. What is the average return per year to the nearest hundredth of a percent?

2.6 Skewness and the Mean, Median, and Mode

105. The median age of the U.S. population in 1980 was 30.0 years. In 1991, the median age was 33.1 years.

a. What does it mean for the median age to rise?
b. Give two reasons why the median age could rise.
c. For the median age to rise, is the actual number of children less in 1991 than it was in 1980? Why or why not?

2.7 Measures of the Spread of the Data

Use the following information to answer the next nine exercises: The population parameters below describe the full-time equivalent number of students (FTES) each year at Lake Tahoe Community College from 1976–1977 through 2004–2005.

- μ = 1000 FTES
- median = 1,014 FTES
- σ = 474 FTES
- first quartile = 528.5 FTES
- third quartile = 1,447.5 FTES
- n = 29 years

106. A sample of 11 years is taken. About how many are expected to have a FTES of 1014 or above? Explain how you determined your answer.

107. 75% of all years have an FTES:

a. at or below: _____

b. at or above: _____

108. The population standard deviation = _____

109. What percent of the FTES were from 528.5 to 1447.5? How do you know?

110. What is the *IQR*? What does the *IQR* represent?

111. How many standard deviations away from the mean is the median?

Additional Information: The population FTES for 2005–2006 through 2010–2011 was given in an updated report. The data are reported here.

Table 2.77

Year	2005–06	2006–07	2007–08	2008–09	2009–10	2010–11
Total FTES	1,585	1,690	1,735	1,935	2,021	1,890

112. Calculate the mean, median, standard deviation, the first quartile, the third quartile and the *IQR*. Round to one decimal place.

113. Compare the *IQR* for the FTES for 1976–77 through 2004–2005 with the *IQR* for the FTES for 2005-2006 through 2010–2011. Why do you suppose the *IQR*s are so different?

114. Three students were applying to the same graduate school. They came from schools with different grading systems. Which student had the best GPA when compared to other students at his school? Explain how you determined your answer.

Table 2.78

Student	GPA	School Average GPA	School Standard Deviation
Thuy	2.7	3.2	0.8

Student	GPA	School Average GPA	School Standard Deviation
Vichet	87	75	20
Kamala	8.6	8	0.4

115. A music school has budgeted to purchase three musical instruments. They plan to purchase a piano costing $3,000, a guitar costing $550, and a drum set costing $600. The mean cost for a piano is $4,000 with a standard deviation of $2,500. The mean cost for a guitar is $500 with a standard deviation of $200. The mean cost for drums is $700 with a standard deviation of $100. Which cost is the lowest, when compared to other instruments of the same type? Which cost is the highest when compared to other instruments of the same type. Justify your answer.

116. An elementary school class ran one mile with a mean of 11 minutes and a standard deviation of three minutes. Rachel, a student in the class, ran one mile in eight minutes. A junior high school class ran one mile with a mean of nine minutes and a standard deviation of two minutes. Kenji, a student in the class, ran 1 mile in 8.5 minutes. A high school class ran one mile with a mean of seven minutes and a standard deviation of four minutes. Nedda, a student in the class, ran one mile in eight minutes.

a. Why is Kenji considered a better runner than Nedda, even though Nedda ran faster than he?

b. Who is the fastest runner with respect to his or her class? Explain why.

117. The most obese countries in the world have obesity rates that range from 11.4% to 74.6%. This data is summarized in Table 14.

Table 2.79

Percent of population obese	Number of countries
11.4–20.45	29
20.45–29.45	13
29.45–38.45	4
38.45–47.45	0
47.45–56.45	2
56.45–65.45	1
65.45–74.45	0
74.45–83.45	1

What is the best estimate of the average obesity percentage for these countries? What is the standard deviation for the listed obesity rates? The United States has an average obesity rate of 33.9%. Is this rate above average or below? How "unusual" is the United States' obesity rate compared to the average rate? Explain.

118. Table 2.80 gives the percent of children under five considered to be underweight.

Table 2.80

Percent of underweight children	Number of countries
16–21.45	23
21.45–26.9	4
26.9–32.35	9
32.35–37.8	7
37.8–43.25	6
43.25–48.7	1

What is the best estimate for the mean percentage of underweight children? What is the standard deviation? Which interval(s) could be considered unusual? Explain.

Bringing it Together: Homework

119. Javier and Ercilia are supervisors at a shopping mall. Each was given the task of estimating the mean distance that shoppers live from the mall. They each randomly surveyed 100 shoppers. The samples yielded the following information.

Table 2.81

	Javier	Ercilia
$\bar{x}$	6.0 miles	6.0 miles
s	4.0 miles	7.0 miles

a. How can you determine which survey was correct ?
b. Explain what the difference in the results of the surveys implies about the data.
c. If the two histograms depict the distribution of values for each supervisor, which one depicts Ercilia's sample? How do you know?

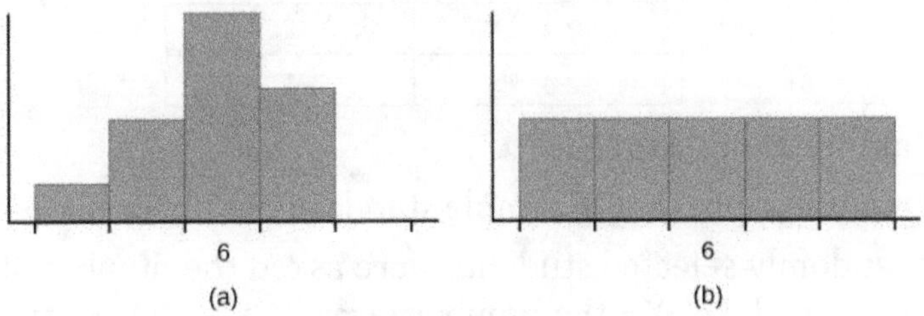

Figure 2.24

Use the following information to answer the next three exercises: We are interested in the number of years students in a particular elementary statistics class have lived in California. The information in the following table is from the entire section.

Table 2.82

Number of years	Frequency	Number of years	Frequency
7	1	22	1
14	3	23	1
15	1	26	1
18	1	40	2
19	4	42	2
20	3		
			Total = 20

120. What is the *IQR*?

a. 8
b. 11
c. 15
d. 35

121. What is the mode?

a. 19
b. 19.5
c. 14 and 20
d. 22.65

122. Is this a sample or the entire population?

a. sample
b. entire population
c. neither

123. Twenty-five randomly selected students were asked the number of movies they watched the previous week. The results are as follows:

Table 2.83

# of movies	Frequency
0	5
1	9
2	6
3	4
4	1

a. Find the sample mean x¯x^-.
b. Find the approximate sample standard deviation, *s*.

124. Forty randomly selected students were asked the number of pairs of sneakers they owned. Let *X* = the number of pairs of sneakers owned. The results are as follows:

Table 2.84

X	Frequency
1	2
2	5

X	Frequency
3	8
4	12
5	12
6	0
7	1

a. Find the sample mean x¯x^{-}
b. Find the sample standard deviation, *s*
c. Construct a histogram of the data.
d. Complete the columns of the chart.
e. Find the first quartile.
f. Find the median.
g. Find the third quartile.
h. What percent of the students owned at least five pairs?
i. Find the 40th percentile.
j. Find the 90th percentile.
k. Construct a line graph of the data
l. Construct a stemplot of the data

125. Following are the published weights (in pounds) of all of the team members of the San Francisco 49ers from a previous year.
177; 205; 210; 210; 232; 205; 185; 185; 178; 210; 206; 212; 184; 174; 185; 242; 188; 212; 215; 247; 241; 223; 220; 260; 245; 259; 278; 270; 280; 295; 275; 285; 290; 272; 273; 280; 285; 286; 200; 215; 185; 230; 250; 241; 190; 260; 250; 302; 265; 290; 276; 228; 265

a. Organize the data from smallest to largest value.
b. Find the median.
c. Find the first quartile.
d. Find the third quartile.
e. The middle 50% of the weights are from _______ to _______.
f. If our population were all professional football players, would the above data be a sample of weights or the population of weights? Why?
g. Assume the population was the San Francisco 49ers. Find:
   i. the population mean, μ.
   ii. the population standard deviation, σ.
   iii. the weight that is two standard deviations below the mean.

iv. When Steve Young, quarterback, played football, he weighed 205 pounds. How many standard deviations above or below the mean was he?

h. That same year, the mean weight for the Dallas Cowboys was 240.08 pounds with a standard deviation of 44.38 pounds. Emmit Smith weighed in at 209 pounds. With respect to his team, who was lighter, Smith or Young? How did you determine your answer?

126. One hundred teachers attended a seminar on mathematical problem solving. The attitudes of a representative sample of 12 of the teachers were measured before and after the seminar. A positive number for change in attitude indicates that a teacher's attitude toward math became more positive. The 12 change scores are as follows:

3; 8; –1; 2; 0; 5; –3; 1; –1; 6; 5; –2

a. What is the mean change score?

b. What is the standard deviation for this population?

c. What is the median change score?

d. Find the change score that is 2.2 standard deviations below the mean.

127. Refer to Figure 2.25 determine which of the following are true and which are false. Explain your solution to each part in complete sentences.

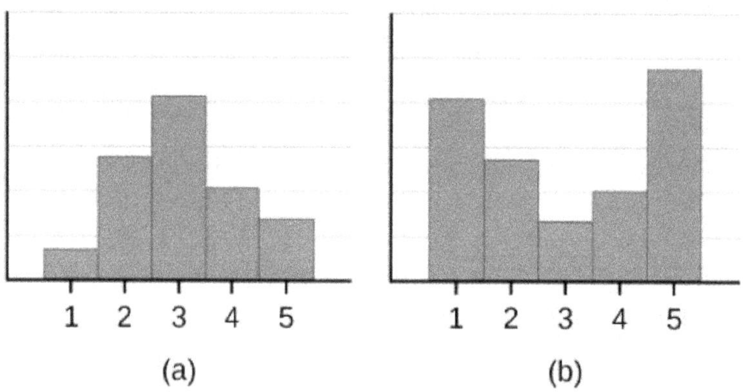

Figure 2.25

a. The medians for both graphs are the same.

b. We cannot determine if any of the means for both graphs is different.

c. The standard deviation for graph b is larger than the standard deviation for graph a.

d. We cannot determine if any of the third quartiles for both graphs is different.

128. In a recent issue of the *IEEE Spectrum*, 84 engineering conferences were announced. Four conferences lasted two days. Thirty-six lasted three days.

Eighteen lasted four days. Nineteen lasted five days. Four lasted six days. One lasted seven days. One lasted eight days. One lasted nine days. Let X = the length (in days) of an engineering conference.

a. Organize the data in a chart.
b. Find the median, the first quartile, and the third quartile.
c. Find the 65th percentile.
d. Find the 10th percentile.
e. The middle 50% of the conferences last from _______ days to _______ days.
f. Calculate the sample mean of days of engineering conferences.
g. Calculate the sample standard deviation of days of engineering conferences.
h. Find the mode.
i. If you were planning an engineering conference, which would you choose as the length of the conference: mean; median; or mode? Explain why you made that choice.
j. Give two reasons why you think that three to five days seem to be popular lengths of engineering conferences.

129. A survey of enrollment at 35 community colleges across the United States yielded the following figures:
6414; 1550; 2109; 9350; 21828; 4300; 5944; 5722; 2825; 2044; 5481; 5200; 5853; 2750; 10012; 6357; 27000; 9414; 7681; 3200; 17500; 9200; 7380; 18314; 6557; 13713; 17768; 7493; 2771; 2861; 1263; 7285; 28165; 5080; 11622

a. Organize the data into a chart with five intervals of equal width. Label the two columns "Enrollment" and "Frequency."
b. Construct a histogram of the data.
c. If you were to build a new community college, which piece of information would be more valuable: the mode or the mean?
d. Calculate the sample mean.
e. Calculate the sample standard deviation.
f. A school with an enrollment of 8000 would be how many standard deviations away from the mean?

Use the following information to answer the next two exercises. X = the number of days per week that 100 clients use a particular exercise facility.

Table 2.85

x	Frequency
0	3
1	12

x	Frequency
2	33
3	28
4	11
5	9
6	4

130. The 80th percentile is _____

a. 5

b. 80

c. 3

d. 4

131. The number that is 1.5 standard deviations BELOW the mean is approximately _____

a. 0.7

b. 4.8

c. –2.8

d. Cannot be determined

132. Suppose that a publisher conducted a survey asking adult consumers the number of fiction paperback books they had purchased in the previous month. The results are summarized in the Table 2.86.

Table 2.86

# of books	Freq.	Rel. Freq.
0	18	
1	24	
2	24	
3	22	
4	15	
5	10	
7	5	
9	1	

a. Are there any outliers in the data? Use an appropriate numerical test involving the *IQR* to identify outliers, if any, and clearly state your conclusion.

b. If a data value is identified as an outlier, what should be done about it?

c. Are any data values further than two standard deviations away from the mean? In some situations, statisticians may use this criteria to identify data values that are unusual, compared to the other data values. (Note that this criteria is most appropriate to use for data that is mound-shaped and symmetric, rather than for skewed data.)

d. Do parts a and c of this problem give the same answer?

e. Examine the shape of the data. Which part, a or c, of this question gives a more appropriate result for this data?

f. Based on the shape of the data which is the most appropriate measure of center for this data: mean, median or mode?

REFERENCES

2.1 Display Data

Burbary, Ken. *Facebook Demographics Revisited – 2001 Statistics,* 2011. Available online at http://www.kenburbary.com/2011/03/facebook-demographics-revisited-2011-statistics-2/ (accessed August 21, 2013).

"9th Annual AP Report to the Nation." CollegeBoard, 2013. Available online at http://apreport.collegeboard.org/goals-and-findings/promoting-equity (accessed September 13, 2013).

"Overweight and Obesity: Adult Obesity Facts." Centers for Disease Control and Prevention. Available online at http://www.cdc.gov/obesity/data/adult.html (accessed September 13, 2013).

Data on annual homicides in Detroit, 1961–73, from Gunst & Mason's book 'Regression Analysis and its Application', Marcel Dekker

"Timeline: Guide to the U.S. Presidents: Information on every president's birthplace, political party, term of office, and more." Scholastic, 2013. Available online at http://www.scholastic.com/teachers/article/timeline-guide-us-presidents (accessed April 3, 2013).

"Presidents." Fact Monster. Pearson Education, 2007. Available online at http://www.factmonster.com/ipka/A0194030.html (accessed April 3, 2013).

"Food Security Statistics." Food and Agriculture Organization of the United Nations. Available online at http://www.fao.org/economic/ess/ess-fs/en/ (accessed April 3, 2013).

"Consumer Price Index." United States Department of Labor: Bureau of Labor Statistics. Available online at http://data.bls.gov/pdq/SurveyOutputServlet (accessed April 3, 2013).

"CO2 emissions (kt)." The World Bank, 2013. Available online at http://databank.worldbank.org/data/home.aspx (accessed April 3, 2013).

"Births Time Series Data." General Register Office For Scotland, 2013. Available online at http://www.gro-scotland.gov.uk/statistics/theme/vital-events/births/time-series.html (accessed April 3, 2013).

"Demographics: Children under the age of 5 years underweight." Indexmundi. Available online at http://www.indexmundi.com/g/r.aspx?t=50&v=2224&aml=en (accessed April 3, 2013).

Gunst, Richard, Robert Mason. *Regression Analysis and Its Application: A Data-Oriented Approach*. CRC Press: 1980.

"Overweight and Obesity: Adult Obesity Facts." Centers for Disease Control and Prevention. Available online at http://www.cdc.gov/obesity/data/adult.html (accessed September 13, 2013).

2.2 Measures of the Location of the Data

Cauchon, Dennis, Paul Overberg. "Census data shows minorities now a majority of U.S. births." USA Today, 2012. Available online at http://usatoday30.usatoday.com/news/nation/story/2012-05-17/minority-birthscensus/55029100/1 (accessed April 3, 2013).

Data from the United States Department of Commerce: United States Census Bureau. Available online at http://www.census.gov/ (accessed April 3, 2013).

"1990 Census." United States Department of Commerce: United States Census Bureau. Available online at http://www.census.gov/main/www/cen1990.html (accessed April 3, 2013).

Data from *San Jose Mercury News*.

Data from *Time Magazine*; survey by Yankelovich Partners, Inc.

2.3 Measures of the Center of the Data

Data from The World Bank, available online at http://www.worldbank.org (accessed April 3, 2013).

"Demographics: Obesity – adult prevalence rate." Indexmundi. Available online at http://www.indexmundi.com/g/r.aspx?t=50&v=2228&l=en (accessed April 3, 2013).

2.7 Measures of the Spread of the Data

Data from Microsoft Bookshelf.

King, Bill."Graphically Speaking." Institutional Research, Lake Tahoe Community College. Available online at http://www.ltcc.edu/web/about/institutional-research (accessed April 3, 2013).

Solutions

1.

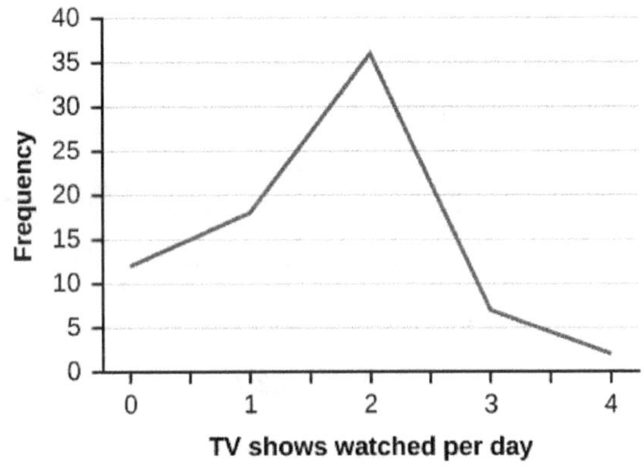

Figure 2.26

3.

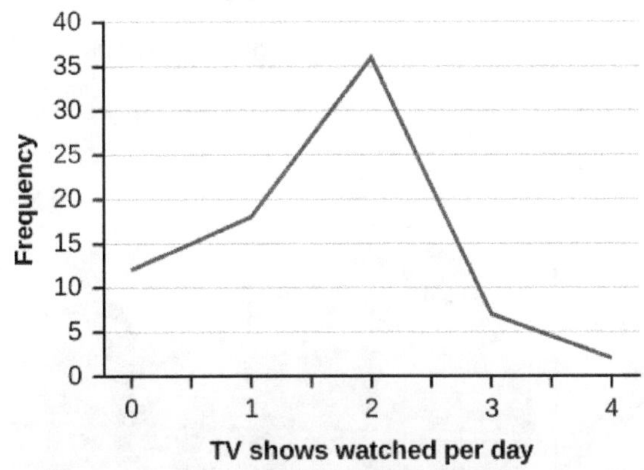

Figure 2.27

5.

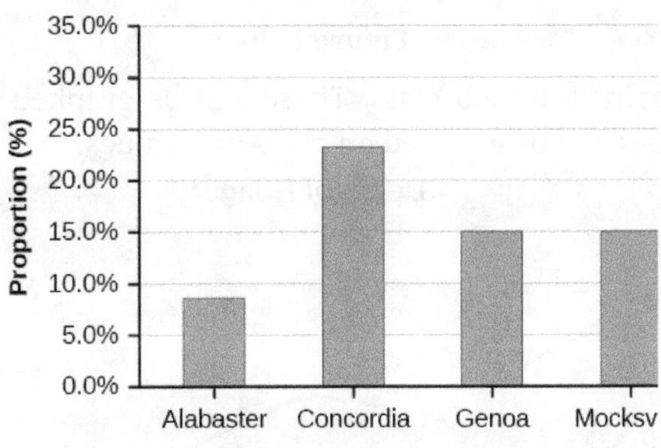

Figure 2.28

7.

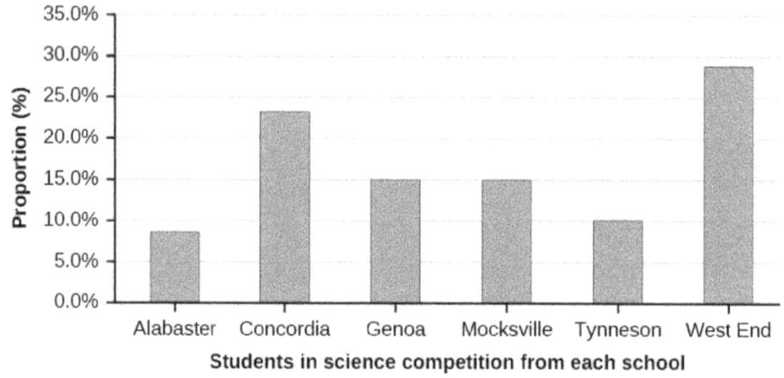

Figure 2.29

9. 65

11. The relative frequency shows the *proportion* of data points that have each value. The frequency tells the *number* of data points that have each value.

13. Answers will vary. One possible histogram is shown:

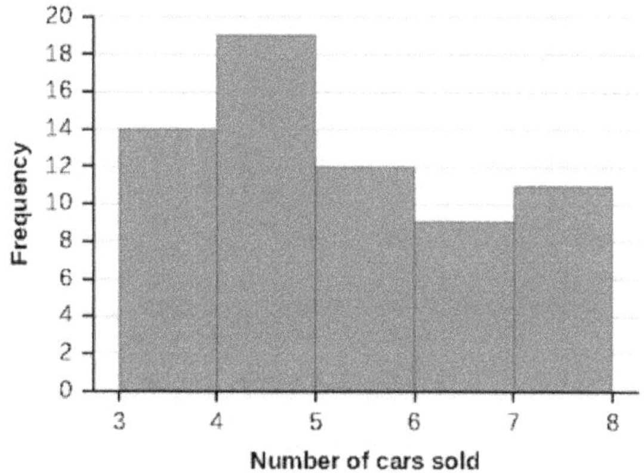

Figure 2.30

15. Find the midpoint for each class. These will be graphed on the *x*-axis. The frequency values will be graphed on the *y*-axis values.

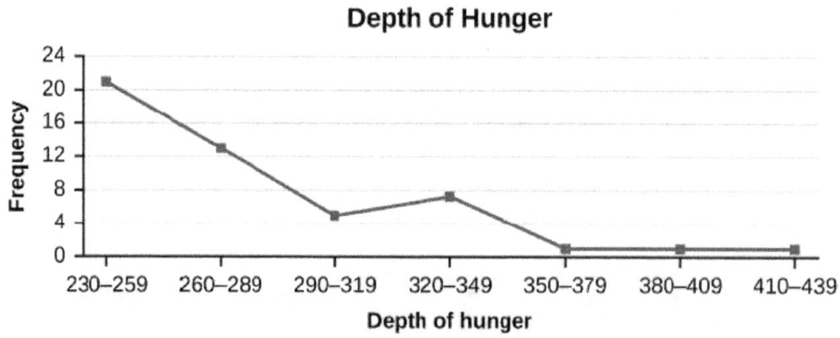

Figure 2.31

17.

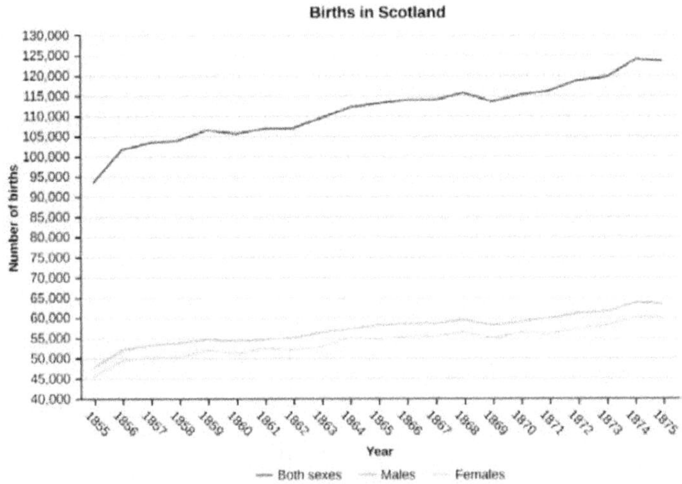

Figure 2.32

19. a. The 40th percentile is 37 years.
b. The 78th percentile is 70 years.

21. Jesse graduated 37th out of a class of 180 students. There are 180 – 37 = 143 students ranked below Jesse. There is one rank of 37. $x = 143$ $y = 1$. $\frac{x + 0.5y}{n}(100) = \frac{143 + 0.5(1)}{180}(100) = 79.72$. Jesse's rank of 37 puts him at the 80th percentile.

23. a. For runners in a race it is more desirable to have a high percentile for speed. A high percentile means a higher speed which is faster.
b. 40% of runners ran at speeds of 7.5 miles per hour or less (slower). 60% of runners ran at speeds of 7.5 miles per hour or more (faster).

25. When waiting in line at the DMV, the 85th percentile would be a long wait time compared to the other people waiting. 85% of people had shorter wait times than Mina. In this context, Mina would prefer a wait time corresponding to a lower percentile. 85% of people at the DMV waited 32 minutes or less. 15% of people at the DMV waited 32 minutes or longer.

27. The manufacturer and the consumer would be upset. This is a large repair cost for the damages, compared to the other cars in the sample. INTERPRETATION: 90% of the crash tested cars had damage repair costs of $1700 or less; only 10% had damage repair costs of $1700 or more.

29. You can afford 34% of houses. 66% of the houses are too expensive for your budget. INTERPRETATION: 34% of houses cost $240,000 or less. 66% of houses cost $240,000 or more.

31. 4

33. $6 - 4 = 2$

35. 6

37. Mean: 16 + 17 + 19 + 20 + 20 + 21 + 23 + 24 + 25 + 25 + 25 + 26 + 26 + 27 + 27 + 27 + 28 + 29 + 30 + 32 + 33 + 33 + 34 + 35 + 37 + 39 + 40 = 738; $\frac{738}{27} = 27.33$

39. The most frequent lengths are 25 and 27, which occur three times. Mode = 25, 27

41. 4

44. 39.48 in.

45. $21,574

46. 15.98 ounces

47. 81.56

48. 4 hours

49. 2.01 inches

50. 18.25

51. 10

52. 14.15

53. 14

54. 14.78

55. 44%

56. 100%

57. 6%

58. 33%

59. The data are symmetrical. The median is 3 and the mean is 2.85. They are close, and the mode lies close to the middle of the data, so the data are symmetrical.

61. The data are skewed right. The median is 87.5 and the mean is 88.2. Even though they are close, the mode lies to the left of the middle of the data, and there are many more instances of 87 than any other number, so the data are skewed right.

63. When the data are symmetrical, the mean and median are close or the same.

65. The distribution is skewed right because it looks pulled out to the right.

67. The mean is 4.1 and is slightly greater than the median, which is four.

69. The mode and the median are the same. In this case, they are both five.

71. The distribution is skewed left because it looks pulled out to the left.

73. The mean and the median are both six.

75. The mode is 12, the median is 12.5, and the mean is 15.1. The mean is the largest.

77. The mean tends to reflect skewing the most because it is affected the most by outliers.

79. $s = 34.5$

81. For Fredo: $z = \frac{0.158 - 0.166}{0.012} = -0.67$For Karl $z = \frac{0.177 - 0.189}{0.015} = -0.8$

Fredo's *z*-score of −0.67 is higher than Karl's *z*-score of −0.8. For batting average, higher values are better, so Fredo has a better batting average compared to his team.

83. a. $s_x = \sqrt{\frac{\sum fm^2}{n} - \bar{x}^2} = \sqrt{\frac{193157.45}{30} - 79.5^2} = 10.88$

b. $s_x = \sqrt{\frac{\sum fm^2}{n} - \bar{x}^2} = \sqrt{\frac{380945.3}{101} - 60.94^2} = 7.62$

c. $s_x = \sqrt{\frac{\sum fm^2}{n} - \bar{x}^2} = \sqrt{\frac{440051.5}{86} - 70.66^2} = 11.14$

84. a. Example solution for using the random number generator for the TI-84+ to generate a simple random sample of 8 states. Instructions are as follows.

Number the entries in the table 1–51 (Includes Washington, DC; Numbered vertically)

Press MATH

Arrow over to PRB

Press 5:randInt(

Enter 51,1,8)

Eight numbers are generated (use the right arrow key to scroll through the numbers). The numbers correspond to the numbered states (for this example: {47 21 9 23 51 13 25 4}. If any numbers are repeated, generate a different number by using 5:randInt(51,1)). Here, the states (and Washington DC) are {Arkansas, Washington DC, Idaho, Maryland, Michigan, Mississippi, Virginia, Wyoming}.

Corresponding percents are {30.1, 22.2, 26.5, 27.1, 30.9, 34.0, 26.0, 25.1}.

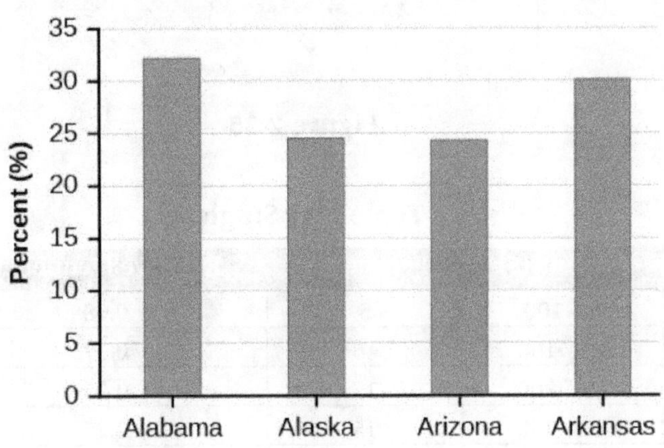

Figure 2.33

b.

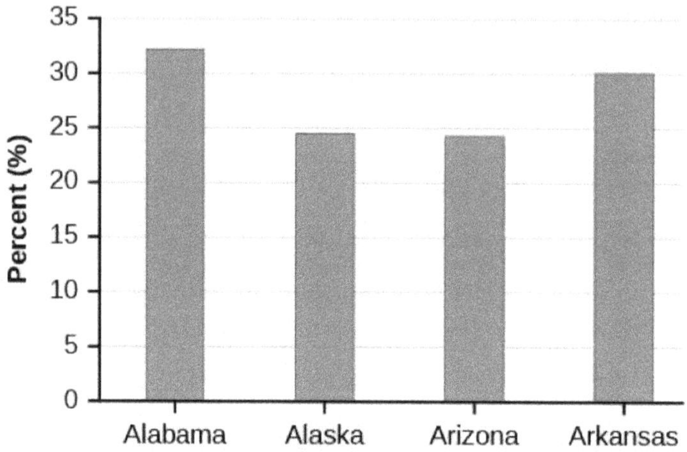

Figure 2.34

c.

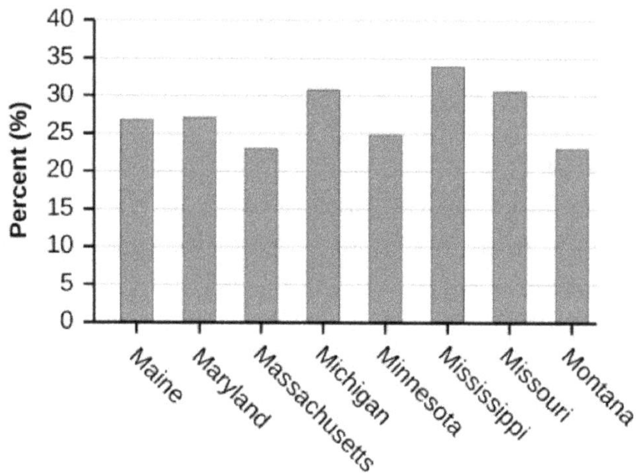

Figure 2.35

86.

Table 2.87 Singles

Amount($)	Frequency	Relative frequency
51–100	5	0.08
101–150	10	0.17
151–200	15	0.25
201–250	15	0.25
251–300	10	0.17
301–350	5	0.08

Table 2.88 Couples

Amount($)	Frequency	Relative frequency
100–150	5	0.07
201–250	5	0.07
251–300	5	0.07
301–350	5	0.07
351–400	10	0.14
401–450	10	0.14
451–500	10	0.14
501–550	10	0.14
551–600	5	0.07
601–650	5	0.07

a. See Table 2.87 and Table 2.88.

b. In the following histogram data values that fall on the right boundary are counted in the class interval, while values that fall on the left boundary are not counted (with the exception of the first interval where both boundary values are included).

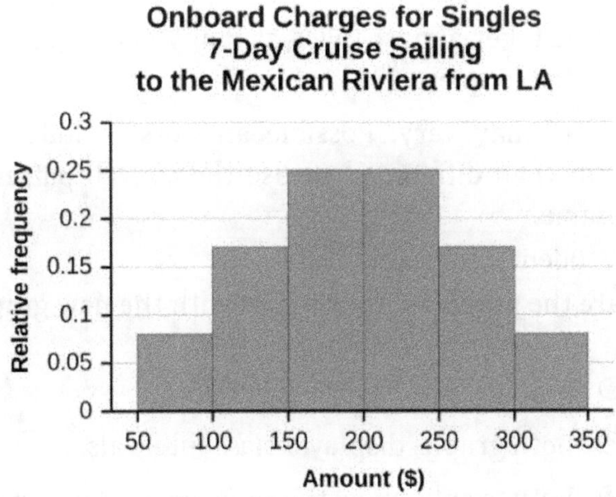

Figure 2.36

c. In the following histogram, the data values that fall on the right boundary are counted in the class interval, while values that fall on the left boundary are not counted (with the exception of the first interval where values on both boundaries are included).

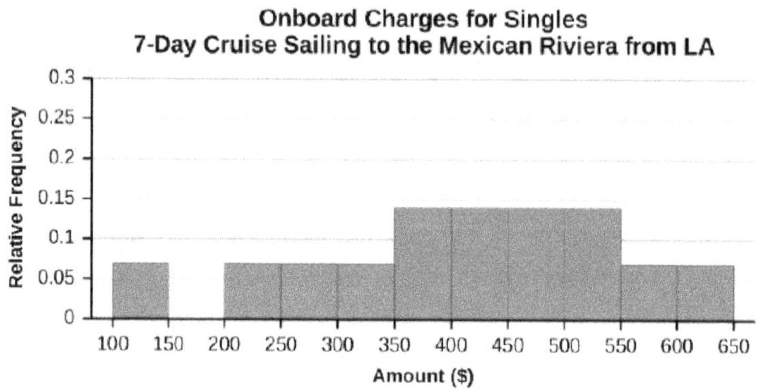

Figure 2.37

d. Compare the two graphs:
   i. Answers may vary. Possible answers include:
      - Both graphs have a single peak.
      - Both graphs use class intervals with width equal to $50.
   ii. Answers may vary. Possible answers include:
      - The couples graph has a class interval with no values.
      - It takes almost twice as many class intervals to display the data for couples.
   iii. Answers may vary. Possible answers include: The graphs are more similar than different because the overall patterns for the graphs are the same.
e. Check student's solution.
f. Compare the graph for the Singles with the new graph for the Couples:
   i.
      - Both graphs have a single peak.
      - Both graphs display 6 class intervals.
      - Both graphs show the same general pattern.
   ii. Answers may vary. Possible answers include: Although the width of the class intervals for couples is double that of the class intervals for singles, the graphs are more similar than they are different.
g. Answers may vary. Possible answers include: You are able to compare the graphs interval by interval. It is easier to compare the overall patterns with the new scale on the Couples graph. Because a couple represents two individuals, the new scale leads to a more accurate comparison.
h. Answers may vary. Possible answers include: Based on the histograms, it seems that spending does not vary much from singles to individuals who are

part of a couple. The overall patterns are the same. The range of spending for couples is approximately double the range for individuals.

88. c

90. Answers will vary.

92. a. 1 – (0.02+0.09+0.19+0.26+0.18+0.17+0.02+0.01) = 0.06

b. 0.19+0.26+0.18 = 0.63

c. Check student's solution.

d. 40th percentile will fall between 30,000 and 40,000

e. 80th percentile will fall between 50,000 and 75,000

f. Check student's solution.

94. The mean percentage, $\bar{x} = \frac{1328.65}{50} = 26.75$

95. a. Yes

b. The sample is 0.5 higher.

96.

a. 20

b. No

97. 51

98. a. 42

b. 99

99. $10.19

100. 17%

101. $30,772.48

102. 4.4%

103. 7.24%

104. -1.27%

106. The median value is the middle value in the ordered list of data values. The median value of a set of 11 will be the 6th number in order. Six years will have totals at or below the median.

108. 474 FTES

110. 919

112.

- mean = 1,809.3
- median = 1,812.5
- standard deviation = 151.2
- first quartile = 1,690
- third quartile = 1,935
- IQR = 245

113. Hint: Think about the number of years covered by each time period and what happened to higher education during those periods.

115. For pianos, the cost of the piano is 0.4 standard deviations BELOW the mean. For guitars, the cost of the guitar is 0.25 standard deviations ABOVE the mean. For drums, the cost of the drum set is 1.0 standard deviations BELOW the mean. Of the three, the drums cost the lowest in comparison to the cost of other instruments of the same type. The guitar costs the most in comparison to the cost of other instruments of the same type.

117.

- $\bar{x} = 23.32$
- Using the TI 83/84, we obtain a standard deviation of: s_x=12.95.
- The obesity rate of the United States is 10.58% higher than the average obesity rate.
- Since the standard deviation is 12.95, we see that 23.32 + 12.95 = 36.27 is the obesity percentage that is one standard deviation from the mean. The United States obesity rate is slightly less than one standard deviation from the mean. Therefore, we can assume that the United States, while 34% obese, does not have an unusually high percentage of obese people.

120. a

122. b

123. a. 1.48
b. 1.12

125. a. 174; 177; 178; 184; 185; 185; 185; 185; 188; 190; 200; 205; 205; 206; 210; 210; 210; 212; 212; 215; 215; 220; 223; 228; 230; 232; 241; 241; 242; 245; 247; 250; 250; 259; 260; 260; 265; 265; 270; 272; 273; 275; 276; 278; 280; 280; 285; 285; 286; 290; 290; 295; 302

b. 241 c. 205.5
d. 272.5 e. 205.5, 272.5
f. Sample g. Population
h.

i. 236.34
ii. 37.50
iii. 161.34
iv. 0.84 std. dev. below the mean

i. Young

127. a. True b. True
c. True d. False

129. a.

Table 2.89

Enrollment	Frequency
1000-5000	10
5000-10000	16
10000-15000	3
15000-20000	3
20000-25000	1
25000-30000	2

a. Check student's solution.

b. mode

c. 8628.74

d. 6943.88

e. –0.09

131. a

3 Chapter PROBABILITY

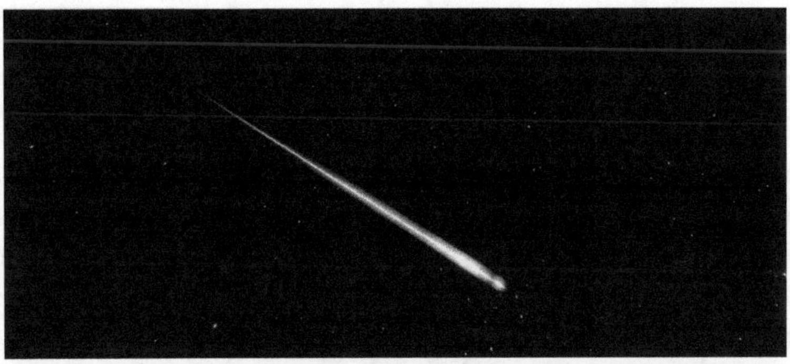

Figure 3.1 Meteor showers are rare, but the probability of them occurring can be calculated. (credit: Navicore/flickr)

INTRODUCTION: AN OVERVIEW

It is frequently essential to hypothesize about the result of an event to facilitate decision-making. Politicians analyze polls to assess their probability of electoral success. Educators select a certain curriculum depending on their assessment of students' comprehension abilities. Physicians select therapies for different diseases based on their evaluation of probable outcomes. You may have frequented a casino where individuals engage in games selected based on the perception of favorable odds for winning. Your selection of a field of study may have been influenced by the anticipated job prospects. It is highly probable that you have utilized probability. You likely possess an inherent understanding of probability. Probability pertains to the likelihood of an event transpiring. When you assess the likelihood of completing your schoolwork or preparing for an examination, you are use probability. This chapter will instruct you on systematically solving probability questions.

3.1 TERMINOLOGY

Probability is a measure that is associated with how certain we are of outcomes of a particular experiment or activity. An experiment is a planned operation carried out under controlled conditions. If the result is not predetermined, then the experiment is said to be a chance experiment. Flipping one fair coin twice is an example of an experiment.

A result of an experiment is called an outcome. The sample space of an experiment is the set of all possible outcomes. Three ways to represent a sample space are: to list the possible outcomes, to create a tree diagram, or to create a Venn diagram. The uppercase letter S is used to denote the sample space. For example, if you flip one fair coin, $S = \{H, T\}$ where H = heads and T = tails are the outcomes.

An event is any combination of outcomes. Upper case letters like A and B represent events. For example, if the experiment is to flip one fair coin, event A might be getting at most one head. The probability of an event A is written $P(A)$.

The probability of any outcome is the long-term relative frequency of that outcome. Probabilities are between zero and one, inclusive (that is, zero and one and all numbers between these values). $P(A) = 0$ means the event A can never happen. $P(A) = 1$ means the event A always happens. $P(A) = 0.5$ means the event A is equally likely to occur or not to occur. For example, if you flip one fair coin repeatedly (from 20 to 2,000 to 20,000 times) the relative frequency of heads approaches 0.5 (the probability of heads).

Equally likely means that each outcome of an experiment occurs with equal probability. For example, if you toss a fair, six-sided die, each face (1, 2, 3, 4, 5, or 6) is as likely to occur as any other face. If you toss a fair coin, a Head (H) and a Tail (T) are equally likely to occur. If you randomly guess the answer to a true/false question on an exam, you are equally likely to select a correct answer or an incorrect answer.

To calculate the probability of an event A when all outcomes in the sample space are equally likely, count the number of outcomes for event A and divide by the total number of outcomes in the sample space. For example, if you toss a fair dime and a fair nickel, the sample space is $\{HH, TH, HT, TT\}$ where T = tails and H = heads. The sample space has four outcomes. A = getting one head. There are two outcomes that meet this condition $\{HT, TH\}$, so $P(A) = \frac{2}{4} = 0.5$.

Suppose you roll one fair six-sided die, with the numbers $\{1, 2, 3, 4, 5, 6\}$ on its faces. Let event E = rolling a number that is at least five. There are two outcomes $\{5, 6\}$. $P(E) = \frac{2}{6}$. If you were to roll the die only a few times, you would not be surprised if your observed results did not match the probability. If you were to roll the die a very large number of times, you would expect that, overall, $\frac{2}{6}$

of the rolls would result in an outcome of "at least five". You would not expect exactly $\frac{2}{6}$. The long-term relative frequency of obtaining this result would approach the theoretical probability of $\frac{2}{6}$ as the number of repetitions grows larger and larger.

This important characteristic of probability experiments is known as the **law of large numbers** which states that as the number of repetitions of an experiment is increased, the relative frequency obtained in the experiment tends to become closer and closer to the theoretical probability. Even though the outcomes do not happen according to any set pattern or order, overall, the long-term observed relative frequency will approach the theoretical probability. (The word **empirical** is often used instead of the word observed.)

It is important to realize that in many situations, the outcomes are not equally likely. A coin or die may be **unfair**, or **biased.** Two math professors in Europe had their statistics students test the Belgian one Euro coin and discovered that in 250 trials, a head was obtained 56% of the time and a tail was obtained 44% of the time. The data seem to show that the coin is not a fair coin; more repetitions would be helpful to draw a more accurate conclusion about such bias. Some dice may be biased. Look at the dice in a game you have at home; the spots on each face are usually small holes carved out and then painted to make the spots visible. Your dice may or may not be biased; it is possible that the outcomes may be affected by the slight weight differences due to the different numbers of holes in the faces. Gambling casinos make a lot of money depending on outcomes from rolling dice, so casino dice are made differently to eliminate bias. Casino dice have flat faces; the holes are completely filled with paint having the same density as the material that the dice are made out of so that each face is equally likely to occur. Later we will learn techniques to use to work with probabilities for events that are not equally likely.

"∪" Event: The Union

An outcome is in the event $A \cup\cup B$ if the outcome is in A or is in B or is in both A and B. For example, let $A = \{1, 2, 3, 4, 5\}$ and $B = \{4, 5, 6, 7, 8\}$. $A \cup\cup B = \{1, 2, 3, 4, 5, 6, 7, 8\}$. Notice that 4 and 5 are NOT listed twice.

"∩" Event: The Intersection

An outcome is in the event $A \cap\cap B$ if the outcome is in both A and B at the same time. For example, let A and B be $\{1, 2, 3, 4, 5\}$ and $\{4, 5, 6, 7, 8\}$, respectively. Then $A \cap\cap B = \{4, 5\}$.

The **complement** of event A is denoted A' (read "A prime"). A' consists of all outcomes that are **NOT** in A. Notice that $P(A) + P(A') = 1$. For example, let S

Let A = {1, 2, 3, 4}. Then, $A' = \{5, 6\}$. $P(A) = \frac{4}{6}$, $P(A') = \frac{2}{6}$, and $P(A) + P(A') = \frac{4}{6} + \frac{2}{6} = 1$

The conditional probability of A given B is written P(A||B). P(A||B) is the probability that event A will occur given that the event B has already occurred. A conditional reduces the sample space. We calculate the probability of A from the reduced sample space B. The formula to $P(A \mid B)$ is $P(A \mid B) = \frac{P(A \cap B)}{P(B)}$ where P(B) is greater than zero.

For example, suppose we toss one fair, six-sided die. The sample space *S* = {1, 2, 3, 4, 5, 6}. Let *A* = face is 2 or 3 and *B* = face is even (2, 4, 6). To calculate *P*(*A* | | *B*), we count the number of outcomes 2 or 3 in the sample space *B* = {2, 4, 6}. Then we divide that by the number of outcomes *B* (rather than *S*).

We get the same result by using the formula. Remember that *S* has six outcomes.

$$P(A \mid B) = \frac{P(A \cap B)}{P(B)} = \frac{\frac{\text{(the number of outcomes that are 2 or 3 and even in} S)}{6}}{\frac{\text{(the number of outcomes that are even in} S)}{6}} = \frac{\frac{1}{6}}{\frac{3}{6}} = \frac{1}{3}$$

Odds

The odds of an event presents the probability as a ratio of success to failure. This is common in various gambling formats. Mathematically, the odds of an event can be defined as:

$$\frac{P(A)}{1 - P(A)}$$

where *P*(*A*) is the probability of success and of course 1 - *P*(*A*) is the probability of failure. Odds are always quoted as "numerator to denominator," e.g. 2 to 1. Here the probability of winning is twice that of losing; thus, the probability of winning is 0.66. A probability of winning of 0.60 would generate odds in favor of winning of 3 to 2. While the calculation of odds can be useful in gambling venues in determining payoff amounts, it is not helpful for understanding probability or statistical theory.

Understanding Terminology and Symbols

It is important to read each problem carefully to think about and understand what the events are. Understanding the wording is the first very important step in solving probability problems. Reread the problem several times if necessary. Clearly identify the event of interest. Determine whether there is a condition stated in the wording that would indicate that the probability is conditional; carefully identify the condition, if any.

EXAMPLE 3.1

The sample space S is the whole numbers starting at one and less than 20.

a. S = ______________________________

Let event A = the even numbers and event B = numbers greater than 13.

b. A = ________________________, B = ____________________

c. $P(A)$ = ______________, $P(B)$ = ________________

d. $A \cap\cap B$ = ______________________, A OR B = ________________

e. $P(A \cap\cap B)$ = __________, $P(A \cup\cup B)$ = _____________

f. A' = ______________, $P(A')$ = _____________

g. $P(A) + P(A')$ = ____________

h. $P(A||B)$ = ___________, $P(B||A)$ = _____________; are the probabilities equal?

Solution 3.1

a. $S = \{1, 2, 3, 4, 5, 6, 7, 8, 9, 10, 11, 12, 13, 14, 15, 16, 17, 18, 19\}$

b. $A = \{2, 4, 6, 8, 10, 12, 14, 16, 18\}$, $B = \{14, 15, 16, 17, 18, 19\}$

c. $P(A) = \frac{9}{19}$, $P(B) = \frac{6}{19}$

d. $A \cap B = \{14,16,18\}$, A OR $B = \{2, 4, 6, 8, 10, 12, 14, 15, 16, 17, 18, 19\}$

e. $P(A \cap B) = \frac{3}{19}$, $P(A \cup B) = \frac{12}{19}$

f. $A' = 1, 3, 5, 7, 9, 11, 13, 15, 17, 19$; $P(A') = \frac{10}{19}$

g. $P(A) + P(A') = 1$ $(\frac{9}{19} + \frac{10}{19} = 1)$

h. $P(A \mid B) = \frac{P(A \cap B)}{P(B)} = \frac{3}{6}$, $P(B \mid A) = \frac{P(A \cap B)}{P(A)} = \frac{3}{9}$, No

TRY IT 3.1

The sample space S is all the ordered pairs of two whole numbers, the first from one to three and the second from one to four (Example: (1, 4)).

a. S = ______________________________

Let event A = the sum is even and event B = the first number is prime.

b. A = ________________________, B = ____________________

c. $P(A)$ = ______________, $P(B)$ = ________________

d. $A \cap\cap B$ = ______________________, $A \cup\cup B$ = ________________

e. $P(A \cap\cap B)$ = __________, $P(A \cup\cup B)$ = _____________

f. B' = ______________, $P(B')$ = _____________

g. $P(A) + P(A')$ = ____________

h. $P(A||B)$ = ___________, $P(B||A)$ = _____________; are the probabilities equal?

EXAMPLE 3.2

A fair, six-sided die is rolled. Describe the sample space S, identify each of the following events with a subset of S and compute its probability (an outcome is the number of dots that show up).

a. Event T = the outcome is two.
b. Event A = the outcome is an even number.
c. Event B = the outcome is less than four.
d. The complement of A.
e. $A \,||\, B$
f. $B \,||\, A$
g. $A \cap\cap B$
h. $A \cup\cup B$
i. $A \cup\cup B'$
j. Event N = the outcome is a prime number.
k. Event I = the outcome is seven.

Solution 3.2

a. $T = \{2\}$, $P(T) = \frac{1}{6}$
b. $A = \{2, 4, 6\}$, $P(A) = \frac{1}{2}$
c. $B = \{1, 2, 3\}$, $P(B) = \frac{1}{2}$
d. $A' = \{1, 3, 5\}$, $P(A') = \frac{1}{2}$
e. $A \mid B = \{2\}$, $P(A \mid B) = \frac{1}{3}$
f. $B \mid A = \{2\}$, $P(B \mid A) = \frac{1}{3}$
g. $A \cap B = \{2\}$, $P(A \cap B) = \frac{1}{6}$
h. $A \cup B = \{1, 2, 3, 4, 6\}$, $P(A \cup B) = \frac{5}{6}$
i. $A \cup B' = \{2, 4, 5, 6\}$, $P(A \cup B') = \frac{2}{3}$
j. $N = \{2, 3, 5\}$, $P(N) = \frac{1}{2}$
k. A six-sided die does not have seven dots. $P(7) = 0$.

Example 3.3

Table 3.1 describes the distribution of a random sample S of 100 individuals, organized by gender and whether they are right- or left-handed.

Table 3.1

	Right-handed	Left-handed
Males	43	9
Females	44	4

Let's denote the events M = the subject is male, F = the subject is female, R = the subject is right-handed, L = the subject is left-handed. Compute the following probabilities:

a. $P(M)$
b. $P(F)$
c. $P(R)$
d. $P(L)$
e. $P(M \cap R)$
f. $P(F \cap L)$
g. $P(M \cup F)$
h. $P(M \cup R)$
i. $P(F \cup L)$
j. $P(M')$
k. $P(R|M)$
l. $P(F|L)$
m. $P(L|F)$

Solution 3.3

a. $P(M) = 0.52$
b. $P(F) = 0.48$
c. $P(R) = 0.87$
d. $P(L) = 0.13$
e. $P(M \cap R) = 0.43$
f. $P(F \cap L) = 0.04$
g. $P(M \cup F) = 1$
h. $P(M \cup R) = 0.96$
i. $P(F \cup L) = 0.57$
j. $P(M') = 0.48$
k. $P(R \mid M) = 0.8269$ (rounded to four decimal places)
l. $P(F \mid L) = 0.3077$ (rounded to four decimal places)
m. $P(L \mid F) = 0.0833$

3.2 INDEPENDENT AND MUTUALLY EXCLUSIVE EVENTS

Independent and mutually exclusive do **not** mean the same thing.

Independent Events

Two events are independent if one of the following are true:

- P(A|B)=P(A)$P(A|B)=P(A)$
- P(B|A)=P(B)$P(B|A)=P(B)$
- P(A∩B)=P(A)P(B)$P(A\cap B)=P(A)P(B)$

Two events A and B are independent if the knowledge that one occurred does not affect the chance the other occurs. For example, the outcomes of two roles of a fair die are independent events. The outcome of the first roll does not change the probability for the outcome of the second roll. To show two events are independent, you must show only one of the above conditions. If two events are NOT independent, then we say that they are dependent.

Sampling may be done with replacement or without replacement.

- **With replacement**: If each member of a population is replaced after it is picked, then that member has the possibility of being chosen more than once. When sampling is done with replacement, then events are considered to be independent, meaning the result of the first pick will not change the probabilities for the second pick.
- **Without replacement**: When sampling is done without replacement, each member of a population may be chosen only once. In this case, the probabilities for the second pick are affected by the result of the first pick. The events are considered to be dependent or not independent.

If it is not known whether A and B are independent or dependent, assume they are dependent until you can show otherwise.

EXAMPLE 3.4

You have a fair, well-shuffled deck of 52 cards. It consists of four suits. The suits are clubs, diamonds, hearts and spades. There are 13 cards in each suit consisting of 1, 2, 3, 4, 5, 6, 7, 8, 9, 10, J (jack), Q (queen), K (king) of that suit.

a. Sampling with replacement: Suppose you pick three cards with replacement. The first card you pick out of the 52 cards is the Q of spades. You put this card back, reshuffle the cards and pick a second card from the 52-card deck. It is the ten of clubs. You put this card back, reshuffle the cards and pick a third card from the 52-card deck. This time, the card is the Q of spades again. Your picks are {Q of spades, ten of clubs, Q of spades}. You have picked the Q of spades twice. You pick each card from the 52-card deck.
b. Sampling without replacement: Suppose you pick three cards without replacement. The first card you pick out of the 52 cards is the K of hearts. You put this card aside and pick the second card from the 51 cards remaining in the deck. It is the three of diamonds. You put this card aside and pick the third card from the remaining 50 cards in the deck. The third card is the J of spades. Your picks are {K of hearts, three of diamonds, J of spades}. Because you have picked the cards without replacement, you cannot pick the same card twice. The probability of picking the three of diamonds is called a conditional probability because it is conditioned on what was picked first. This is true also of the probability of picking the J of spades. The probability of picking the J of spades is actually conditioned on *both* the previous picks.

TRY IT 3.4

You have a fair, well-shuffled deck of 52 cards. It consists of four suits. The suits are clubs, diamonds, hearts and spades. There are 13 cards in each suit consisting of 1, 2, 3, 4, 5, 6, 7, 8, 9, 10, *J* (jack), *Q* (queen), *K* (king) of that suit. Three cards are picked at random.

a. Suppose you know that the picked cards are *Q* of spades, *K* of hearts and *Q* of spades. Can you decide if the sampling was with or without replacement?
b. Suppose you know that the picked cards are *Q* of spades, *K* of hearts, and *J* of spades. Can you decide if the sampling was with or without replacement?

EXAMPLE 3.5

You have a fair, well-shuffled deck of 52 cards. It consists of four suits. The suits are clubs, diamonds, hearts, and spades. There are 13 cards in each suit consisting of 1, 2, 3, 4, 5, 6, 7, 8, 9, 10, *J* (jack), *Q* (queen), and *K* (king) of that suit. *S* = spades, *H* = Hearts, *D* = Diamonds, *C* = Clubs.

a. Suppose you pick four cards, but do not put any cards back into the deck. Your cards are *QS*, 1*D*, 1*C*, *QD*.
b. Suppose you pick four cards and put each card back before you pick the next card. Your cards are *KH*, 7*D*, 6*D*, *KH*.

Which of a. or b. did you sample with replacement and which did you sample without replacement?

Solution 3.5

a. Without replacement; b. With replacement

TRY IT 3.5

You have a fair, well-shuffled deck of 52 cards. It consists of four suits. The suits are clubs, diamonds, hearts, and spades. There are 13 cards in each suit consisting of 1, 2, 3, 4, 5, 6, 7, 8, 9, 10, *J* (jack), *Q* (queen), and *K* (king) of that suit. *S* = spades, *H* = Hearts, *D* = Diamonds, *C* = Clubs. Suppose that you sample four cards without replacement. Which of the following outcomes are possible? Answer the same question for sampling with replacement.

a. *QS*, 1*D*, 1*C*, *QD*
b. *KH*, 7*D*, 6*D*, *KH*
c. *QS*, 7*D*, 6*D*, *KS*

Mutually Exclusive Events

A and *B* are mutually exclusive events if they cannot occur at the same time. Said another way, If *A* occurred then *B* cannot occur and vise-a-versa. This means that *A* and *B* do not share any outcomes and $P(A \cap B)=0$.

For example, suppose the sample space $S = \{1, 2, 3, 4, 5, 6, 7, 8, 9, 10\}$. Let $A = \{1, 2, 3, 4, 5\}$, $B = \{4, 5, 6, 7, 8\}$, and $C = \{7, 9\}$. A ∩ B = {4, 5 $P(A \cap B) = \frac{2}{10}$ and is not equal to zero. Therefore, A and B are not mutually exclusive. A and C do not have any numbers in common so P(A∩C)=0$P(A \cap C)=0$. Therefore, A and C are mutually exclusive.

If it is not known whether A and B are mutually exclusive, assume they are not until you can show otherwise. The following examples illustrate these definitions and terms.

EXAMPLE 3.6

Flip two fair coins. (This is an experiment.)

The sample space is {*HH*, *HT*, *TH*, *TT*} where T = tails and H = heads. The outcomes are *HH*, *HT*, *TH*, and *TT*. The outcomes HT and TH are different. The *HT* means that the first coin showed heads and the second coin showed tails. The *TH* means that the first coin showed tails and the second coin showed heads.

- Let A = the event of getting at most one tail. (At most one tail means zero or one tail.) Then A can be written as {*HH*, *HT*, *TH*}. The outcome *HH* shows zero tails. *HT* and *TH* each show one tail.
- Let B = the event of getting all tails. B can be written as {*TT*}. B is the complement of A, so $B = A'$. Also, $P(A) + P(B) = P(A) + P(A') = 1$.
- The probabilities for A and for B $P(A) = \frac{3}{4}$ and $P(B) = \frac{1}{4}$.
- Let C=the event of getting all heads. C={*HH*}. Since B={*TT*}, P(B∩C)=0$P(B \cap C)=0$. B and C are mutually exclusive. (B and C have no members in common because you cannot have all tails and all heads at the same time.)
- Let D = event of getting more than one tail. $D = \{TT\}$. $P(D) = \frac{1}{4}$
- Let E = event of getting a head on the first roll. (This implies you can get either a head or tail on the second roll.) $E = \{HT, HH\}$. $P(E) = \frac{2}{4}$
- Find the probability of getting at least one (one or two) tail in two flips. Let F = event of getting at least one tail in two flips. $F = \{HT, TH, TT\}$. $P(F) = \frac{3}{4}$

TRY IT 3.6

Draw two cards from a standard 52-card deck with replacement. Find the probability of getting at least one black card.

EXAMPLE 3.7

Flip two fair coins. Find the probabilities of the events.

a. Let F = the event of getting at most one tail (zero or one tail).

b. Let G = the event of getting two faces that are the same.
c. Let H = the event of getting a head on the first flip followed by a head or tail on the second flip.
d. Are F and G mutually exclusive?
e. Let J = the event of getting all tails. Are *J* and *H* mutually exclusive?

Solution 3.7

Look at the sample space in Example 3.6.

a. Zero (0) or one (1) tails occur when the outcomes *HH*, *TH*, *HT* show up. $P(F) = \frac{3}{4}$
b. Two faces are the same if *HH* or *TT* show up. $P(G) = \frac{2}{4}$
c. A head on the first flip followed by a head or tail on the second flip occurs when *HH* or *HT* show up. $P(H) = \frac{2}{4}$
d. *F* and *G* share *HH* so $P(F \cap G)$ is not equal to zero (0). *F* and *G* are not mutually exclusive.
e. Getting all tails occurs when tails shows up on both coins (*TT*). *H*'s outcomes are *HH* and *HT*.

J and *H* have nothing in common so $P(J \cap H) = 0$. *J* and *H* are mutually exclusive.

TRY IT 3.7

A box has two balls, one white and one red. We select one ball, put it back in the box, and select a second ball (sampling with replacement). Find the probability of the following events:

a. Let *F* = the event of getting the white ball twice.
b. Let *G* = the event of getting two balls of different colors.
c. Let *H* = the event of getting white on the first pick.
d. Are F and G mutually exclusive?
e. Are G and *H* mutually exclusive?

EXAMPLE 3.8

Roll one fair, six-sided die. The sample space is {1, 2, 3, 4, 5, 6}. Let event *A* = a face is odd. Then *A* = {1, 3, 5}. Let event *B* = a face is even. Then *B* = {2, 4, 6}.

- Find the complement of *A*, *A′*. The complement of *A*, *A′*, is *B* because *A* and *B* together make up the sample space. $P(A) + P(B) = P(A) + P(A') = 1$. Also, $P(A) = \frac{3}{6}$ and $P(B) = \frac{3}{6}$.

- Let event *C* = odd faces larger than two. Then *C* = {3, 5}. Let event *D* = all even faces smaller than five. Then *D* = {2, 4}. $P(C \cap D)=0$ because you cannot have an odd and even face at the same time. Therefore, *C* and *D* are mutually exclusive events.

- Let event *E* = all faces less than five. *E* = {1, 2, 3, 4}.

Are *C* and *E* mutually exclusive events? (Answer yes or no.) Why or why not?

Problem

- Find P(C|A)*P*(*C*|*A*). This is a conditional probability. Recall that the event *C* is {3, 5} and event *A* is {1, 3, 5}. To find P(C|A)*P*(*C*|*A*), find the probability of *C* using the sample space *A*. You have reduced the sample space from the original sample space {1, 2, 3, 4, 5, 6} to {1, 3, 5}. So, P(C||A)=23*P*(*C*|*A*)=23.

Solution 3.8

No. C = {3, 5} and $E = \{1, 2, 3, 4\}$. $P(C \cap E) = \frac{1}{6}$. To be mutually exclusive, P(C ∩ E) must be zero.

- Find P(C|A) . This is a conditional probability. Recall that the event C is {3, 5} and event A is {1, 3, 5}. To find P(C|A) , find the probability of C using the sample space A. You have reduced the sample space from the original sample space {1, 2, 3, 4, 5, 6} to {1, 3, 5}. So, $P(C|A) = \frac{2}{3}$.

TRY IT 3.8

Let event *A* = learning Spanish. Let event *B* = learning German. Then A∩B=*A*∩*B* = learning Spanish and German. Suppose P(A)=0.4*P*(*A*)=0.4 and P(B)=0.2*P*(*B*)=0.2. P(A∩B)=0.08*P*(*A*∩*B*)=0.08. Are events *A* and *B* independent? Hint: You must show ONE of the following:

- P(A|B)=P(A)
- P(B|A)=P(B)
- P(A∩B)=P(A)P(B)

EXAMPLE 3.9

Let event *G* = taking a math class. Let event *H* = taking a science class. Then, G ∩∩ H = taking a math class and a science class. Suppose P(G)=0.6,P(H)=0.5, and P(G∩H)=0.3. Are *G* and *H* independent?

If *G* and *H* are independent, then you must show ONE of the following:

- P(G|H)=P(G)
- P(H|G)=P(H)
- P(G∩H)=P(G)P(H)

NOTE

The choice you make depends on the information you have. You could choose any of the methods here because you have the necessary information.

a. Show that P(G|H)=P(G)

Solution 3.9

$$P(G|H) = \frac{P(G \cap H)}{P(H)} = \frac{0.3}{0.5} = 0.6 = P(G)$$

b. Show $P(G \cap H) = P(G)P(H)$.

Solution 3.9

$P(G)P(H) = (0.6)(0.5) = 0.3 = P(G \cap H)$

Since *G* and *H* are independent, knowing that a person is taking a science class does not change the chance that he or she is taking a math class. If the two events had not been independent (that is, they are dependent) then knowing that a person is taking a science class would change the chance he or she is taking math. For practice, show that P(H|G)=P(H) to show that *G* and *H* are independent events.

TRY IT 3.9

In a bag, there are six red marbles and four green marbles. The red marbles are marked with the numbers 1, 2, 3, 4, 5, and 6. The green marbles are marked with the numbers 1, 2, 3, and 4.

- *R* = a red marble
- *G* = a green marble
- *O* = an odd-numbered marble
- The sample space is *S* = {*R*1, *R*2, *R*3, *R*4, *R*5, *R*6, *G*1, *G*2, *G*3, *G*4}.

S has ten outcomes. What is P(G∩O)?

EXAMPLE 3.10

Problem

Let event *C* = taking an English class. Let event *D* = taking a speech class.
Suppose P(C)=0.75, P(D)=0.3, P(C|D)=0.75 and P(C∩D)=0.225.
Justify your answers to the following questions numerically.

a. Are *C* and *D* independent?
b. Are *C* and *D* mutually exclusive?
c. What is P(D|C)*P*(*D*|*C*)?

Solution 3.10

a. Yes, because $P(C|D) = P(C)$.
b. No, because $P(C \cap D)$ is not equal to zero.
c $P(D|C) = \frac{P(C \cap D)}{P(C)} = \frac{0.225}{0.75} = 0.3$

TRY IT 3.10

A student goes to the library. Let events *B* = the student checks out a book and *D* = the student checks out a DVD. Suppose that P(B)=0.40$P(B)$=0.40, P(D)=0.30 and P(B∩D)=0.20.

a. Find P(B|D).
b. Find P(D|B).
c. Are *B* and *D* independent?
d. Are *B* and *D* mutually exclusive?

EXAMPLE 3.11

In a box there are three red cards and five blue cards. The red cards are marked with the numbers 1, 2, and 3, and the blue cards are marked with the numbers 1, 2, 3, 4, and 5. The cards are well-shuffled. You reach into the box (you cannot see into it) and draw one card.

Let *R* = red card is drawn, *B* = blue card is drawn, *E* = even-numbered card is drawn.

The sample space *S* = *R*1, *R*2, *R*3, *B*1, *B*2, *B*3, *B*4, *B*5. *S* has eight outcomes.

- $P(R)=\frac{3}{8}.P(B)=\frac{5}{8}.P(R\cap B)=0$. (You cannot draw one card that is both red and blue.)
- $P(E)=\frac{3}{8}$. (There are three even-numbered cards, *R*2, *B*2, and *B*4.)
- $P(E|B)=\frac{2}{5}$. (There are five blue cards: *B*1, *B*2, *B*3, *B*4, and *B*5. Out of the blue cards, there are two even cards; *B*2 and *B*4.)
- $P(B|E)=\frac{2}{3}$. (There are three even-numbered cards: *R*2, *B*2, and *B*4. Out of the even-numbered cards, to are blue; *B*2 and *B*4.)
- The events *R* and *B* are mutually exclusive because P(R∩B)=0$P(R\cap B)$=0.
- Let *G* = card with a number greater than 3. $G=\{B4, B5\}$. $P(G)=\frac{2}{8}$. Let *H* = blue card numbered between one and four, inclusive. *H* = {*B*1, *B*2, *B*3, *B*4} $P(G|H)=\frac{1}{4}$.. (The only card in *H* that has a number greater than three is *B*4.) Since $\frac{2}{8}=\frac{1}{4}$, $P(G)=P(G|H)$, which means that *G* and *H* are independent.

TRY IT 3.11

In a basketball arena,

- 70% of the fans are rooting for the home team.
- 25% of the fans are wearing blue.
- 20% of the fans are wearing blue and are rooting for the away team.
- Of the fans rooting for the away team, 67% are wearing blue.

Let *A* be the event that a fan is rooting for the away team.
Let *B* be the event that a fan is wearing blue.

Are the events of rooting for the away team and wearing blue independent? Are they mutually exclusive?

EXAMPLE 3.12

In a particular college class, 60% of the students are female. Fifty percent of all students in the class have long hair. Forty-five percent of the students are female and have long hair. Of the female students, 75% have long hair. Let *F* be the event that a student is female. Let *L* be the event that a student has long hair. One student is picked randomly. Are the events of being female and having long hair independent?

- The following probabilities are given in this example:
- $P(F)=0.60;P(L)=0.50$
- $P(F \cap L)=0.45$
- $P(L|F)=0.75$

NOTE

The choice you make depends on the information you have. You could use the first or last condition on the list for this example. You do not know $P(F|L)$ yet, so you cannot use the second condition.

Solution 1

Check whether $P(F \cap L)=P(F)P(L)$. We are given that $P(F \cap L)=0.45$, but $P(F)P(L)=(0.60)(0.50)=0.30$. The events of being female and having long hair are not independent because $P(F \cap L)$ does not equal $P(F)P(L)$.

Check whether $P(L|F)$ equals $P(L)$. We are given that $P(L|F)=0.75$, but $P(L)=0.50$; they are not equal. The events of being female and having long hair are not independent.

Interpretation of Results

The events of being female and having long hair are not independent; knowing that a student is female changes the probability that a student has long hair.

TRY IT 3.12

Mark is deciding which route to take to work. His choices are *I* = the Interstate and *F* = Fifth Street.

- $P(I)=0.44$ and $P(F)=0.56$
- $P(I \cap F)=0$ because Mark will take only one route to work.

What is the probability of $P(I \cup F)$?

EXAMPLE 3.13

Problem

a. Toss one fair coin (the coin has two sides, *H* and *T*). The outcomes are ________. Count the outcomes. There are ____ outcomes.
b. Toss one fair, six-sided die (the die has 1, 2, 3, 4, 5 or 6 dots on a side). The outcomes are ________________. Count the outcomes. There are ___ outcomes.
c. Multiply the two numbers of outcomes. The answer is _______.
d. If you flip one fair coin and follow it with the toss of one fair, six-sided die, the answer to c is the number of outcomes (size of the sample space). What are the outcomes? (Hint: Two of the outcomes are H1 and *T*6.)
e. Event *A* = heads (*H*) on the coin followed by an even number (2, 4, 6) on the die. *A* = {_________________}. Find *P*(*A*).
f. Event *B* = heads on the coin followed by a three on the die. *B* = {________}. Find *P*(*B*).
g. Are *A* and *B* mutually exclusive? (Hint: What is P(A∩B)? If P(A∩B)=0$P(A\cap B)=0$, then *A* and *B* are mutually exclusive.)
h. Are *A* and *B* independent? (Hint: Is P(A∩B)=P(A)P(B)? If P(A∩B)=P(A)P(B), then *A* and *B* are independent. If not, then they are dependent).

Solution 3.13

a. *H* and *T*; 2
b. 1, 2, 3, 4, 5, 6; 6
c. 2(6) = 12
d. *T*1, *T*2, *T*3, *T*4, *T*5, *T*6, *H*1, *H*2, *H*3, *H*4, *H*5, *H*6
e. $A = \{H2, H4, H6\}$; $P(A) = \frac{3}{12}$
f. $B = \{H3\}$; $P(B) = \frac{1}{12}$
g. $P(A \cap B) = 0.P(A)P(B) = (\frac{3}{12}).P(A \cap B)$does not $P(A)P(B)$,, so *A* and *B* are dependent.

TRY IT 3.13

A box has two balls, one white and one red. We select one ball, put it back in the box, and select a second ball (sampling with replacement). Let *T* be the event of getting the white ball twice, *F* the event of picking the white ball first, *S* the event of picking the white ball in the second drawing.

a. Compute P(T).
b. Compute P(T|F).

c. Are *T* and *F* independent?.
d. Are *F* and *S* mutually exclusive?
e. Are *F* and *S* independent?

3.3 TWO BASIC RULES OF PROBABILITY

When calculating probability, there are two rules to consider when determining if two events are independent or dependent and if they are mutually exclusive or not.

The Multiplication Rule

If A and B are two events defined on a sample space, then: P(A∩B)=P(B)P(A|B). We can think of the intersection symbol as substituting for the word "and".

This rule may also be written as: $P\left(A\middle|B\right)=\frac{P(A\cap B)}{P(B)}$

This equation is read as the probability of *A* given *B* equals the probability of *A* and *B* divided by the probability of *B*.

If *A* and *B* are independent, then P(A|B)=P(A). Then P(A∩B)=P(A) (B) becomes P(A|B)=P(A) if *A* and *B* are independent.

One easy way to remember the multiplication rule is that the word "and" means that the event has to satisfy two conditions. For example the name drawn from the class roster is to be both a female and a sophomore. It is harder to satisfy two conditions than only one and of course when we multiply fractions the result is always smaller. This reflects the increasing difficulty of satisfying two conditions.

The Addition Rule

If *A* and *B* are defined on a sample space, then: P(A∪B)=P(A)+P(B)-P(A∩B). We can think of the union symbol substituting for the word "or". The reason we subtract the intersection of *A* and *B* is to keep from double counting elements that are in both *A* and *B*.

If *A* and *B* are **mutually exclusive**, then P(A∩B)=0. Then P(A∪B)=P(A)+P(B)-P(A∩B) becomes P(A∪B)=P(A)+P(B).

EXAMPLE 3.14

Klaus is trying to choose where to go on vacation. His two choices are: *A* = New Zealand and *B* = Alaska

- Klaus can only afford one vacation. The probability that he chooses *A* is *P*(*A*) = 0.6 and the probability that he chooses *B* is *P*(*B*) = 0.35.
- P(A∩B)=0$P(A\cap B)=0$ because Klaus can only afford to take one vacation
- Therefore, the probability that he chooses either New Zealand or Alaska is

P(A∪B)=P(A)+P(B)=0.6+0.35=0.95. Note that the probability that he does not choose to go anywhere on vacation must be 0.05.

EXAMPLE 3.15

Carlos plays college soccer. He makes a goal 65% of the time he shoots. Carlos is going to attempt two goals in a row in the next game. A = the event Carlos is successful on his first attempt. $P(A) = 0.65$. B = the event Carlos is successful on his second attempt. $P(B) = 0.65$. Carlos tends to shoot in streaks. The probability that he makes the second goal | that he made the first goal is 0.90.

a. What is the probability that he makes both goals?

Solution 3.15

a. The problem is asking you to find $P(A \cap B) = P(B \cap A)$. Since $P(B \mid A) = 0.90$:
   $P(B \cap A) = P(B \mid A)\ P(A) = (0.90)(0.65) = 0.585$
   Carlos makes the first and second goals with probability 0.585.

Solution 3.15

b. The problem is asking you to find $P(A \cup B)$.
   $P(A \cup B) = P(A) + P(B) - P(A \cap B) = 0.65 + 0.65 - 0.585 = 0.715$
   Carlos makes either the first goal or the second goal with probability 0.715.
c. Are A and B independent?

Solution 3.15

c. No, they are not, because $P(B \cap A) = 0.585$.
   $P(B)P(A) = (0.65)(0.65) = 0.423$
   $0.423 \neq 0.585 = P(B \cap A)$
   So, $P(B \cap A)$ is **not** equal to $P(B)P(A)$.
d. Are A and B mutually exclusive?

Solution 3.15

d. No, they are not because $P(A \cap B) = 0.585$.
   To be mutually exclusive, $P(A \cap B)$ must equal zero.

TRY IT 3.15

Helen plays basketball. For free throws, she makes the shot 75% of the time. Helen must now attempt two free throws. C = the event that Helen makes the first shot. $P(C) = 0.75$. D = the event Helen makes the second shot. $P(D) = 0.75$. The probability that Helen makes the second free throw given that she made the first is 0.85. What is the probability that Helen makes both free throws?

EXAMPLE 3.16

A community swim team has 150 members. Seventy-five of the members are advanced swimmers. Forty-seven of the members are intermediate swimmers. The remainder are novice swimmers. Forty of the advanced swimmers practice four times a week. Thirty of the intermediate swimmers practice four times a week. Ten of the novice swimmers practice four times a week. Suppose one member of the swim team is chosen randomly. What is the probability that the member is a novice swimmer?

Solution 3.16

a. $\frac{28}{150}$

b. What is the probability that the member practices four times a week?

Solution 3.16

b. $\frac{80}{150}$

c. What is the probability that the member is an advanced swimmer and practices four times a week?

Solution 3.16

c. $\frac{40}{150}$

d. What is the probability that a member is an advanced swimmer and an intermediate swimmer? Are being an advanced swimmer and an intermediate swimmer mutually exclusive? Why or why not?

Solution 3.16

d. P(advanced ∩ intermediate) = 0, so these are mutually exclusive events. A swimmer cannot be an advanced swimmer and an intermediate swimmer at the same time.

e. Are being a novice swimmer and practicing four times a week independent events? Why or why not?

Solution 3.16

e. No, these are not independent events.
P(novice ∩ practices four times per week) = 0.0667
P(novice)P(practices four times per week) = 0.0996
0.0667 ≠ 0.0996

TRY IT 3.16

A school has 200 seniors of whom 140 will be going to college next year. Forty will be going directly to work. The remainder are taking a gap year. Fifty of the seniors going to college play sports. Thirty of the seniors going directly to work play sports. Five of the seniors taking a gap year play sports. What is the probability that a senior is taking a gap year?

EXAMPLE 3.17

Felicity attends Modesto JC in Modesto, CA. The probability that Felicity enrolls in a math class is 0.2 and the probability that she enrolls in a speech class is 0.65. The probability that she enrolls in a math class || that she enrolls in speech class is 0.25.

Let: M = math class, S = speech class, $M||S$ = math given speech

a. What is the probability that Felicity enrolls in math and speech? Find $P(M \cap S) = P(M|S)P(S)$.
b. What is the probability that Felicity enrolls in math or speech classes? Find $P(M \cup S) = P(M) + P(S) - P(M \cap S)$.
c. Are M and S independent? Is $P(M|S) = P(M)$?
d. Are M and S mutually exclusive? Is $P(M \cap S) = 0$?

Solution 3.17

a. 0.1625, b. 0.6875, c. No, d. No

TRY IT 3.17

A student goes to the library. Let events B = the student checks out a book and D = the student checks out a DVD.

Suppose that $P(B) = 0.40$, $P(D) = 0.30$ and $P(D|B) = 0.5$.

a. Find $P(B \cap D)$.
b. Find $P(B \cup D)$.

EXAMPLE 3.18

Studies show that about one woman in seven (approximately 14.3%) who live to be 90 will develop breast cancer. Suppose that of those women who develop breast cancer, a test is negative 2% of the time. Also suppose that in the general population of women, the test for breast cancer is negative about 85% of the time. Let B = woman develops breast cancer and let N = tests negative. Suppose one woman is selected at random.

a. What is the probability that the woman develops breast cancer? What is the probability that woman tests negative?

Solution 3.18

a. $P(B) = 0.143$; $P(N) = 0.85$
b. Given that the woman has breast cancer, what is the probability that she tests negative?

Solution 3.18

b. $P(N \mid B) = 0.02$

c. What is the probability that the woman has breast cancer AND tests negative?

Solution 3.18

c. $P(B \cap N) = P(B)P(N \mid B) = (0.143)(0.02) = 0.0029$

d. What is the probability that the woman has breast cancer or tests negative?

Solution 3.18

d. $P(B \cup N) = P(B) + P(N) - P(B \cap N) = 0.143 + 0.85 - 0.0029 = 0.9901$

e. Are having breast cancer and testing negative independent events?

Solution 3.18

e. No. $P(N) = 0.85$; $P(N \mid B) = 0.02$. So, $P(N \mid B)$ does not equal $P(N)$.

f. Are having breast cancer and testing negative mutually exclusive?

Solution 3.18

f. No. $P(B \cap N) = 0.0029$. For B and N to be mutually exclusive, $P(B \cap N)$ must be zero.

TRY IT 3.18

A school has 200 seniors of whom 140 will be going to college next year. Forty will be going directly to work. The remainder are taking a gap year. Fifty of the seniors going to college play sports. Thirty of the seniors going directly to work play sports. Five of the seniors taking a gap year play sports. What is the probability that a senior is going to college and plays sports?

EXAMPLE 3.19

Refer to the information in Example 3.18. P = tests positive.

a. Given that a woman develops breast cancer, what is the probability that she tests positive. Find $P(P|B) = 1 - P(N|B)$.
b. What is the probability that a woman develops breast cancer and tests positive. Find $P(B \cap P) = P(P|B)P(B)$.
c. What is the probability that a woman does not develop breast cancer. Find $P(B') = 1 - P(B)$.
d. What is the probability that a woman tests positive for breast cancer. Find $P(P) = 1 - P(N)$.

Solution 3.19

a. 0.98; b. 0.1401; c. 0.857; d. 0.15

TRY IT 3.19

A student goes to the library. Let events B = the student checks out a book and D = the student checks out a DVD. Suppose that $P(B) = 0.40$, $P(D) = 0.30$ and $P(D|B) = 0.5$.

a. Find $P(B')$.
b. Find $P(D \cap B)$.
c. Find $P(B|D)$.
d. Find $P(D \cap B')$.
e. Find $P(D|B')$.

3.4 CONTINGENCY TABLES AND PROBABILITY TREES

Contingency Tables

A **contingency table** provides a way of portraying data that can facilitate calculating probabilities. The table helps in determining conditional probabilities quite easily. The table displays sample values in relation to two different variables that may be dependent or contingent on one another. Later on, we will use contingency tables again, but in another manner.

EXAMPLE 3.20

Suppose a study of speeding violations and drivers who use cell phones produced the following fictional data:

Table 3.2

	Speeding violation in the last year	No speeding violation in the last year	Total
Uses cell phone while driving	25	280	305
Does not use cell phone while driving	45	405	450
Total	70	685	755

The total number of people in the sample is 755. The row totals are 305 and 450. The column totals are 70 and 685. Notice that 305 + 450 = 755 and 70 + 685 = 755.

Calculate the following probabilities using the table.

a. Find P(Driver is a cell phone user).

Solution 3.20

a. $\frac{\text{number of cell phone users}}{\text{total number in study}} = \frac{305}{755}$

b. Find P(Driver had no violation in the last year).

Solution 3.20

a. $\frac{\text{number that had no violation}}{\text{total number in study}} = \frac{685}{755}$

b. Find P(Driver had no violation in the last year ∩ was a cell phone user).

Solution 3.20

c. $\frac{280}{755}$

b. Find P(Driver is a cell phone user ∪ driver had no violation in the last year).

Solution 3.20

d. $\left(\frac{305}{755} + \frac{685}{755}\right) - \frac{280}{755} = \frac{710}{755}$

e. Find P(Driver is a cell phone user | driver had a violation in the last year).

Solution 3.20

e. $\frac{25}{70}$ (The sample space is reduced to the number of drivers who had a violation.)

f. Find P(Driver had no violation last year | driver was not a cell phone user)

Solution 3.20

f. $\frac{405}{450}$ (The sample space is reduced to the number of drivers who were not cell phone users.)

TRY IT 3.20

Table 3.3 shows the number of athletes who stretch before exercising and how many had injuries within the past year.

Table 3.3

	Injury in last year	No injury in last year	Total
Stretches	55	295	350
Does not stretch	231	219	450
Total	286	514	800

a. What is P(athlete stretches before exercising)?

b. What is P(athlete stretches before exercising||no injury in the last year)?

EXAMPLE 3.21

Table 3.4 shows a random sample of 100 hikers and the areas of hiking they prefer.

Table 3.4 Hiking Area Preference

Sex	The coastline	Near lakes and streams	On mountain peaks	Total
Female	18	16	___	45
Male	___	___	14	55
Total	___	41	___	___

a. Complete the table.

Solution 3.21

a.

Table 3.5 Hiking Area Preference

Sex	The coastline	Near lakes and streams	On mountain peaks	Total
Female	18	16	11	45
Male	16	25	14	55
Total	34	41	25	100

b. Are the events "being female" and "preferring the coastline" independent events?

Let F = being female and let C = preferring the coastline.

1. Find P(F∩C)$P(F \cap C)$.
2. Find $P(F)P(C)$

Are these two numbers the same? If they are, then F and C are independent. If they are not, then F and C are not independent.

Solution 3.21

b.

1. $P(F \cap C) = \frac{18}{100} = 0.18$
2. $P(F)P(C) = \left(\frac{45}{100}\right)\left(\frac{34}{100}\right) = (0.45)(0.34) = 0.153$

$P(F \cap C) \neq P(F)P(C)$, so the events F and C are not independent.

c. Find the probability that a person is male given that the person prefers hiking near lakes and streams. Let M =being male, and let L = prefers hiking near lakes and streams.

1. What word tells you this is a conditional?
2. Fill in the blanks and calculate the probability: P (___ | ___) = ___.
3. Is the sample space for this problem all 100 hikers? If not, what is it?

Solution 3.21

c.

1. $P(M \mid L) = \frac{25}{41}$ ven' tells you that this is a conditional.
2.
3. No, the sample space for this problem is the 41 hikers who prefer lakes and streams.

d. Find the probability that a person is female or prefers hiking on mountain peaks. Let F = being female, and let P = prefers mountain peaks.

1. Find P(F).
2. Find P(P).

3. Find $P(F \cap P)$.
4. Find $P(F \cup P)$.

Solution 3.21

d.

1. $P(F) = \frac{45}{100}$
2. $P(P) = \frac{25}{100}$
3. $P(F \cap P) = \frac{11}{100}$
4. $P(F \cup P) = \frac{45}{100} + \frac{25}{100} - \frac{11}{100} = \frac{59}{100}$

TRY IT 3.21

Table 3.6 shows a random sample of 200 cyclists and the routes they prefer. Let M = males and H = hilly path.

Table 3.6

Gender	Lake path	Hilly path	Wooded path	Total
Female	45	38	27	110
Male	26	52	12	90
Total	71	90	39	200

a. Out of the males, what is the probability that the cyclist prefers a hilly path?
b. Are the events “being male” and “preferring the hilly path” independent events?

EXAMPLE 3.22

Muddy Mouse lives in a cage with three doors. If Muddy goes out the first door, the probability that he gets caught by Alissa the cat is $\frac{1}{5}$ and the probability he is not caught is $\frac{4}{5}$. If he goes out the second door, the probability he gets caught by Alissa is $\frac{1}{4}$ and the probability he is not caught is $\frac{3}{4}$. The probability that Alissa catches Muddy coming out of the third door is $\frac{1}{2}$ and the probability she does not catch Muddy is $\frac{1}{2}$. It is equally likely that Muddy will choose any of the three doors so the probability of choosing each door is $\frac{1}{3}$.

Table 3.7 Door Choice

Caught or not	Door one	Door two	Door three	Total
Caught	$\frac{1}{15}$	$\frac{1}{12}$	$\frac{1}{6}$	----
Not caught	$\frac{4}{15}$	$\frac{3}{12}$	$\frac{1}{6}$	----
Total	----	----	----	1

- The first entry $\frac{1}{15} = \left(\frac{1}{5}\right)\left(\frac{1}{3}\right)$ is P(Door One ∩ Caught)
- The entry $\frac{4}{15} = \left(\frac{4}{5}\right)\left(\frac{1}{3}\right)$ is P(Door One ∩ Not Caught)

Verify the remaining entries.

a. Complete the probability contingency table. Calculate the entries for the totals. Verify that the lower-right corner entry is 1.

Solution 3.22

a.

Table 3.8 Door Choice

Caught or not	Door one	Door two	Door three	Total
Caught	$\frac{1}{15}$	$\frac{1}{12}$	$\frac{1}{6}$	$\mathbf{\frac{19}{60}}$
Not caught	$\frac{4}{15}$	$\frac{3}{12}$	$\frac{1}{6}$	$\mathbf{\frac{41}{60}}$
Total	$\mathbf{\frac{5}{15}}$	$\mathbf{\frac{4}{12}}$	$\mathbf{\frac{2}{6}}$	1

b. What is the probability that Alissa does not catch Muddy?

Solution 3.22

b. $\frac{41}{60}$

c. What is the probability that Muddy chooses Door One ∪ Door Two given that Muddy is caught by Alissa?

Solution 3.22

c. $\frac{9}{19}$

EXAMPLE 3.23

Table 3.9 contains the number of crimes per 100,000 inhabitants from 2008 to 2011 in the U.S.

Table 3.9 United States Crime Index Rates Per 100,000 Inhabitants 2008–2011

Year	Robbery	Burglary	Rape	Vehicle	Total
2008	145.7	732.1	29.7	314.7	
2009	133.1	717.7	29.1	259.2	
2010	119.3	701	27.7	239.1	
2011	113.7	702.2	26.8	229.6	
Total					

TOTAL each column and each row. Total data = 4,520.7

a. Find P(2009 ∩ Robbery).
b. Find P(2010 ∩ Burglary).
c. Find P(2010 ∪ Burglary).
d. Find P(2011|Rape).
e. Find P(Vehicle|2008).

Solution 3.23

a. 0.0294, b. 0.1551, c. 0.7165, d. 0.2365, e. 0.2575

TRY IT 3.23

Table 3.10 relates the weights and heights of a group of individuals participating in an observational study.

Table 3.10

Weight/height	Tall	Medium	Short	Totals
Obese	18	28	14	
Normal	20	51	28	
Underweight	12	25	9	
Totals				

a. Find the total for each row and column
b. Find the probability that a randomly chosen individual from this group is Tall.
c. Find the probability that a randomly chosen individual from this group is Obese and Tall.
d. Find the probability that a randomly chosen individual from this group is Tall given that the individual is Obese.
e. Find the probability that a randomly chosen individual from this group is Obese given that the individual is Tall.
f. Find the probability a randomly chosen individual from this group is Tall and Underweight.
g. Are the events Obese and Tall independent?

Tree Diagrams

Sometimes, when the probability problems are complex, it can be helpful to graph the situation. Tree diagrams can be used to visualize and solve conditional probabilities.

Tree Diagrams

A **tree diagram** is a special type of graph used to determine the outcomes of an experiment. It consists of "branches" that are labeled with either frequencies or probabilities. Tree diagrams can make some probability problems easier to visualize and solve. The following example illustrates how to use a tree diagram.

EXAMPLE 3.24

In an urn, there are 11 balls. Three balls are red (*R*) and eight balls are blue (*B*). Draw two balls, one at a time, with replacement. "With replacement" means that you put the first ball back in the urn before you select the second ball. The tree diagram using frequencies that show all the possible outcomes follows.

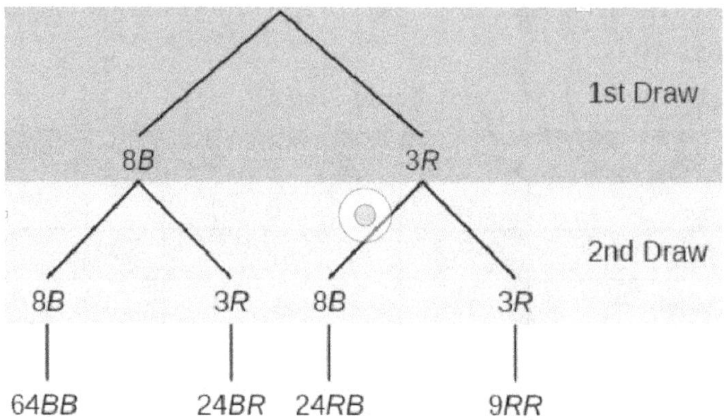

Figure 3.2 Total = 64 + 24 + 24 + 9 = 121

The first set of branches represents the first draw. The second set of branches represents the second draw. Each of the outcomes is distinct. In fact, we can list each red ball as *R*1, *R*2, and *R*3 and each blue ball as *B*1, *B*2, *B*3, *B*4, *B*5, *B*6, *B*7, and *B*8. Then the nine *RR* outcomes can be written as:

*R*1*R*1; *R*1*R*2; *R*1*R*3; *R*2*R*1; *R*2*R*2; *R*2*R*3; *R*3*R*1; *R*3*R*2; *R*3*R*3

The other outcomes are similar.

There are a total of 11 balls in the urn. Draw two balls, one at a time, with replacement. There are 11(11) = 121 outcomes, the size of the sample space.

a List the 24 *BR* outcomes: *B*1*R*1, *B*1*R*2, *B*1*R*3, ...

Solution 3.24

a. *B*1*R*1; *B*1*R*2; *B*1*R*3; *B*2*R*1; *B*2*R*2; *B*2*R*3; *B*3*R*1; *B*3*R*2; *B*3*R*3; *B*4*R*1; *B*4*R*2; *B*4*R*3; *B*5*R*1; *B*5*R*2; *B*5*R*3; *B*6*R*1; *B*6*R*2; *B*6*R*3; *B*7*R*1; *B*7*R*2; *B*7*R*3; *B*8*R*1; *B*8*R*2; *B*8*R*3

b. Using the tree diagram, calculate *P*(*RR*).

Solution 3.24

b. $P(RR) = \left(\frac{3}{11}\right)\left(\frac{3}{11}\right) = \frac{9}{121}$

c. Using the tree diagram, calculate $P(RB \cup BR)$.

Solution 3.24

c. $P(RB \cup BR) = \left(\frac{3}{11}\right)\left(\frac{8}{11}\right) + \left(\frac{8}{11}\right)\left(\frac{3}{11}\right) = \frac{48}{121}$

d. Using the tree diagram, calculate $P(R$ on 1st draw $\cap$ B on 2nd draw) .

Solution 3.24

d. $P(R$ on 1st draw $\cap$ B on 2nd draw) = $\left(\frac{3}{11}\right)\left(\frac{8}{11}\right) = \frac{24}{121}$

e. Using the tree diagram, calculate $P(R$ on 2nd draw | B on 1st draw).

Solution 3.24

e. $P(R$ on 2nd draw | B on 1st draw) = $P(R$ on 2nd | B on 1st)= $\frac{24}{88} = \frac{3}{11}$

This problem is a conditional one. The sample space has been reduced to those outcomes that already have a blue on the first draw. There are 24 + 64 = 88 possible outcomes (24 *BR* and 64 *BB*). Twenty-four of the 88 possible outcomes are *BR*. $\frac{24}{88} = \frac{3}{11}$.

f. Using the tree diagram, calculate $P(BB)$.

Solution 3.24

f. $P(BB) = \frac{64}{121}$

g. Using the tree diagram, calculate $P(B$ on the 2nd draw | R on the first draw).

Solution 3.24

g. $P(B$ on 2nd draw | R on 1st draw) $= \frac{8}{11}$

There are 9 + 24 outcomes that have *R* on the first draw (9 *RR* and 24 *RB*). The sample space is then 9 + 24 = 33. 24 of the 33 outcomes have *B* on the second draw. The probability is then $\frac{24}{33}$.

TRY IT 3.24

In a standard deck, there are 52 cards. 12 cards are face cards (event *F*) and 40 cards are not face cards (event *N*). Draw two cards, one at a time, with replacement. All possible outcomes are shown in the tree diagram as frequencies. Using the tree diagram, calculate $P(FF)$.

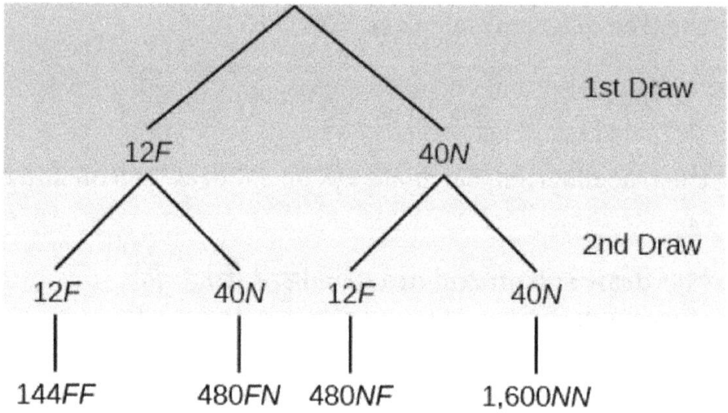

Figure 3.33

EXAMPLE 3.25

An urn has three red marbles and eight blue marbles in it. Draw two marbles, one at a time, this time without replacement, from the urn. **"Without replacement"** means that you do not put the first ball back before you select the second marble. Following is a tree diagram for this situation. The branches are labeled with probabilities instead of frequencies. The numbers at the ends of the branches are calculated by multiplying the numbers on the two corresponding branches, for example, $\left(\frac{3}{11}\right)\left(\frac{2}{10}\right)=\frac{6}{110}$.

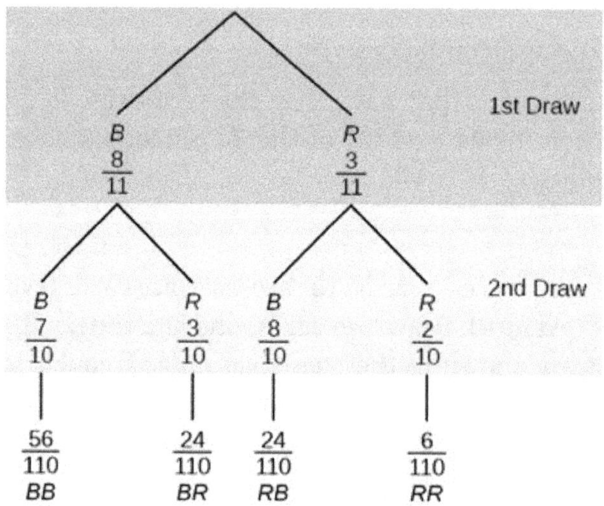

$$\text{Total} = \frac{56+24+24+6}{110} = \frac{110}{110} = 1$$

Figure 3.4

NOTE

If you draw a red on the first draw from the three red possibilities, there are two red marbles left to draw on the second draw. You do not put back or replace the first marble after you have drawn it. You draw **without replacement**, so that on the second draw there are ten marbles left in the urn.

Calculate the following probabilities using the tree diagram.

a. $P(RR)$ = ________

Solution 3.25

a. $P(RR) = \left(\frac{3}{11}\right)\left(\frac{2}{10}\right) = \frac{6}{110}$

b. Fill in the blanks:

$P(RB \cup BR) = \left(\frac{3}{11}\right)\left(\frac{8}{10}\right) + (\_\_)(\_\_) = \frac{48}{110}$

Solution 3.25

b. $P(RB \cup BR) = \left(\frac{3}{11}\right)\left(\frac{8}{10}\right) + \left(\frac{8}{11}\right)\left(\frac{3}{10}\right) = \frac{48}{110}$

c. $P(R \text{ on 2nd} \mid B \text{ on 1st})$ =

Solution 3.25

c. $P(R \text{ on 2nd} \mid B \text{ on 1st}) = \frac{3}{10}$

d. Fill in the blanks.

$P(R \text{ on 1st} \cap B \text{ on 2nd}) = (\_\_)(\_\_) = \frac{24}{100}$

Solution 3.25

e. $P(BB) = \left(\frac{8}{11}\right)\left(\frac{7}{10}\right)$

f. Find $P(B \text{ on 2nd} \mid R \text{ on 1st})$.

Solution 3.25

f. Using the tree diagram, $P(B \text{ on 2nd} \mid R \text{ on 1st}) = P(R \mid B) = \frac{8}{10}$.

If we are using probabilities, we can label the tree in the following general way.

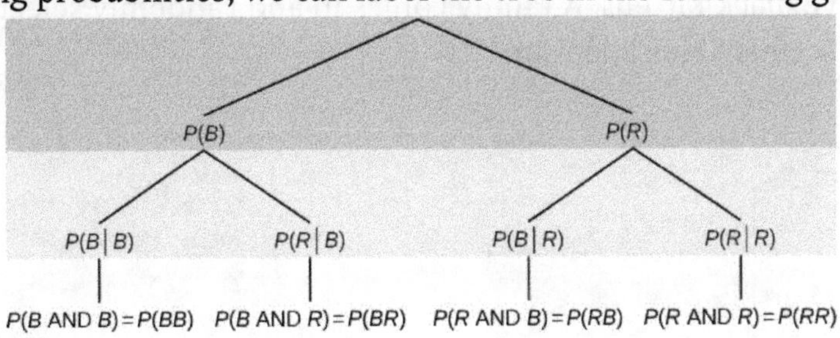

- $P(R||R)$ here means $P(R \text{ on 2nd}|R \text{ on 1st})$
- $P(B||R)$ here means $P(B \text{ on 2nd}|R \text{ on 1st})$
- $P(R||B)$ here means $P(R \text{ on 2nd}|B \text{ on 1st})$
- $P(B||B)$ here means $P(B \text{ on 2nd}|B \text{ on 1st})$

TRY IT 3.25

In a standard deck, there are 52 cards. Twelve cards are face cards (*F*) and 40 cards are not face cards (*N*). Draw two cards, one at a time, without replacement. The tree diagram is labeled with all possible probabilities.

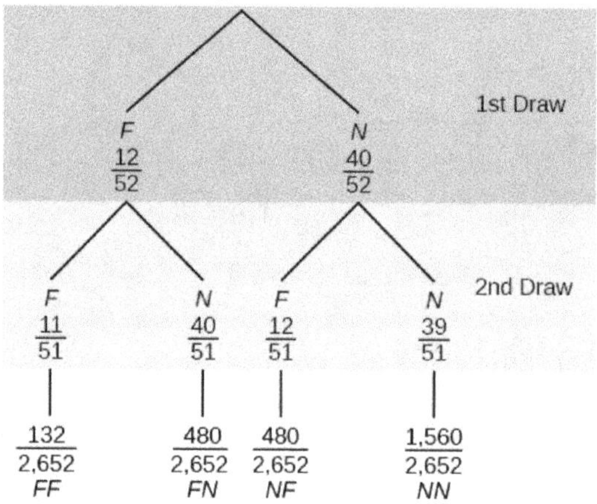

Figure 3.5

a. Find P(FN∪NF).
b. Find $P(N||F)$.
c. Find *P*(at most one face card).
   Hint: "At most one face card" means zero or one face card.
d. Find *P*(at least one face card).
   Hint: "At least one face card" means one or two face cards.

EXAMPLE 3.26

A litter of kittens available for adoption at the Humane Society has four tabby kittens and five black kittens. A family comes in and randomly selects two kittens (without replacement) for adoption.

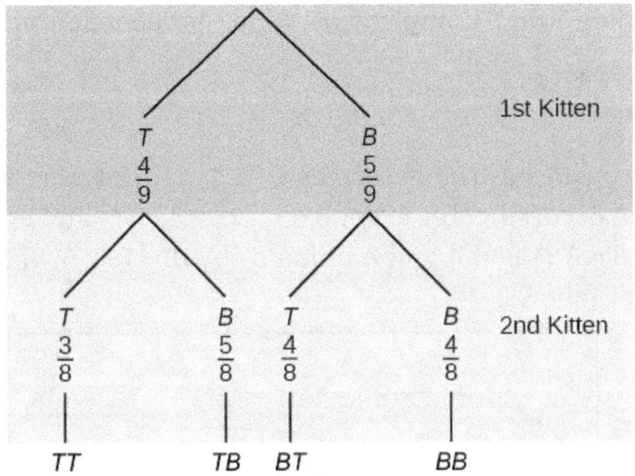

a. What is the probability that both kittens are tabby?

a. $\left(\frac{1}{2}\right)\left(\frac{1}{2}\right)$ b. $\left(\frac{4}{9}\right)\left(\frac{4}{9}\right)$ c. $\left(\frac{4}{9}\right)\left(\frac{3}{8}\right)$ d. $\left(\frac{4}{9}\right)\left(\frac{5}{9}\right)$

b. What is the probability that one kitten of each coloring is selected?

a. $\left(\frac{4}{9}\right)\left(\frac{5}{9}\right)$ b. $\left(\frac{4}{9}\right)\left(\frac{5}{8}\right)$ c. $\left(\frac{4}{9}\right)\left(\frac{5}{9}\right)+\left(\frac{5}{9}\right)\left(\frac{4}{9}\right)$ d. $\left(\frac{4}{9}\right)\left(\frac{5}{8}\right)+\left(\frac{5}{9}\right)\left(\frac{4}{8}\right)$

c. What is the probability that a tabby is chosen as the second kitten when a black kitten was chosen as the first?

d. What is the probability of choosing two kittens of the same color?

Solution 3.26

a. c, b. d, c. $\frac{4}{8}$, d. $\frac{32}{72}$

TRY IT 3.26

Suppose there are four red balls and three yellow balls in a box. Two balls are drawn from the box without replacement. What is the probability that one ball of each coloring is selected?

3.5 VENN DIAGRAMS

A Venn diagram is a picture that represents the outcomes of an experiment. It generally consists of a box that represents the sample space S together with circles or ovals. The circles or ovals represent events. Venn diagrams also help us to convert common English words into mathematical terms that help add precision.

Venn diagrams are named for their inventor, John Venn, a mathematics professor at Cambridge and an Anglican minister. His main work was conducted during the late 1870's and gave rise to a whole branch of mathematics and a new way to approach issues of logic. We will develop the probability rules just covered using this powerful way to demonstrate the probability postulates including the Addition

Rule, Multiplication Rule, Complement Rule, Independence, and Conditional Probability.

EXAMPLE 3.27

Suppose an experiment has the outcomes 1, 2, 3, ... , 12 where each outcome has an equal chance of occurring. Let event A = {1, 2, 3, 4, 5, 6} and event B = {6, 7, 8, 9}. Then A intersect B = A ∩ B ={6} and A union B = AUB={1, 2, 3, 4, 5, 6, 7, 8, 9}.... The Venn diagram is as follows:

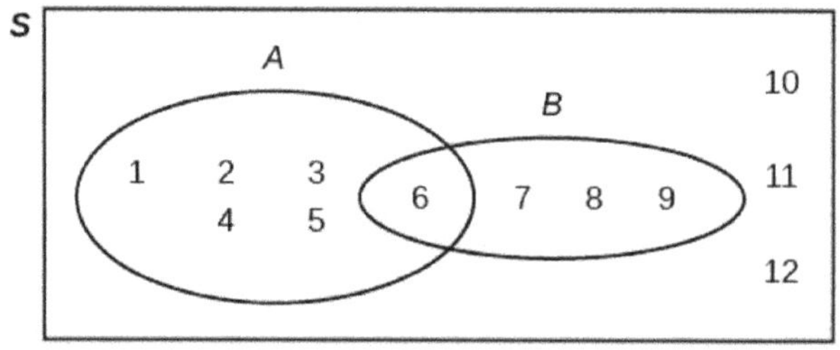

Figure 3.6

Figure 3.6 shows the most basic relationship among these numbers. First, the numbers are in groups called sets; set A and set B. Some number are in both sets; we say in set A ∩∩ in set B. The English word "and" means inclusive, meaning having the characteristics of both A and B, or in this case, being a part of both A and B. This condition is called the INTERSECTION of the two sets. All members that are part of both sets constitute the intersection of the two sets. The intersection is written as A ∩ B where ∩ is the mathematical symbol for intersection. The statement A ∩ B is read as "A intersect B." You can remember this by thinking of the intersection of two streets.

There are also those numbers that form a group that, for membership, the number must be in either one or the other group. The number does not have to be in BOTH groups, but instead only in either one of the two. These numbers are called the UNION of the two sets and in this case they are the numbers 1-5 (from A exclusively), 7-9 (from set B exclusively) and also 6, which is in both sets A and B. The symbol for the UNION is ∪∪, thus A∪B=$A \cup B$= numbers 1-9, but excludes number 10, 11, and 12. The values 10, 11, and 12 are part of the universe, but are not in either of the two sets.

Translating the English word "AND" into the mathematical logic symbol ∩∩, intersection, and the word "OR" into the mathematical symbol ∪∪, union, provides a very precise way to discuss the issues of probability and logic. The general terminology for the three areas of the Venn diagram in Figure 3.6 is shown in Figure 3.7.

TRY IT 3.27

Suppose an experiment has outcomes black, white, red, orange, yellow, green, blue, and purple, where each outcome has an equal chance of occurring. Let event *C* = {green, blue, purple} and event *P* = {red, yellow, blue}. Then C ∩ P={blue} and C∪P={green, blue, purple, red, yellow}. Draw a Venn diagram representing this situation.

EXAMPLE 3.28

Flip two fair coins. Let *A* = tails on the first coin. Let *B* = tails on the second coin. Then *A* = {*TT*, *TH*} and *B* = {*TT*, *HT*}. Therefore, A∩B={TT}. A∪B={TH, TT, HT}.

The sample space when you flip two fair coins is *X* = {*HH*, *HT*, *TH*, *TT*}. The outcome *HH* is in NEITHER *A* NOR *B*. The Venn diagram is as follows:

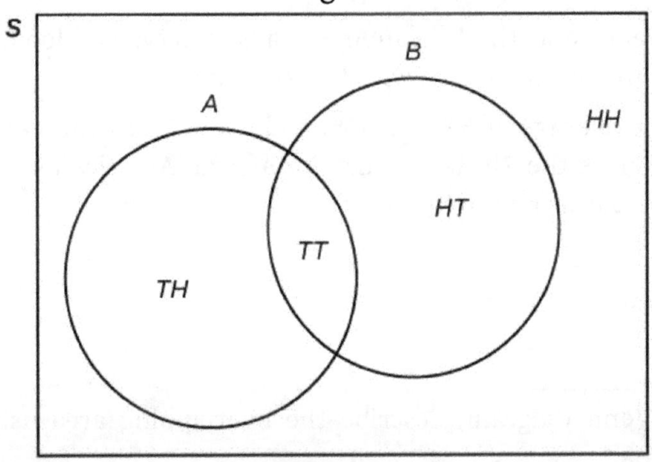

Figure 3.7

TRY IT 3.28

Roll a fair, six-sided die. Let *A* = a prime number of dots is rolled. Let *B* = an odd number of dots is rolled. Then *A* = {2, 3, 5} and *B* = {1, 3, 5}. Therefore, A∩B={3, 5}. A∪B={1, 2, 3, 5}. The sample space for rolling a fair die is *S* = {1, 2, 3, 4, 5, 6}. Draw a Venn diagram representing this situation.

EXAMPLE 3.29

A person with type O blood and a negative Rh factor (Rh-) can donate blood to any person with any blood type. Four percent of African Americans have type O blood and a negative RH factor, 5-10% of African Americans have the Rh- factor, and 51% have type O blood.

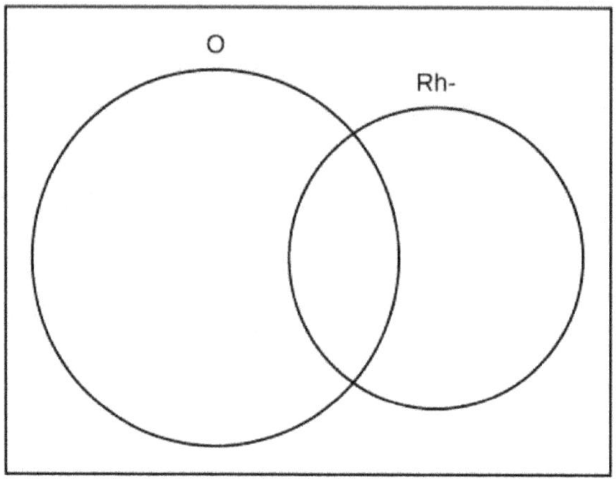

Figure 3.8

The "O" circle represents the African Americans with type O blood. The "Rh-" oval represents the African Americans with the Rh- factor.

We will take the average of 5% and 10% and use 7.5% as the percent of African Americans who have the Rh- factor. Let O = African American with Type O blood and R = African American with Rh- factor.

a. $P(O)$ = ___________
b. $P(R)$ = ___________
c. $P(O \cap R)$= ___________
d. $P(O \cup R)$= ____________
e. In the Venn Diagram, describe the overlapping area using a complete sentence.
f. In the Venn Diagram, describe the area in the rectangle but outside both the circle and the oval using a complete sentence.

EXAMPLE 3.30

Forty percent of the students at a local college belong to a club and **50%** work part time. **Five percent** of the students work part time and belong to a club. Draw a Venn diagram showing the relationships. Let C = student belongs to a club and PT = student works part time.

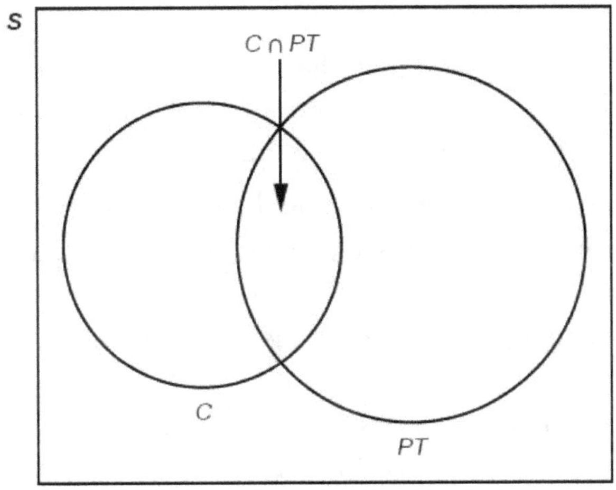

Figure 3.9

If a student is selected at random, find

- the probability that the student belongs to a club. $P(C) = 0.40$
- the probability that the student works part time. $P(PT) = 0.50$
- the probability that the student belongs to a club AND works part time. $P(C \cap PT)=0.05$
- the probability that the student belongs to a club given that the student works part time. $P(C|PT) = \frac{P(C \cap PT)}{P(PT)} = \frac{0.05}{0.50} = 0.1$
- the probability that the student belongs to a club OR works part time. $P(C \cup PT)=P(C)+P(PT)-P(C \cap PT)=0.40+0.50-0.05=0.85$

In order to solve Example 3.30 we had to draw upon the concept of conditional probability from the previous section. There we used tree diagrams to track the changes in the probabilities, because the sample space changed as we drew without replacement. In short, conditional probability is the chance that something will happen given that some other event has already happened. Put another way, the probability that something will happen conditioned upon the situation that something else is also true. In Example 3.30 the probability $P(C||PT)$ is the conditional probability that the randomly drawn student is a member of the club, conditioned upon the fact that the student also is working part time. This allows us to see the relationship between Venn diagrams and the probability postulates.

TRY IT 3.30

Fifty percent of the workers at a factory work a second job, 25% have a spouse who also works, 5% work a second job and have a spouse who also works. Draw a Venn diagram showing the relationships. Let W = works a second job and S = spouse also works.

TRY IT 3.30

In a bookstore, the probability that the customer buys a novel is 0.6, and the probability that the customer buys a non-fiction book is 0.4. Suppose that the probability that the customer buys both is 0.2.

a. Draw a Venn diagram representing the situation.
b. Find the probability that the customer buys either a novel or a non-fiction book.
c. In the Venn diagram, describe the overlapping area using a complete sentence.
d. Suppose that some customers buy only compact disks. Draw an oval in your Venn diagram representing this event.

EXAMPLE 3.31

A set of 20 German Shepherd dogs is observed. 12 are male, 8 are female, 10 have some brown coloring, and 5 have some white sections of fur. Answer the following using Venn Diagrams.

Draw a Venn diagram simply showing the sets of male and female dogs.

Solution 3.31

The Venn diagram below demonstrates the situation of mutually exclusive events where the outcomes are independent events. If a dog cannot be both male and female, then there is no intersection. Being male precludes being female and being female precludes being male: in this case, the characteristic gender is therefore mutually exclusive. A Venn diagram shows this as two sets with no intersection. The intersection is said to be the null set using the mathematical symbol ∅.

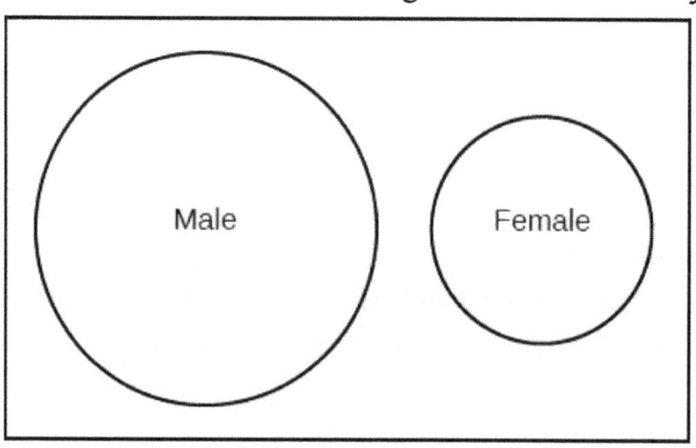

Figure 3.10

Draw a second Venn diagram illustrating that 10 of the male dogs have brown coloring.

Solution 3.31

The Venn diagram below shows the overlap between male and brown where the number 10 is placed in it. This represents Male ∩ Brown : both male and brown. This is the intersection of these two characteristics. To get the union of Male and Brown, then it is simply the two circled areas minus the overlap. In proper terms, Male ∪ Brown = Male + Brown - Male ∩ Brown will give us the number of dogs in the union of these two sets. If we did not subtract the intersection, we would have double counted some of the dogs.

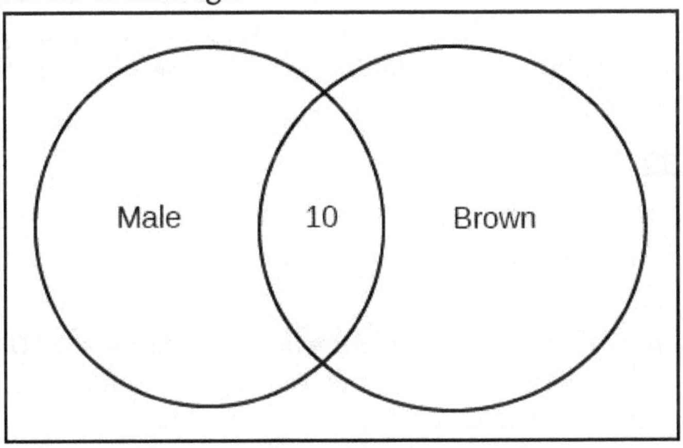

Figure 3.11

Now draw a situation depicting a scenario in which the non-shaded region represents "No white fur and female," or *White fur'* ∩∩ *Female.* the prime above "fur" indicates "not white fur." The prime above a set means not in that set, e.g. A'A' means not AA. Sometimes, the notation used is a line above the letter. For example, $\bar{A}$ = A'.

Solution 3.31

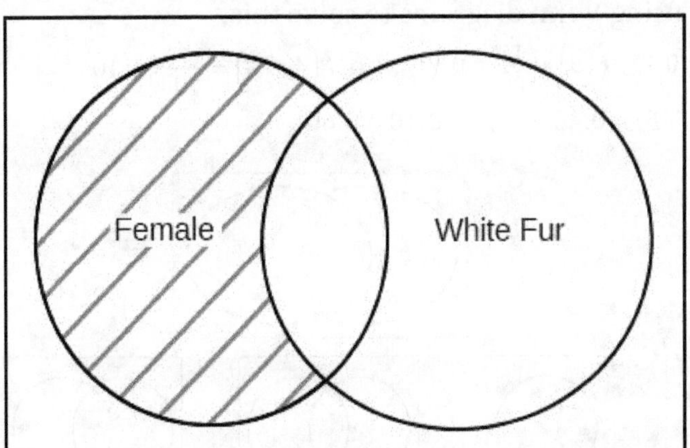

Figure 3.12

The Addition Rule of Probability

We met the addition rule earlier but without the help of Venn diagrams. Venn diagrams help visualize the counting process that is inherent in the calculation of probability. To restate the Addition Rule of Probability:

$$P(A \cup B) = P(A) + P(B) - P(A \cap B)$$

Remember that probability is simply the proportion of the objects we are interested in relative to the total number of objects. This is why we can see the usefulness of the Venn diagrams. Example 3.31 shows how we can use Venn diagrams to count the number of dogs in the union of brown and male by reminding us to subtract the intersection of brown and male. We can see the effect of this directly on probabilities in the addition rule.

EXAMPLE 3.32

Let's sample 50 students who are in a statistics class. 20 are freshmen and 30 are sophomores. 15 students get a "B" in the course, and 5 students both get a "B" and are freshmen.

Find the probability of selecting a student who either earns a "B" OR is a freshmen. We are translating the word OR to the mathematical symbol for the addition rule, which is the union of the two sets.

Solution 3.32

We know that there are 50 students in our sample, so we know the denominator of our fraction to give us probability. We need only to find the number of students that meet the characteristics we are interested in, i.e. any freshman and any student who earned a grade of "B." With the Addition Rule of probability, we can skip directly to probabilities.

Let "A" = the number of freshmen, and let "B" = the grade of "B." Below we can see the process for using Venn diagrams to solve this.

The $P(A) = \frac{20}{50} = 0.40$, $P(B) = \frac{15}{50} = 0.30$, and $P(A \cap B) = \frac{5}{50} = 0.10$.

Therefore, $P(A \cap B) = 0.40 + 0.30 - 0.10 = 0.60$.

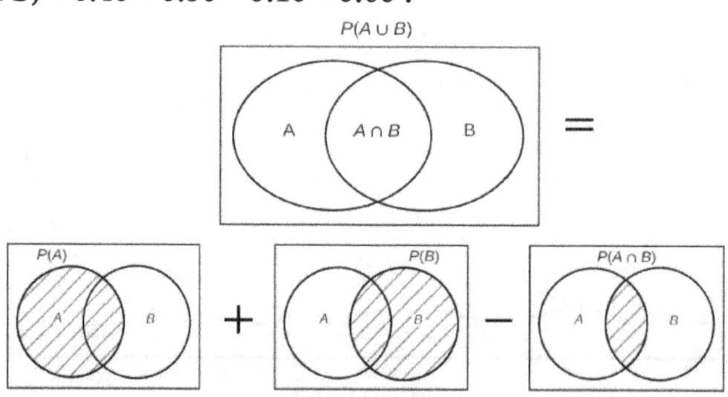

Figure 3.13

If two events are mutually exclusive, then, like the example where we diagram the male and female dogs, the addition rule is simplified to just $P(A \cup B) = P(A) + P(B) - 0$. This is true because, as we saw earlier, the union of mutually exclusive events is the null set, $\emptyset$. The diagrams below demonstrate this.

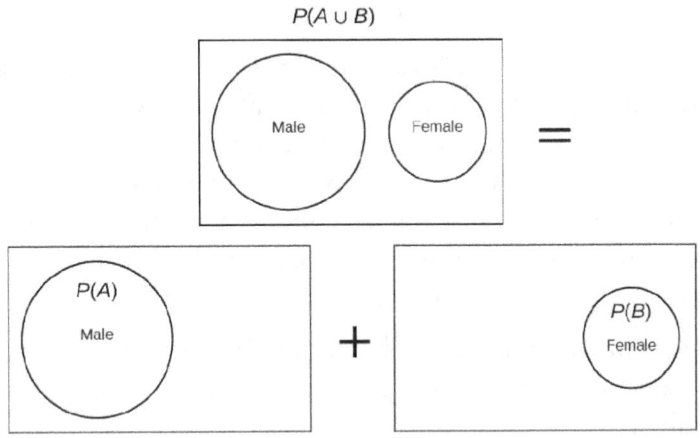

Figure 3.14

The Multiplication Rule of Probability

Restating the Multiplication Rule of Probability using the notation of Venn diagrams, we have:

$$P(A \cap B) = P(A|B) \cdot P(B)$$

The multiplication rule can be modified with a bit of algebra into the following conditional rule. Then Venn diagrams can then be used to demonstrate the process. The conditional rule: $P(A|B) = \frac{P(A \cap B)}{P(B)}$

Using the same facts from Example 3.32 above, find the probability that someone will earn a "B" if they are a "freshman."

$$P(A|B) = \frac{0.10}{0.30} = \frac{1}{3}$$

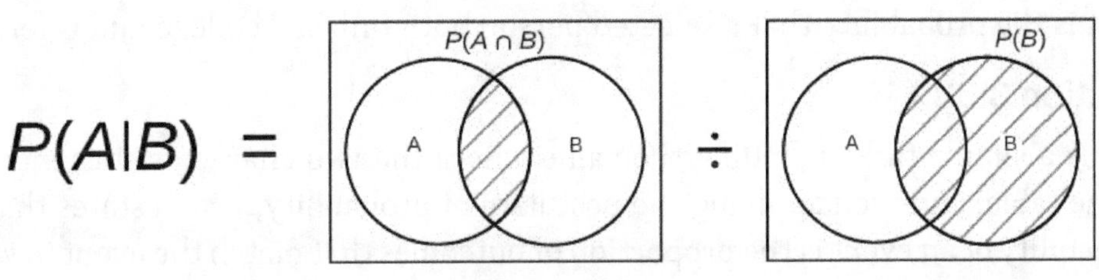

Figure 3.15

The multiplication rule must also be altered if the two events are independent. Independent events are defined as a situation where the conditional probability is simply the probability of the event of interest. Formally, independence of events is defined as P(A|B)=P(A) or P(B|A)=P(B). When flipping coins, the outcome of the second flip is independent of the outcome of the first flip; coins do not have memory. The Multiplication Rule of Probability for independent events thus becomes:

$$P(A \cap B) = P(A) \cdot P(B)$$

One easy way to remember this is to consider what we mean by the word "and." We see that the Multiplication Rule has translated the word "and" to the Venn notation for intersection. Therefore, the outcome must meet the two conditions of freshmen and grade of "B" in the above example. It is harder, less probable, to meet two conditions than just one or some other one. We can attempt to see the logic of the Multiplication Rule of probability due to the fact that fractions multiplied times each other become smaller.

The development of the Rules of Probability with the use of Venn diagrams can be shown to help as we wish to calculate probabilities from data arranged in a contingency table.

EXAMPLE 3.33

Table 3.11 is from a sample of 200 people who were asked how much education they completed. The columns represent the highest education they completed, and the rows separate the individuals by male and female.

Table 3.11

	Less than high school grad	High school grad	Some college	College grad	Total
Male	5	15	40	60	120
Female	8	12	30	30	80
Total	13	27	70	90	200

Now, we can use this table to answer probability questions. The following examples are designed to help understand the format above while connecting the knowledge to both Venn diagrams and the probability rules.

What is the probability that a selected person both finished college and is female?

Solution 3.33

This is a simple task of finding the value where the two characteristics intersect on the table, and then applying the postulate of probability, which states that the probability of an event is the proportion of outcomes that match the event in which we are interested as a proportion of all total possible outcomes.

$$P(College\ Grad \cap Female) = \frac{30}{200} = 0.15$$

What is the probability of selecting either a female or someone who finished college?

Solution 3.33

This task involves the use of the addition rule to solve for this probability.

$$P(College\ Grad \cup Female) = P(F) + P(CG) - P(F \cap CG)$$

$$P(College\ Grad \cup Female) = \frac{80}{200} + \frac{90}{200} - \frac{30}{200} = \frac{140}{200} = 0.70$$

What is the probability of selecting a high school graduate if we only select from the group of males?

Solution 3.33

Here we must use the conditional probability rule (the modified multiplication rule) to solve for this probability.

$$P(HS\ Grad \mid Male = \frac{P(\text{HS Grad} \cap \text{Male})}{P(\text{Male})} = \frac{\left(\frac{15}{200}\right)}{\left(\frac{120}{200}\right)} = \frac{15}{120} = 0.125$$

Can we conclude that the level of education attained by these 200 people is independent of the gender of the person?

Solution 3.33

There are two ways to approach this test. The first method seeks to test if the intersection of two events equals the product of the events separately remembering that if two events are independent than $P(A)*P(B) = P(A \cap B)$.

Does P(College Grad ∩ Female) = P(CG) . P(F)?

$$\frac{30}{200} \neq \frac{90}{200} \cdot \frac{80}{200} \text{ because } 0.15 \neq 0.18.$$

Therefore, gender and education here are **not** independent.

The second method is to test if the conditional probability of *A given B* is equal to the probability of *A*. Again for simplicity, we can use an already calculated value from above.

Does P(HS Grad | Male) = P(HS Grad)?

$$\frac{15}{120} \neq \frac{27}{200} \text{ because } 0.125 \neq 0.135.$$

Therefore, again gender and education here are not independent.

Key Terms

Conditional Probability the likelihood that an event will occur given that another event has already occurred

Contingency Table the method of displaying a frequency distribution as a table with rows and columns to show how two variables may be dependent (contingent) upon each other; the table provides an easy way to calculate conditional probabilities.

Dependent Events If two events are NOT independent, then we say that they are dependent.

Equally Likely Each outcome of an experiment has the same probability.

Event a subset of the set of all outcomes of an experiment; the set of all outcomes of an experiment is called a sample space and is usually denoted by *S*. An event is an arbitrary subset in *S*. It can contain one outcome, two outcomes, no outcomes (empty subset), the entire sample space, and the like. Standard notations for events are capital letters such as *A*, *B*, *C*, and so on.

Experiment a planned activity carried out under controlled conditions

Independent Events The occurrence of one event has no effect on the probability of the occurrence of another event. Events *A* and *B* are independent if one of the following is true:

1. $P(A|B) = P(A)$
2. $P(B|A) = P(B)$
3. $P(A \cap B) = P(A)P(B)$

Mutually Exclusive Two events are mutually exclusive if the probability that they both happen at the same time is zero. If events *A* and *B* are mutually exclusive, then $P(A \cap B) = 0$.

Outcome a particular result of an experiment

Probability a number between zero and one, inclusive, that gives the likelihood that a specific event will occur; the foundation of statistics is given by the following 3 axioms (by A.N. Kolmogorov, 1930's): Let *S* denote the sample space and *A* and *B* are two events in *S*. Then:

- $0 \leq P(A) \leq 1$
- If *A* and *B* are any two mutually exclusive events, then $P(A \cup\cup B) = P(A) + P(B)$.
- $P(S) = 1$

Sample Space the set of all possible outcomes of an experiment

Sampling with Replacement If each member of a population is replaced after it is picked, then that member has the possibility of being chosen more than once.

Sampling without Replacement When sampling is done without replacement, each member of a population may be chosen only once.

The Complement Event The complement of event *A* consists of all outcomes that are NOT in *A*.

The Conditional Probability of A | B $P(A|B)$ is the probability that event *A* will occur given that the event *B* has already occurred.

The Intersection: the ∩ Event An outcome is in the event $A \cap B$ if the outcome is in both *A* ∩∩ *B* at the same time.

The Union: the ∪ Event An outcome is in the event $A \cup B$ if the outcome is in *A* or is in *B* or is in both *A* and *B*.

Tree Diagram the useful visual representation of a sample space and events in the form of a "tree" with branches marked by possible outcomes together with associated probabilities (frequencies, relative frequencies)

Venn Diagram the visual representation of a sample space and events in the form of circles or ovals showing their intersections

CHAPTER REVIEW

3.1 Terminology

In this module we learned the basic terminology of probability. The set of all possible outcomes of an experiment is called the sample space. Events are subsets of the sample space, and they are assigned a probability that is a number between zero and one, inclusive.

3.2 Independent and Mutually Exclusive Events

Two events *A* and *B* are independent if the knowledge that one occurred does not affect the chance the other occurs. If two events are not independent, then we say that they are dependent.

In sampling with replacement, each member of a population is replaced after it is picked, so that member has the possibility of being chosen more than once, and the events are considered to be independent. In sampling without replacement, each member of a population may be chosen only once, and the events are considered not to be independent. When events do not share outcomes, they are mutually exclusive of each other.

3.3 Two Basic Rules of Probability

The multiplication rule and the addition rule are used for computing the probability of *A* and *B*, as well as the probability of *A* or *B* for two given events *A*, *B* defined on the sample space. In sampling with replacement each member of a population is replaced after it is picked, so that member has the possibility of being chosen more than once, and the events are considered to be independent. In sampling without replacement, each member of a population may be chosen only once, and the events are considered to be not independent. The events *A* and *B* are mutually exclusive events when they do not have any outcomes in common.

3.4 Contingency Tables and Probability Trees

There are several tools you can use to help organize and sort data when calculating probabilities. Contingency tables help display data and are particularly useful when calculating probabilities that have multiple dependent variables.

A tree diagram use branches to show the different outcomes of experiments and makes complex probability questions easy to visualize.

3.5 Venn Diagrams

A Venn diagram is a picture that represents the outcomes of an experiment. It generally consists of a box that represents the sample space S or universe of the objects of interest together with circles or ovals. The circles or ovals represent groups of events called sets. A Venn diagram is especially helpful for visualizing the ∪∪ event, the ∩∩ event, and the complement of an event and for understanding conditional probabilities. A Venn diagram is especially helpful for visualizing an Intersection of two events, a Union of two events, or a Complement of one event. A system of Venn diagrams can also help to understand Conditional probabilities. Venn diagrams connect the brain and eyes by matching the literal arithmetic to a picture. It is important to note that more than one Venn diagram is needed to solve the probability rule formulas introduced in Section 3.3.

FORMULA REVIEW

3.1 Terminology

A and *B* are events

$P(S) = 1$ where *S* is the sample space

$0 \leq P(A) \leq 1$

$$P(A \mid B) = \frac{P(A \cap B)}{P(B)}$$

3.2 Independent and Mutually Exclusive Events

If AA and BB are independent, $P(A \cap B) = P(A)P(B)$. $P(A|B) = P(A) \text{and} P(B|A) = P(B)$.

If AA and BB are mutually exclusive, $P(A \cup B) = P(A) + P(B) \text{and} P(A \cap B) = 0$.

3.3 Two Basic Rules of Probability

The multiplication rule: $P(A \cap B) = P(A|B)P(B)$

The addition rule: $P(A \cup B) = P(A) + P(B) - P(A \cap B)$

PRACTICE

3.1 Terminology

1. In a particular college class, there are male and female students. Some students have long hair and some students have short hair. Write the **symbols** for the probabilities of the events for parts a through j. (Note that you cannot find numerical answers here. You were not given enough information to find any probability values yet; concentrate on understanding the symbols.)
   - Let F be the event that a student is female.
   - Let M be the event that a student is male.
   - Let S be the event that a student has short hair.
   - Let L be the event that a student has long hair.
     a. The probability that a student does not have long hair.
     b. The probability that a student is male or has short hair.
     c. The probability that a student is a female and has long hair.
     d. The probability that a student is male, given that the student has long hair.
     e. The probability that a student has long hair, given that the student is male.
     f. Of all the female students, the probability that a student has short hair.
     g. Of all students with long hair, the probability that a student is female.
     h. The probability that a student is female or has long hair.
     i. The probability that a randomly selected student is a male student with short hair.
     j. The probability that a student is female.

Use the following information to answer the next four exercises. A box is filled with several party favors. It contains 12 hats, 15 noisemakers, ten finger traps, and five bags of confetti. One party favor is chosen from the box at random.

Let H = the event of getting a hat.
Let N = the event of getting a noisemaker.
Let F = the event of getting a finger trap.
Let C = the event of getting a bag of confetti.

2. Find $P(H)$.
3. Find $P(N)$.
4. Find $P(F)$.
5. Find $P(C)$.

Use the following information to answer the next six exercises. A jar of 150 jelly beans contains 22 red jelly beans, 38 yellow, 20 green, 28 purple, 26 blue, and the rest are orange. One jelly bean is chosen from the box at random.

Let B = the event of getting a blue jelly bean
Let G = the event of getting a green jelly bean.
Let O = the event of getting an orange jelly bean.
Let P = the event of getting a purple jelly bean.
Let R = the event of getting a red jelly bean.
Let Y = the event of getting a yellow jelly bean.

6. Find $P(B)$.

7. Find $P(G)$.

8. Find $P(P)$.

9. Find $P(R)$.

10. Find $P(Y)$.

11. Find $P(O)$.

Use the following information to answer the next six exercises. There are 23 countries in North America, 12 countries in South America, 47 countries in Europe, 44 countries in Asia, 54 countries in Africa, and 14 in Oceania (Pacific Ocean region).

Let A = the event that a country is in Asia.
Let E = the event that a country is in Europe.
Let F = the event that a country is in Africa.
Let N = the event that a country is in North America.
Let O = the event that a country is in Oceania.
Let S = the event that a country is in South America.

12. Find $P(A)$.

13. Find $P(E)$.

14. Find $P(F)$.

15. Find $P(N)$.

16. Find $P(O)$.

17. Find $P(S)$.

18. What is the probability of drawing a red card in a standard deck of 52 cards?

19. What is the probability of drawing a club in a standard deck of 52 cards?

20. What is the probability of rolling an even number of dots with a fair, six-sided die numbered one through six?

21. What is the probability of rolling a prime number of dots with a fair, six-sided die numbered one through six?

Use the following information to answer the next two exercises. You see a game at a local fair. You have to throw a dart at a color wheel. Each section on the color wheel is equal in area.

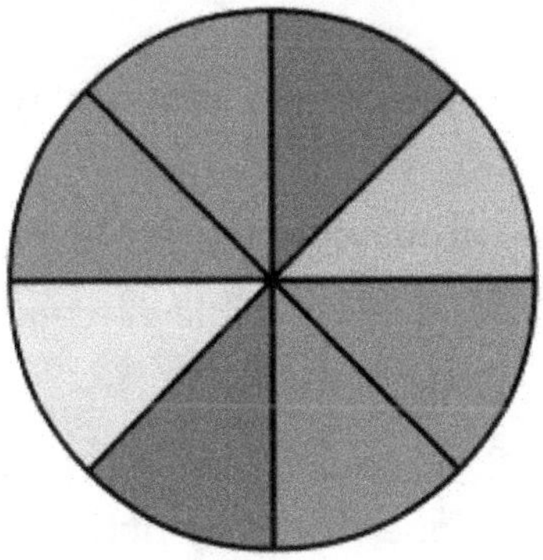

Figure 3.16

Let B = the event of landing on blue.
Let R = the event of landing on red.
Let G = the event of landing on green.
Let Y = the event of landing on yellow.

22. If you land on Y, you get the biggest prize. Find $P(Y)$.

23. If you land on red, you don't get a prize. What is $P(R)$?

Use the following information to answer the next ten exercises. On a baseball team, there are infielders and outfielders. Some players are great hitters, and some players are not great hitters.

Let I = the event that a player in an infielder.
Let O = the event that a player is an outfielder.
Let H = the event that a player is a great hitter.
Let N = the event that a player is not a great hitter.

24. Write the symbols for the probability that a player is not an outfielder.

25. Write the symbols for the probability that a player is an outfielder or is a great hitter.

26. Write the symbols for the probability that a player is an infielder and is not a great hitter.

27. Write the symbols for the probability that a player is a great hitter, given that the player is an infielder.

28. Write the symbols for the probability that a player is an infielder, given that the player is a great hitter.

29. Write the symbols for the probability that of all the outfielders, a player is not a great hitter.
30. Write the symbols for the probability that of all the great hitters, a player is an outfielder.
31. Write the symbols for the probability that a player is an infielder or is not a great hitter.
32. Write the symbols for the probability that a player is an outfielder and is a great hitter.
33. Write the symbols for the probability that a player is an infielder.
34. What is the word for the set of all possible outcomes?
35. What is conditional probability?
36. A shelf holds 12 books. Eight are fiction and the rest are nonfiction. Each is a different book with a unique title. The fiction books are numbered one to eight. The nonfiction books are numbered one to four. Randomly select one book

    Let F = event that book is fiction
    Let N = event that book is nonfiction
    What is the sample space?
37. What is the sum of the probabilities of an event and its complement?

Use the following information to answer the next two exercises. You are rolling a fair, six-sided number cube. Let E = the event that it lands on an even number. Let M = the event that it lands on a multiple of three.

38. What does $P(E|M)$ mean in words?
39. What does $P(E \cup M)$ mean in words?

3.2 Independent and Mutually Exclusive Events

40. E and F are mutually exclusive events. P(E)=0.4; $P(F)$=0.5. Find $P(E|F)$.
41. J and K are independent events. P(J|K)=0.3. Find P(J).$P(J)$.
42. U and V are mutually exclusive events. $P(U)$ = 0.26; $P(V)$ = 0.37. Find:

    P(U∩V)=
    P(U|V)=
    P(U∪V)=
43. Q and R are independent events. P(Q)=0.4 and P(Q∩R)=0.1. Find P(R).$P(R)$.

3.3 Two Basic Rules of Probability

Use the following information to answer the next ten exercises. Forty-eight percent of all Californians registered voters prefer life in prison without parole over the death penalty for a person convicted of first degree murder. Among Latino California registered voters, 55% prefer life in prison without parole over the death penalty for a person convicted of first degree murder. 37.6% of all Californians are Latino.

In this problem, let:

- C = Californians (registered voters) preferring life in prison without parole over the death penalty for a person convicted of first degree murder.
- L = Latino Californians

Suppose that one Californian is randomly selected.

44. Find $P(C)$.

45. Find $P(L)$.

46. Find $P(C|L)$.

47. In words, what is $C|L$?

48. Find $P(L \cap C)$.

49. In words, what is $L \cap C$?

50. Are L and C independent events? Show why or why not.

51. ind $P(L \cup C)$.

52. In words, what is $L \cup C$?

53. Are L and C mutually exclusive events? Show why or why not.

3.5 Venn Diagrams

Use the following information to answer the next four exercises. Table 3.12 shows a random sample of musicians and how they learned to play their instruments.

Table 3.12

Gender	Self-taught	Studied in school	Private instruction	Total
Female	12	38	22	72
Male	19	24	15	58
Total	31	62	37	130

54. Find P(musician is a female).

55. Find P(musician is a male $\cap$ had private instruction).

56. Find P(musician is a female $\cup$ is self taught).

57. Are the events "being a female musician" and "learning music in school" mutually exclusive events?

58. The probability that a man develops some form of cancer in his lifetime is 0.4567. The probability that a man has at least one false positive test result (meaning the test comes back for cancer when the man does not have it) is 0.51. Let: C = a man develops cancer in his lifetime; P = man has at least one false positive. Construct a tree diagram of the situation.

Bringing It Together: Practice

Use the following information to answer the next seven exercises. An article in the *New England Journal of Medicine*, reported about a study of smokers in California and Hawaii. In one part of the report, the self-reported ethnicity and smoking levels per day were given. Of the people smoking at most ten cigarettes per day, there were 9,886 African Americans, 2,745

Native Hawaiians, 12,831 Latinos, 8,378 Japanese Americans, and 7,650 Whites. Of the people smoking 11 to 20 cigarettes per day, there were 6,514 African Americans, 3,062 Native Hawaiians, 4,932 Latinos, 10,680 Japanese Americans, and 9,877 Whites. Of the people smoking 21 to 30 cigarettes per day, there were 1,671 African Americans, 1,419 Native Hawaiians, 1,406 Latinos, 4,715 Japanese Americans, and 6,062 Whites. Of the people smoking at least 31 cigarettes per day, there were 759 African Americans, 788 Native Hawaiians, 800 Latinos, 2,305 Japanese Americans, and 3,970 Whites.

59. Complete the table using the data provided. Suppose that one person from the study is randomly selected. Find the probability that person smoked 11 to 20 cigarettes per day.

Table 3.13 Smoking Levels by Ethnicity

Smoking level	African American	Native Hawaiian	Latino	Japanese American	White	TOTALS
1–10						
11–20						
21–30						
31+						
TOTALS						

60. Suppose that one person from the study is randomly selected. Find the probability that person smoked 11 to 20 cigarettes per day.

61. Find the probability that the person was Latino.

62. In words, explain what it means to pick one person from the study who is "Japanese American **AND** smokes 21 to 30 cigarettes per day." Also, find the probability.

63. In words, explain what it means to pick one person from the study who is "Japanese American UU smokes 21 to 30 cigarettes per day." Also, find the probability.

64. In words, explain what it means to pick one person from the study who is "Japanese American | that person smokes 21 to 30 cigarettes per day." Also, find the probability.

65. Prove that smoking level/day and ethnicity are dependent events.

Use the following information to answer the next two exercises. Suppose that you have eight cards. Five are green and three are yellow. The cards are well shuffled.

66. Suppose that you randomly draw two cards, one at a time, **with replacement.**

    Let G_1 = first card is green

    Let G_2 = second card is green

    a. Draw a tree diagram of the situation.
    b. Find $P(G_1 \cap G_2)$.
    c. Find P(at least one green).
    d. Find $P(G_2|G_1)$.
    e. Are G_2 and G_1 independent events? Explain why or why not.

67. Suppose that you randomly draw two cards, one at a time, **without replacement.**

    G_1 = first card is green.

    G_2 = second card is green

    a. Draw a tree diagram of the situation.
    b. Find $P(G_1 \cap G_2)$.
    c. Find P(at least one green).
    d. Find $P(G_2 \mid G_1)$.

Are G_2 and G_1 independent events? Explain why or why not.

Use the following information to answer the next two exercises. The percent of licensed U.S. drivers (from a recent year) that are female is 48.60. Of the females, 5.03% are age 19 and under; 81.36% are age 20–64; 13.61% are age 65 or over. Of the licensed U.S. male drivers, 5.04% are age 19 and under; 81.43% are age 20–64; 13.53% are age 65 or over.

68. Complete the following.

a. Construct a table or a tree diagram of the situation.
b. Find *P*(driver is female).
c. Find *P*(driver is age 65 or over || driver is female).
d. Find *P*(driver is age 65 or over ∩∩ female).
e. In words, explain the difference between the probabilities in part c and part d.
f. Find *P*(driver is age 65 or over).
g. Are being age 65 or over and being female mutually exclusive events? How do you know?

69. Suppose that 10,000 U.S. licensed drivers are randomly selected.

a. How many would you expect to be male?
b. Using the table or tree diagram, construct a contingency table of gender versus age group.
c. Using the contingency table, find the probability that out of the age 20–64 group, a randomly selected driver is female.

70. Approximately 86.5% of Americans commute to work by car, truck, or van. Out of that group, 84.6% drive alone and 15.4% drive in a carpool. Approximately 3.9% walk to work and approximately 5.3% take public transportation.

a. Construct a table or a tree diagram of the situation. Include a branch for all other modes of transportation to work.
b. Assuming that the walkers walk alone, what percent of all commuters travel alone to work?
c. Suppose that 1,000 workers are randomly selected. How many would you expect to travel alone to work?
d. Suppose that 1,000 workers are randomly selected. How many would you expect to drive in a carpool?

71. When the Euro coin was introduced in 2002, two math professors had their statistics students test whether the Belgian one Euro coin was a fair coin. They spun the coin rather than tossing it and found that out of 250 spins, 140 showed a head (event *H*) while 110 showed a tail (event *T*). On that basis, they claimed that it is not a fair coin.

a. Based on the given data, find *P*(*H*) and *P*(*T*).
b. Use a tree to find the probabilities of each possible outcome for the experiment of tossing the coin twice.

c. Use the tree to find the probability of obtaining exactly one head in two tosses of the coin.

d. Use the tree to find the probability of obtaining at least one head.

HOMEWORK

3.1 Terminology

72.

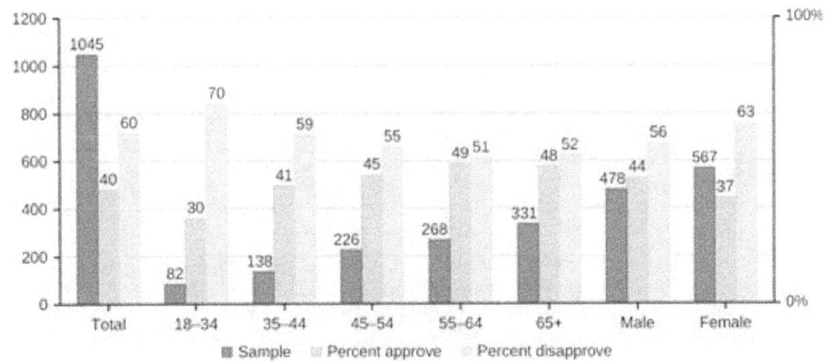

Figure 3.17

The graph in**Figure 3.17** displays the sample sizes and percentages of people in different age and gender groups who were polled concerning their approval of Mayor Ford's actions in office. The total number in the sample of all the age groups is 1,045.

a. Define three events in the graph.
b. Describe in words what the entry 40 means.
c. Describe in words the complement of the entry in question 2.
d. Describe in words what the entry 30 means.
e. Out of the males and females, what percent are males?
f. Out of the females, what percent disapprove of Mayor Ford?
g. Out of all the age groups, what percent approve of Mayor Ford?
h. Find*P*(Approve||Male).
i. Out of the age groups, what percent are more than 44 years old?
j. Find*P*(Approve||Age < 35).

73. Explain what is wrong with the following statements. Use complete sentences.

a. If there is a 60% chance of rain on Saturday and a 70% chance of rain on Sunday, then there is a 130% chance of rain over the weekend.
b. The probability that a baseball player hits a home run is greater than the probability that he gets a successful hit.

3.2Independent and Mutually Exclusive Events

Use the following information to answer the next 12 exercises.From January through December 2012, Gallup interviewed more than 170,000 employed Americans 18 years of age or order. The graph shown is compiled from data collected by Gallup. The Emotional Health Index Scores are the sample space. We randomly sample one Emotional Health Index Score.

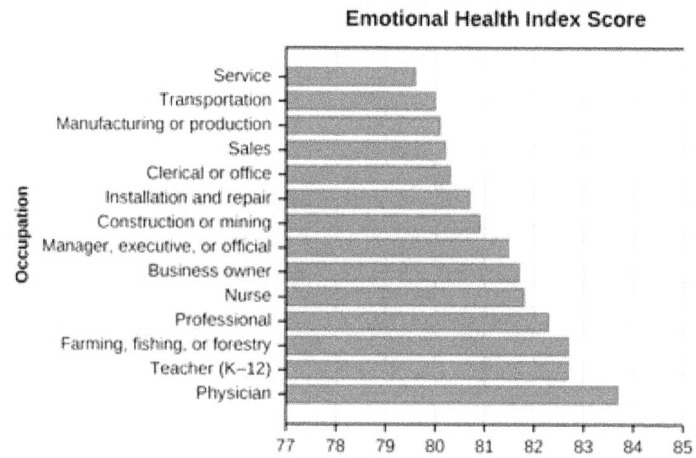

Figure 3.18

74. Find the probability that the selected Emotional Health Index Score is 82.7.

75. Find the probability that the selected Emotional Health Index Score is 81.0.

76. Find the probability that the selected Emotional Health Index Score is more than 81?

77. Find the probability that the selected Emotional Health Index Score is between 80.5 and 82?

78. If we know the selected Emotional Health Index Score is 81.5 or more, what is the probability that it is 82.7?

79. What is the probability that the selected Emotional Health Index Score is 80.7 or 82.7?

80. What is the probability that the selected Emotional Health Index Score is less than 80.2 given that it is already less than 81.

81. What occupation has the highest emotional index score?

82. What occupation has the lowest emotional index score?

83. What is the range of the data?

84. Compute the average EHIS.

85. If all occupations are equally likely for a certain individual, what is the probability that he or she will have an occupation with lower than average EHIS?

3.3 Two Basic Rules of Probability

86. On February 28, 2013, a Field Poll Survey reported that 61% of California registered voters approved of allowing two people of the same gender to marry and have regular marriage laws apply to them. Among 18 to 39 year olds (California registered voters), the approval rating was 78%. Six in ten California registered voters said that the upcoming Supreme Court's ruling about the constitutionality of California's Proposition 8 was either very or somewhat important to them. Out of those CA registered voters who support same-sex marriage, 75% say the ruling is important to them.

In this problem, let:

- C= California registered voters who support same-sex marriage.
- B= California registered voters who say the Supreme Court's ruling about the constitutionality of California's Proposition 8 is very or somewhat important to them
- A= California registered voters who are 18 to 39 years old.

a. Find$P(C)$.
b. Find$P(B)$.
c. Find$P(C||A)$.
d. Find$P(B|C)$.
e. In words, what is$C||A$?
f. In words, what is$B||C$?
g. Find$P(C\cap\cap B)$.
h. In words, what is$C\cap\cap B$?
i. Find$P(C\cup\cup B)$.

Are C and B mutually exclusive events? Show why or why not.

87. After Rob Ford, the mayor of Toronto, announced his plans to cut budget costs in late 2011, the Forum Research polled 1,046 people to measure the mayor's popularity. Everyone polled expressed either approval or disapproval. These are the results their poll produced:

- In early 2011, 60 percent of the population approved of Mayor Ford's actions in office.
- In mid-2011, 57 percent of the population approved of his actions.
- In late 2011, the percentage of popular approval was measured at 42 percent.

a. What is the sample size for this study?
b. What proportion in the poll disapproved of Mayor Ford, according to the results from late 2011?
c. How many people polled responded that they approved of Mayor Ford in late 2011?
d. What is the probability that a person supported Mayor Ford, based on the data collected in mid-2011?
e. What is the probability that a person supported Mayor Ford, based on the data collected in early 2011?

Use the following information to answer the next three exercises. The casino game, roulette, allows the gambler to bet on the probability of a ball, which spins in the roulette wheel, landing on a particular color, number, or range of numbers. The table used to place bets contains of 38 numbers, and each number is assigned to a color and a range.

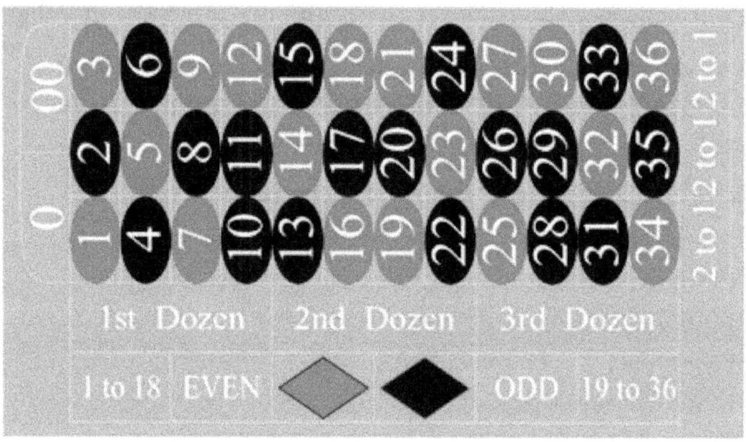

Figure3.19(credit: film8ker/wikibooks)

88. a. List the sample space of the 38 possible outcomes in roulette.
b. You bet on red. Find*P*(red).
c. You bet on -1st 12- (1st Dozen). Find*P*(-1st 12-).
d. You bet on an even number. Find*P*(even number).
e. Is getting an odd number the complement of getting an even number? Why?
f. Find two mutually exclusive events.
g. Are the events Even and 1st Dozen independent?

89. Compute the probability of winning the following types of bets:

a. Betting on two lines that touch each other on the table as in 1-2-3-4-5-6
b. Betting on three numbers in a line, as in 1-2-3

c. Betting on one number
d. Betting on four numbers that touch each other to form a square, as in 10-11-13-14
e. Betting on two numbers that touch each other on the table, as in 10-11 or 10-13
f. Betting on 0-00-1-2-3
g. Betting on 0-1-2; or 0-00-2; or 00-2-3

90. Compute the probability of winning the following types of bets:

a. Betting on a color
b. Betting on one of the dozen groups
c. Betting on the range of numbers from 1 to 18
d. Betting on the range of numbers 19–36
e. Betting on one of the columns
f. Betting on an even or odd number (excluding zero)

91. Suppose that you have eight cards. Five are green and three are yellow. The five green cards are numbered 1, 2, 3, 4, and 5. The three yellow cards are numbered 1, 2, and 3. The cards are well shuffled. You randomly draw one card.

- *G*= card drawn is green
- *E*= card drawn is even-numbered

a. List the sample space.
b. $P(G)$ = _____
c. $P(G||E)$ = _____
d. $P(G∩∩E)$ = _____
e. $P(G∪∪E)$ = _____
f. Are *G* and *E* mutually exclusive? Justify your answer numerically.

92. Roll two fair dice separately. Each die has six faces.

a. List the sample space.
b. Let *A* be the event that either a three or four is rolled first, followed by an even number. Find$P(A)$.
c. Let *B* be the event that the sum of the two rolls is at most seven. Find$P(B)$.
d. In words, explain what "$P(A||B)$" represents. Find$P(A||B)$.
e. Are *A* and *B* mutually exclusive events? Explain your answer in one to three complete sentences, including numerical justification.
f. Are *A* and *B* independent events? Explain your answer in one to three complete sentences, including numerical justification.

93. A special deck of cards has ten cards. Four are green, three are blue, and three are red. When a card is picked, its color of it is recorded. An experiment consists of first picking a card and then tossing a coin.

a. List the sample space.

b. Let A be the event that a blue card is picked first, followed by landing a head on the coin toss. Find $P(A)$.
c. Let B be the event that a red or green is picked, followed by landing a head on the coin toss. Are the events A and B mutually exclusive? Explain your answer in one to three complete sentences, including numerical justification.
d. Let C be the event that a red or blue is picked, followed by landing a head on the coin toss. Are the events A and C mutually exclusive? Explain your answer in one to three complete sentences, including numerical justification.

94. An experiment consists of first rolling a die and then tossing a coin.

a. List the sample space.
b. Let A be the event that either a three or a four is rolled first, followed by landing a head on the coin toss. Find $P(A)$.
c. Let B be the event that the first and second tosses land on heads. Are the events A and B mutually exclusive? Explain your answer in one to three complete sentences, including numerical justification.

95. An experiment consists of tossing a nickel, a dime, and a quarter. Of interest is the side the coin lands on.

a. List the sample space.
b. Let A be the event that there are at least two tails. Find $P(A)$.
c. Let B be the event that the first and second tosses land on heads. Are the events A and B mutually exclusive? Explain your answer in one to three complete sentences, including justification.

96. Consider the following scenario:

Let $P(C) = 0.4$.

Let $P(D) = 0.5$. Let $P(C||D) = 0.6$.

a. Find $P(C \cap D)$.
b. Are C and D mutually exclusive? Why or why not?
c. Are C and D independent events? Why or why not?
d. Find $P(C \cup D)$.
e. Find $P(D||C)$.

97. Y and Z are independent events.

a. Rewrite the basic Addition Rule $P(Y \cup Z) = P(Y) + P(Z) - P(Y \cap Z)$ using the information that Y and Z are independent events.
b. Use the rewritten rule to find $P(Z)$ if $P(Y \cup Z) = 0.71$ and $P(Y) = 0.42$.

98. G and H are mutually exclusive events. $P(G) = 0.5$ $P(H) = 0.3$

a. Explain why the following statement MUST be false: $P(H||G) = 0.4$.

b. Find $P(H \cup G)$.

c. Are G and H independent or dependent events? Explain in a complete sentence.

99. Approximately 281,000,000 people over age five live in the United States. Of these people, 55,000,000 speak a language other than English at home. Of those who speak another language at home, 62.3% speak Spanish.

Let: E= speaks English at home; E'= speaks another language at home; S= speaks Spanish;

Finish each probability statement by matching the correct answer.

Table3.14

Probability Statements	Answers
a. P(E') =	i. 0.8043
b. P(E) =	ii. 0.623
c. P(S∩E') =	iii. 0.1957
d. P(S|E') =	iv. 0.1219

100. 1994, the U.S. government held a lottery to issue 55,000 Green Cards (permits for non-citizens to work legally in the U.S.). Renate Deutsch, from Germany, was one of approximately 6.5 million people who entered this lottery. Let G= won green card.

a. What was Renate's chance of winning a Green Card? Write your answer as a probability statement.

b. In the summer of 1994, Renate received a letter stating she was one of 110,000 finalists chosen. Once the finalists were chosen, assuming that each finalist had an equal chance to win, what was Renate's chance of winning a Green Card? Write your answer as a conditional probability statement. Let F= was a finalist.

c. Are G and F independent or dependent events? Justify your answer numerically and also explain why.

d. Are G and F mutually exclusive events? Justify your answer numerically and explain why.

101. Three professors at George Washington University did an experiment to determine if economists are more selfish than other people. They dropped 64 stamped, addressed envelopes with $10 cash in different classrooms on the George Washington campus. 44% were returned overall. From the economics classes 56% of the envelopes were returned.
From the business, psychology, and history classes 31% were returned.
Let: R= money returned; E= economics classes; O= other classes

a. Write a probability statement for the overall percent of money returned.
b. Write a probability statement for the percent of money returned out of the economics classes.
c. Write a probability statement for the percent of money returned out of the other classes.
d. Is money being returned independent of the class? Justify your answer numerically and explain it.
e. Based upon this study, do you think that economists are more selfish than other people? Explain why or why not. Include numbers to justify your answer.

102. The following table of data obtained from www.baseball-almanac.com shows hit information for four players. Suppose that one hit from the table is randomly selected.

Table3.15

Name	Single	Double	Triple	Home run	Total hits
Babe Ruth	1,517	506	136	714	2,873
Jackie Robinson	1,054	273	54	137	1,518
Ty Cobb	3,603	174	295	114	4,189
Hank Aaron	2,294	624	98	755	3,771
Total	8,471	1,577	583	1,720	12,351

Are "the hit being made by Hank Aaron" and "the hit being a double" independent events?

a. Yes, because*P*(hit by Hank Aaron||hit is a double) =*P*(hit by Hank Aaron)
b. No, because*P*(hit by Hank Aaron||hit is a double) ≠*P*(hit is a double)
c. No, because*P*(hit is by Hank Aaron||hit is a double) ≠*P*(hit by Hank Aaron)
d. Yes, because*P*(hit is by Hank Aaron||hit is a double) =*P*(hit is a double)

103. United Blood Services is a blood bank that serves more than 500 hospitals in 18 states. According to their website, a person with type O blood and a negative Rh factor (Rh-) can donate blood to any person with any bloodtype. Their data show that 43% of people have type O blood and 15% of people have Rh- factor; 52% of people have type O or Rh- factor.

a. Find the probability that a person has both type O blood and the Rh- factor.
b. Find the probability that a person does NOT have both type O blood and the Rh- factor.

104. At a college, 72% of courses have final exams and 46% of courses require research papers. Suppose that 32% of courses have a research paper and a final exam. Let$F$be the event that a course has a final exam. Let$R$be the event that a course requires a research paper.

a. Find the probability that a course has a final exam or a research project.
b. Find the probability that a course has NEITHER of these two requirements.

105. In a box of assorted cookies, 36% contain chocolate and 12% contain nuts. In the box, 8% contain both chocolate and nuts. Sean is allergic to both chocolate and nuts.

a. Find the probability that a cookie contains chocolate or nuts (he can't eat it).
b. Find the probability that a cookie does not contain chocolate or nuts (he can eat it).

106. A college finds that 10% of students have taken a distance learning class and that 40% of students are part time students. Of the part time students, 20% have taken a distance learning class. Let$D$= event that a student takes a distance learning class and$E$= event that a student is a part time student

a. Find$P(D \cap E)$.
b. Find$P(E|D)$.
c. Find$P(D \cup E)$.
d. Using an appropriate test, show whether$D$and$E$are independent.
e. Using an appropriate test, show whether$D$and$E$are mutually exclusive.

3.5Venn Diagrams

*Use the information in theTable 3.16to answer the next eight exercises.*The table shows the political party affiliation of each of 67 members of the US Senate in June 2012, and when they are up for reelection.

Table3.16

Up for reelection:	Democratic party	Republican party	Other	Total
November 2014	20	13	0	
November 2016	10	24	0	
Total				

107. What is the probability that a randomly selected senator has an "Other" affiliation?

108. What is the probability that a randomly selected senator is up for reelection in November 2016?

109. What is the probability that a randomly selected senator is a Democrat and up for reelection in November 2016?

110. What is the probability that a randomly selected senator is a Republican or is up for reelection in November 2014?

111. Suppose that a member of the US Senate is randomly selected. Given that the randomly selected senator is up for reelection in November 2016, what is the probability that this senator is a Democrat?

112. Suppose that a member of the US Senate is randomly selected. What is the probability that the senator is up for reelection in November 2014, knowing that this senator is a Republican?

113. The events "Republican" and "Up for reelection in 2016" are ________

a. mutually exclusive.
b. independent.
c. both mutually exclusive and independent.
d. neither mutually exclusive nor independent.

114. The events "Other" and "Up for reelection in November 2016" are ________

a. mutually exclusive.
b. independent.
c. both mutually exclusive and independent.
d. neither mutually exclusive nor independent.

115. **Table 3.17**gives the number of participants in the recent National Health Interview Survey who had been treated for cancer in the previous 12 months. The results are sorted by age, race (Black or White), and sex. We are interested in possible relationships between age, race, and sex. We will let suicide victims be our population.

Table3.17

Race and sex	15–24	25–40	41–65	Over 65	TOTALS
White, male	1,165	2,036	3,703		8,395
White, female	1,076	2,242	4,060		9,129
Black, male	142	194	384		824
Black, female	131	290	486		1,061
All others					
TOTALS	2,792	5,279	9,354		21,081

Do not include "all others" for parts f and g.

a. Fill in the column for cancer treatment for individuals over age 65.
b. Fill in the row for all other races.
c. Find the probability that a randomly selected individual was a White male.
d. Find the probability that a randomly selected individual was a Black female.

e. Find the probability that a randomly selected individual was Black
f. Find the probability that a randomly selected individual was male.
g. Out of the individuals over age 65, find the probability that a randomly selected individual was a Black or White male.

Use the following information to answer the next two exercises. The table of data obtained from *www.baseball-almanac.com* shows hit information for four well known baseball players. Suppose that one hit from the table is randomly selected.

Table3.18

Name	Single	Double	Triple	Home run	TOTAL HITS
Babe Ruth	1,517	506	136	714	2,873
Jackie Robinson	1,054	273	54	137	1,518
Ty Cobb	3,603	174	295	114	4,189
Hank Aaron	2,294	624	98	755	3,771
TOTAL	8,471	1,577	583	1,720	12,351

116. Find$P$(hit was made by Babe Ruth).

a. $\frac{1518}{2873}$
b. $\frac{2873}{12351}$
c. $\frac{583}{12351}$
d. $\frac{4189}{12351}$

117. Find$P$(hit was made by Ty Cobb|The hit was a Home Run).

a. $\frac{4189}{12351}$
b. $\frac{114}{1720}$
c. $\frac{1720}{4189}$
d. $\frac{114}{12351}$

118. Table 3.19identifies a group of children by one of four hair colors, and by type of hair.

Table3.19

Hair type	Brown	Blond	Black	Red	Totals
Wavy	20		15	3	43
Straight	80	15		12	
Totals		20			215

a. Complete the table.
b. What is the probability that a randomly selected child will have wavy hair?
c. What is the probability that a randomly selected child will have either brown or blond hair?

d. What is the probability that a randomly selected child will have wavy brown hair?

e. What is the probability that a randomly selected child will have red hair, given that he or she has straight hair?

f. If *B* is the event of a child having brown hair, find the probability of the complement of *B*.

g. In words, what does the complement of *B*represent?

119. In a previous year, the weights of the members of the San Francisco 49ers and the Dallas Cowboyswere published in the *San Jose Mercury News*. The factual data were compiled into the following table.

Table3.20

Shirt #	≤ 210	211–250	251–290	> 290
1–33	21	5	0	0
34–66	6	18	7	4
66–99	6	12	22	5

For the following, suppose that you randomly select one player from the 49ers or Cowboys.

a. Find the probability that his shirt number is from 1 to 33.

b. Find the probability that he weighs at most 210 pounds.

c. Find the probability that his shirt number is from 1 to 33 AND he weighs at most 210 pounds.

d. Find the probability that his shirt number is from 1 to 33 OR he weighs at most 210 pounds.

e. Find the probability that his shirt number is from 1 to 33 GIVEN that he weighs at most 210 pounds.

Use the following information to answer the next two exercises. This tree diagram shows the tossing of an unfair coin followed by drawing one bead from a cup containing three red (*R*), four yellow (*Y*) and five blue (*B*) beads. For the coin, $P(H) = \frac{2}{3}$ and $P(T) = \frac{1}{3}$ where *H* is heads and *T* is tails.

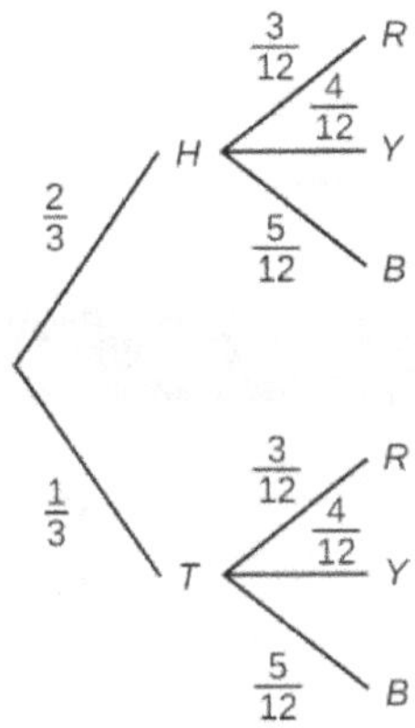

Figure3.20

120. Find P(tossing a Head on the coin AND a Red bead)

a. $\frac{2}{3}$

b. $\frac{5}{15}$

c. $\frac{6}{36}$

d. $\frac{5}{36}$

121. Find P(Blue bead).

a. $\frac{15}{36}$

b. $\frac{10}{36}$

c. $\frac{10}{12}$

d. $\frac{6}{36}$

122. A box of cookies contains three chocolate and seven butter cookies. Miguel randomly selects a cookie and eats it. Then he randomly selects another cookie and eats it. (How many cookies did he take?)

a. Draw the tree that represents the possibilities for the cookie selections. Write the probabilities along each branch of the tree.

b. Are the probabilities for the flavor of the SECOND cookie that Miguel selects independent of his first selection? Explain.

c. For each complete path through the tree, write the event it represents and find the probabilities.

d. Let S be the event that both cookies selected were the same flavor. Find$P(S)$.

e. Let$T$be the event that the cookies selected were different flavors. Find$P(T)$ by two different methods: by using the complement rule and by using the branches of the tree. Your answers should be the same with both methods.

f. Let$U$be the event that the second cookie selected is a butter cookie. Find$P(U)$.

123. A previous year, the weights of the members of the**San Francisco 49ers**and the**Dallas Cowboys**were published in the*San Jose Mercury News*. The factual data are compiled into**Table 3.21.**

Table3.21

Shirt #	≤ 210	211–250	251–290	290≤
1–33	21	5	0	0
34–66	6	18	7	4
66–99	6	12	22	5

For the following, suppose that you randomly select one player from the 49ers or Cowboys.

If having a shirt number from one to 33 and weighing at most 210 pounds were independent events, then what should be true about*P*(Shirt# 1–33|≤ 210 pounds)?

124. The probability that a male develops some form of cancer in his lifetime is 0.4567. The probability that a male has at least one false positive test result (meaning the test comes back for cancer when the man does not have it) is 0.51. Some of the following questions do not have enough information for you to answer them. Write "not enough information" for those answers. Let*C*= a man develops cancer in his lifetime and*P*= man has at least one false positive.

a. P(C)=______

b. P(P|C)=______

c. P(P|C')=______

d. If a test comes up positive, based upon numerical values, can you assume that man has cancer? Justify numerically and explain why or why not.

125. Given events*G*and*H*:*P*(*G*) = 0.43;*P*(*H*) = 0.26;P(H∩G)= 0.14

a. FindP(H∪G).

b. Find the probability of the complement of event(H∩G).

c. Find the probability of the complement of event(H∪G).

126. Given eventsJandK:P(J)=0.18;P(K)=0.37;P(J∪K)=0.45

a. FindP(J∩K).

b. Find the probability of the complement of event(J∩K).

c Find the probability of the complement of event(J∩K).

REFERENCES

3.1 Terminology

"Countries List by Continent." Worldatlas, 2013. Available online at http://www.worldatlas.com/cntycont.htm (accessed May 2, 2013).

3.2 Independent and Mutually Exclusive Events

Lopez, Shane, Preety Sidhu. "U.S. Teachers Love Their Lives, but Struggle in the Workplace." Gallup Wellbeing, 2013.

http://www.gallup.com/poll/161516/teachers-love-lives-struggle-workplace.aspx (accessed May 2, 2013).

Data from Gallup. Available online at www.gallup.com/ (accessed May 2, 2013).

3.3 Two Basic Rules of Probability

DiCamillo, Mark, Mervin Field. "The File Poll." Field Research Corporation. Available online at http://www.field.com/fieldpollonline/subscribers/Rls2443.pdf (accessed May 2, 2013).

Rider, David, "Ford support plummeting, poll suggests," The Star, September 14, 2011. Available online at http://www.thestar.com/news/gta/2011/09/14/ford_support_plummeting_poll_suggests.html (accessed May 2, 2013).

"Mayor's Approval Down." News Release by Forum Research Inc. Available online at http://www.forumresearch.com/forms/News Archives/News Releases/74209_TO_Issues_-

_Mayoral_Approval_%28Forum_Research%29%2820130320%29.pdf (accessed May 2, 2013).

"Roulette." Wikipedia. Available online at http://en.wikipedia.org/wiki/Roulette (accessed May 2, 2013).

Shin, Hyon B., Robert A. Kominski. "Language Use in the United States: 2007." United States Census Bureau. Available online at http://www.census.gov/hhes/socdemo/language/data/acs/ACS-12.pdf (accessed May 2, 2013).

Data from the Baseball-Almanac, 2013. Available online at www.baseball-almanac.com (accessed May 2, 2013).

Data from U.S. Census Bureau.

Data from the Wall Street Journal.

Data from The Roper Center: Public Opinion Archives at the University of Connecticut. Available online at http://www.ropercenter.uconn.edu/ (accessed May 2, 2013).

Data from Field Research Corporation. Available online at www.field.com/fieldpollonline (accessed May 2,2 013).

3.4 Contingency Tables and Probability Trees

"Blood Types." American Red Cross, 2013. Available online at http://www.redcrossblood.org/learn-about-blood/bloodtypes (accessed May 3, 2013).

Data from the National Center for Health Statistics, part of the United States Department of Health and Human Services.

Data from United States Senate. Available online at www.senate.gov (accessed May 2, 2013).

"Human Blood Types." Unite Blood Services, 2011. Available online at http://www.unitedbloodservices.org/ learnMore.aspx (accessed May 2, 2013).

Haiman, Christopher A., Daniel O. Stram, Lynn R. Wilkens, Malcom C. Pike, Laurence N. Kolonel, Brien E. Henderson, and Loīc Le Marchand. "Ethnic and Racial Differences in the Smoking-Related Risk of Lung Cancer." The New England Journal of Medicine, 2013. Available online at http://www.nejm.org/doi/full/10.1056/NEJMoa033250 (accessed May 2, 2013).

Samuel, T. M. "Strange Facts about RH Negative Blood." eHow Health, 2013. Available online at http://www.ehow.com/

facts_5552003_strange-rh-negative-blood.html (accessed May 2, 2013).

"United States: Uniform Crime Report – State Statistics from 1960–2011." The Disaster Center. Available online at http://www.disastercenter.com/crime/ (accessed May 2, 2013).

Data from Clara County Public H.D.

Data from the American Cancer Society.

Data from The Data and Story Library, 1996. Available online at http://lib.stat.cmu.edu/DASL/ (accessed May 2, 2013).

Data from the Federal Highway Administration, part of the United States Department of Transportation.

Data from the United States Census Bureau, part of the United States Department of Commerce.

Data from USA Today.

"Environment." The World Bank, 2013. Available online at http://data.worldbank.org/topic/environment (accessed May 2, 2013).

"Search for Datasets." Roper Center: Public Opinion Archives, University of Connecticut., 2013. Available online at http://www.ropercenter.uconn.edu/data_access/data/search_for_datasets.html (accessed May 2, 2013).

SOLUTIONS

1. a. $P(L') = P(S)$
   b. $P(M \cup S)$
   c. $P(F \cap L)$
   d. $P(M \mid L)$
   e. $P(L \mid M)$
   f. $P(S \mid F)$
   g. $P(F \mid L)$
   h. $P(F \cup L)$
   i. $P(M \cap S)$
   j. $P(F)$

3. $P(N) = \frac{15}{42} = \frac{5}{14} = 0.36$

5. $P(C) = \frac{5}{42} = 0.12$

7. $P(G) = \frac{20}{150} = \frac{2}{15} = 0.13$

9. $P(R) = \frac{22}{150} = \frac{11}{75} = 0.15$

11. $P(O) = \frac{150 - 22 - 38 - 20 - 28 - 26}{150} = \frac{16}{150} = \frac{8}{75} = 0.11$

13. $P(E) = \frac{47}{194} = 0.24$

15. $P(N) = \frac{23}{194} = 0.12$

17. $P(S) = \frac{12}{194} = \frac{6}{97} = 0.06$

19. $\frac{13}{52} = \frac{1}{4} = 0.25$

21. $\frac{3}{6} = \frac{1}{2} = 0.5$

23. $P(R) = \frac{4}{8} = 0.5$

25. $P(O \cup H)$

27. $P(H \mid I)$

29. $P(N \mid O)$

31. $P(I \cup N)$

33. $P(I)$

35. The likelihood that an event will occur given that another event has already occurred.

37. 1

39. the probability of landing on an even number or a multiple of three

41. $P(J) = 0.3$

43. $P(Q \cap R) = P(Q)P(R)$ $0.1 = (0.4)P(R)$ $P(R) = 0.25$

45. 0.376

47. $C \mid L$ means, given the person chosen is a Latino Californian, the person is a registered voter who prefers life in prison without parole for a person convicted of first degree murder.

49. $L \cap C$ is the event that the person chosen is a Latino California registered voter who prefers life without parole over the death penalty for a person convicted of first degree murder.

51. 0.6492

53. No, because $P(L \cap C)$ does not equal 0.

55. $P(\text{musician is a male} \cap \text{had private instruction}) = \frac{15}{130} = \frac{3}{26} = 0.12$

57. $P(\text{being a female musician} \cap \text{learning music in school}) = \frac{38}{130} = \frac{19}{65} = 0.29$
$P(\text{being a female musician})\,P(\text{learning music in school})$
$= \left(\frac{72}{130}\right)\left(\frac{62}{130}\right) = \frac{4,464}{16,900} = \frac{1,116}{4,225} = 0.26$ No, they are not independent because $P(\text{being a female musician} \cap \text{learning music in school})$ is not equal to $P(\text{being a female musician})P(\text{learning music in school})$.

58.

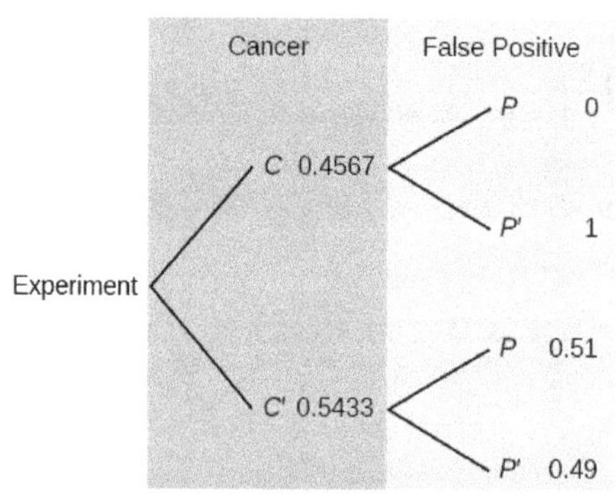

Figure 3.21

60. $\frac{35,065}{100,450}$

62. To pick one person from the study who is Japanese American AND smokes 21 to 30 cigarettes per day means that the person has to meet both criteria: both Japanese American and smokes 21 to 30 cigarettes. The sample space should include everyone in the study. The probability is $\frac{4,715}{100,450}$.

64. To pick one person from the study who is Japanese American given that person smokes 21-30 cigarettes per day, means that the person must fulfill both criteria and the sample space is reduced to those who smoke 21-30 cigarettes per day. The probability is $\frac{4715}{15,273}$.

66.

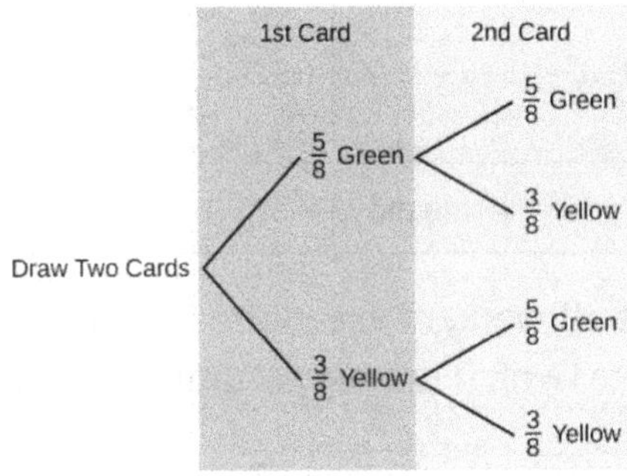

Figure 3.22

b. $P(GG) = \left(\frac{5}{8}\right)\left(\frac{5}{8}\right) = \frac{25}{64}$

c. P(at least one green) $= P(GG) + P(GY) + P(YG) = \frac{25}{64} + \frac{15}{64} + \frac{15}{64} = \frac{55}{64}$

$P(G \mid G) = \frac{5}{8}$

Yes, they are independent because the first card is placed back in the bag before the second card is drawn; the composition of cards in the bag remains the same from draw one to draw two.

68.

Table 3.22

	<20	20–64	>64	Totals
Female	0.0244	0.3954	0.0661	0.486
Male	0.0259	0.4186	0.0695	0.514
Totals	0.0503	0.8140	0.1356	1

b. $P(F) = 0.486$

c. $P(>64|F) = 0.1361$

d. $P(>64 \text{ and} F) = P(F)P(>64|F) = (0.486)(0.1361) = 0.0661$

e. $P(>64|F)$ is the percentage of female drivers who are 65 or older and $P(>64 \cap F)$ is the percentage of drivers who are female and 65 or older.

f. $P(>64) = P(>64 \cap F) + P(>64 \cap M) = 0.1356$

g. No, being female and 65 or older are not mutually exclusive because they can occur at the same time $P(>64 \cap \cap F) = 0.0661$.

70. a.

Table 3.23

	Car, truck or van	Walk	Public transportation	Other	Totals
Alone	0.7318				
Not alone	0.1332				
Totals	0.8650	0.0390	0.0530	0.0430	1

a. If we assume that all walkers are alone and that none from the other two groups travel alone (which is a big assumption) we have:*P*(Alone) = 0.7318 + 0.0390 = 0.7708.

b. Make the same assumptions as in (b) we have: (0.7708)(1,000) = 771

c. (0.1332)(1,000) = 133

73. a. You can't calculate the joint probability knowing the probability of both events occurring, which is not in the information given; the probabilities should be multiplied, not added; and probability is never greater than 100%

b. A home run by definition is a successful hit, so he has to have at least as many successful hits as home runs.

75. 0

77. 0.3571

79. 0.2142

81. Physician (83.7)

83. 83.7 – 79.6 = 4.1

85. P(Occupation < 81.3) = 0.5

87. a, The Forum Research surveyed 1,046 Torontonians.

b. 58%

c. 42% of 1,046 = 439 (rounding to the nearest integer)

d. 0.57

e. 0.60.

89. a. *P*(Betting on two line that touch each other on the table) $= \frac{6}{38}$

b. *P*(Betting on three numbers in a line) $= \frac{3}{38}$

c. *P*(Betting on one number) $= \frac{1}{38}$

d. *P*(Betting on four number that touch each other to form a square) $= \frac{4}{38}$

e. *P*(Betting on two number that touch each other on the table) $= \frac{2}{38}$

f. *P*(Betting on 0-00-1-2-3) $= \frac{5}{38}$

g. *P*(Betting on 0-1-2; or 0-00-2; or 00-2-3) $= \frac{3}{38}$

91. a. {G1,G2,G3,G4,G5,Y1,Y2,Y3}

   b. $\frac{5}{8}$

   c. $\frac{2}{3}$

   d. $\frac{2}{8}$

   e. $\frac{6}{8}$

   f. No, because P(G∩E) does not equal 0.

93.

NOTE

The coin toss is independent of the card picked first.

a. $\{(G,H)\ (G,T)\ (B,H)\ (B,T)\ (R,H)\ (R,T)\}$

b. $P(A) = P(\text{blue})P(\text{head}) = \left(\frac{3}{10}\right)\left(\frac{1}{2}\right) = \frac{3}{20}$

c. Yes, *A* and *B* are mutually exclusive because they cannot happen at the same time; you cannot pick a card that is both blue and also (red or green). $P(A \cap B) = 0$

d. No, *A* and *C* are not mutually exclusive because they can occur at the same time. In fact, *C* includes $P(A \cap C) = P(A) = \frac{3}{20}$ f *A*; if the card chosen is blue it is also (red or blue).

95.

a. $S = \{(HHH), (HHT), (HTH), (HTT), (THH), (THT), (TTH), (TTT)\}$

b. $\frac{4}{8}$

c. Yes, because if *A* has occurred, it is impossible to obtain two tails. In other words, $P(A \cap B) = 0$.

97.

a. If *Y* and *Z* are independent, then $P(Y \cap Z) = P(Y)P(Z)$, so $P(Y \cup\cup Z) = P(Y) + P(Z) - P(Y)P(Z)$.

b. 0.5

99. iii i iv ii

101.

a. $P(R) = 0.44$

b. $P(R||E) = 0.56$

c. $P(R||O) = 0.31$

d. No, whether the money is returned is not independent of which class the money was placed in. There are several ways to justify this mathematically, but one is that the money placed in economics classes is not returned at the same overall rate; $P(R||E) \neq P(R)$.

e. No, this study definitely does not support that notion; *in fact*, it suggests the opposite. The money placed in the economics classrooms was returned at a higher rate than the money place in all classes collectively; $P(R|E) > P(R)$.

103.

a. P(type O ∪∪ Rh-) = P(type O) + P(Rh-) - P(type O ∩ Rh-)
0.52 = 0.43 + 0.15 - P(type O ∩ Rh-); solve to find P(type O ∩ Rh-) = 0.06
6% of people have type O, Rh- blood

b. P(NOT(type O ∩ Rh-)) = 1 - P(type O ∩ Rh-) = 1 - 0.06 = 0.94
94% of people do not have type O, Rh- blood

105.

a. Let C = be the event that the cookie contains chocolate. Let N = the event that the cookie contains nuts.

b. $P(C \cup N) = P(C) + P(N) - P(C \cap N) = 0.36 + 0.12 - 0.08 = 0.40$

c. P(NEITHER chocolate NOR nuts) = 1 - $P(C \cup N)$ = 1 - 0.40 = 0.60

107. 0

109. $\frac{10}{67}$

111. $\frac{10}{34}$

113. d

115. a.

Table 3.24

Race and sex	1–14	15–24	25–64	Over 64	TOTALS
White, male	210	3,356	13,610	4,870	22,050
White, female	80	580	3,380	890	4,930
Black, male	10	460	1,060	140	1,670
Black, female	0	40	270	20	330
All others				100	
TOTALS	310	4,650	18,780	6,020	29,760

b.

Table 3.25

Race and sex	1–14	15–24	25–64	Over 64	TOTALS
White, male	210	3,356	13,610	4,870	22,050
White, female	80	580	3,380	890	4,930
Black, male	10	460	1,060	140	1,670
Black, female	0	40	270	20	330
All others	10	210	460	100	780
TOTALS	310	4,650	18,780	6,020	29,760

c. $\frac{22{,}050}{29{,}760}$

d. $\frac{330}{29{,}760}$

e. $\frac{2{,}000}{29{,}760}$

f. $\frac{23{,}720}{29{,}760}$

g. $\frac{5{,}010}{6{,}020}$

117. b

119. a. $\frac{26}{106}$

b. $\frac{33}{106}$

c. $\frac{21}{106}$

d. $\left(\frac{26}{106}\right) + \left(\frac{33}{106}\right) - \left(\frac{21}{106}\right) = \left(\frac{38}{106}\right)$

e. $\frac{21}{33}$

121. a

124. a. $P(C) = 0.4567$

b. not enough information

c. not enough information

d. No, because over half (0.51) of men have at least one false positive text

126. a. (J∪K)=P(J)+P(K)–P(J∩K); 0.45=0.18+0.37–P(J∩K); solve to find P(J∩K)=0.10

b. P(NOT(J∩K))=1–P(J∩K) =1– 0.10=0.90

c. P(NOT(J∪K))=1–P(J∪K) =1– 0.45=0.55

4 Chapter

DISCRETE RANDOM VARIABLES

Figure 4.1 You can use probability and discrete random variables to calculate the likelihood of lightning striking the ground five times during a half-hour thunderstorm. (Credit: Leszek Leszczynski)

INTRODUCTION: AN OVERVIEW

A student completes a ten-item true-false examination. Due to the student's demanding schedule, he or she was unable to study and resorted to random guessing for each response. What is the likelihood of the student achieving a score of at least 70% on the test? Small enterprises may be concerned with the volume of long-distance telephone calls made by their staff during peak hours. Assume the historical mean is 20 calls. What is the likelihood that employees will make in excess of 20 long-distance phone calls during peak hours? These two examples exemplify distinct categories of probability problems with discrete random variables. Discrete data refers to data that can be enumerated, meaning the random variable can assume only integer values. A random variable articulates the results of a statistical experiment in verbal form. The values of a random variable may fluctuate with each iteration of an experiment, commonly referred to as a trial.

Random Variable Notation

The upper case letter X denotes a random variable. Lower case letters like x or y denote the value of a random variable. If X is a random variable, then X is written in words, and x is given as a number.

For example, let X = the number of heads you get when you toss three fair coins. The sample space for the toss of three fair coins is *TTT*; *THH*; *HTH*; *HHT*; *HTT*; *THT*; *TTH*; *HHH*. Then, $x = 0, 1, 2, 3$. X is in words and x is a number. Notice that for this example, the x values are countable outcomes. Because you can count the possible values as whole numbers that X can take on and the outcomes are random (the x values 0, 1, 2, 3), X is a discrete random variable.

Probability Density Functions (PDF) for a Random Variable

A probability density function or probability distribution function has two characteristics:

1. Each probability is between zero and one, inclusive.
2. The sum of the probabilities is one.

A probability density function is a mathematical formula that calculates probabilities for specific types of events, what we have been calling experiments. There is a sort of magic to a probability density function (Pdf) partially because the same formula often describes very different types of events. For example, the binomial Pdf will calculate probabilities for flipping coins, yes/no questions on an exam, opinions of voters in an up or down opinion poll, indeed any binary event. Other probability density functions will provide probabilities for the time until a part will fail, when a customer will arrive at the turnpike booth, the number of telephone calls arriving at a central switchboard, the growth rate of a bacterium, and on and on. There are whole families of probability density functions that are used in a wide variety of applications, including medicine, business and finance, physics and engineering, among others.

For our needs here we will concentrate on only a few probability density functions as we develop the tools of inferential statistics.

Counting Formulas and the Combinational Formula

To repeat, the probability of event A , P(A), is simply the number of ways the experiment will result in A, relative to the total number of possible outcomes of the experiment.

As an equation this is:

$$P(A) = \frac{\text{number of ways to get A}}{\text{Total number of possible outcomes}}$$

When we looked at the sample space for flipping 3 coins we could easily write the full sample space and thus could easily count the number of events that met our desired result, e.g. x = 1 , where X is the random variable defined as the number of heads.

As we have larger numbers of items in the sample space, such as a full deck of 52 cards, the ability to write out the sample space becomes impossible.

We see that probabilities are nothing more than counting the events in each group we are interested in and dividing by the number of elements in the universe, or sample space. This is easy enough if we are counting sophomores in a Stat class, but in more complicated cases listing all the possible outcomes may take a life time. There are, for example, 36 possible outcomes from throwing just two six-sided dice where the random variable is the sum of the number of spots on the up-facing sides. If there were four dice then the total number of possible outcomes would become 1,296. There are more than 2.5 MILLION possible 5 card poker hands in a standard deck of 52 cards. Obviously keeping track of all these possibilities and counting them to get at a single probability would be tedious at best.

An alternative to listing the complete sample space and counting the number of elements we are interested in, is to skip the step of listing the sample space, and simply figuring out the number of elements in it and doing the appropriate division. If we are after a probability we really do not need to see each and every element in the sample space, we only need to know how many elements are there. Counting formulas were invented to do just this. They tell us the number of unordered subsets of a certain size that can be created from a set of unique elements. By unordered it is meant that, for example, when dealing cards, it does not matter if you got {ace, ace, ace, ace, king} or {king, ace, ace, ace, ace} or {ace, king, ace, ace, ace} and so on. Each of these subsets are the same because they each have 4 aces and one king.

Combinational Formula

$$\binom{n}{x} = {}_nC_x = \frac{n!}{x!(n-x)!}$$

This is the formula that tells the number of unique unordered subsets of size x that can be created from n unique elements. The formula is read "n combinatorial x". Sometimes it is read as "n choose x." The exclamation point "!" is called a factorial and tells us to take all the numbers from 1 through the number before the ! and multiply them together thus 4! is 1*2*3*4=24. By definition 0! = 1. The formula is called the Combinatorial Formula. It is also called the Binomial Coefficient, for reasons that will be clear shortly. While this mathematical concept was understood long before 1653, Blaise Pascal is given major credit for his proof that he published in that year. Further, he developed a generalized method of calculating the values for combinatorials known to us as the Pascal Triangle. Pascal was one of the geniuses of an era of extraordinary intellectual advancement which included the work of Galileo, Rene Descartes, Isaac Newton, William Shakespeare and the refinement of the scientific method, the very rationale for the topic of this text.

Let's find the hard way the total number of combinations of the four aces in a deck of cards if we were going to take them two at a time. The sample space would be:

S={Spade,Heart),(Spade,Diamond),(Spade,Club),(Diamond,Club),(Heart,Diamond),(Heart,Club)}

There are 6 combinations; formally, six unique unordered subsets of size 2 that can be created from 4 unique elements. To use the combinatorial formula we would solve the formula as follows:

$$\binom{4}{2}=\frac{4!}{(4-2)!2!}=\frac{4\cdot 3\cdot 2\cdot 1}{2\cdot 1\cdot 2\cdot 1}=6$$

If we wanted to know the number of unique 5 card poker hands that could be created from a 52 card deck we simply compute:

$$\binom{52}{5}$$

where 52 is the total number of unique elements from which we are drawing and 5 is the size group we are putting them into.

With the combinatorial formula we can count the number of elements in a sample space without having to write each one of them down, truly a lifetime's work for just the number of 5 card hands from a deck of 52 cards. We can now apply this tool to a very important probability density function, the hypergeometric distribution.

Remember, a probability density function computes probabilities for us. We simply put the appropriate numbers in the formula and we get the probability of specific events. However, for these formulas to work they must be applied only to cases for which they were designed.

4.1 HYPERGEOMETRIC DISTRIBUTION

The simplest probability density function is the hypergeometric. This is the most basic one because it is created by combining our knowledge of probabilities from Venn diagrams, the addition and multiplication rules, and the combinatorial counting formula.

To find the number of ways to get 2 aces from the four in the deck we computed:

$$\binom{4}{2}=\frac{4!}{2!(4-2)!}=6$$

And if we did not care what else we had in our hand for the other three cards we would compute:

$$\binom{48}{3} = \frac{48!}{3!45!} = 17{,}296$$

Putting this together, we can compute the probability of getting exactly two aces in a 5 card poker hand as:

$$\frac{\binom{4}{2}\binom{48}{3}}{\binom{52}{5}} = .0399$$

This solution is really just the probability distribution known as the Hypergeometric. The generalized formula is:

$$h(x) = \frac{\binom{A}{x}\binom{N-A}{n-x}}{\binom{N}{n}}$$

where x = the number we are interested in coming from the group with A objects.

h(x) is the probability of x successes, in n attempts, when A successes (aces in this case) are in a population that contains N elements. The hypergeometric distribution is an example of a discrete probability distribution because there is no possibility of partial success, that is, there can be no poker hands with 2 1/2 aces. Said another way, a discrete random variable has to be a whole, or counting, number only. This probability distribution works in cases where the probability of a success changes with each draw. Another way of saying this is that the events are NOT independent. In using a deck of cards, we are sampling WITHOUT replacement. If we put each card back after it was drawn then the hypergeometric distribution be an inappropriate Pdf.

For the hypergeometric to work,

1. the population must be dividable into two and only two independent subsets (aces and non-aces in our example). The random variable X = the number of items from the group of interest.
2. the experiment must have changing probabilities of success with each experiment (the fact that cards are not replaced after the draw in our example makes this true in this case). Another way to say this is that you sample without replacement and therefore each pick is not independent.
3. the random variable must be discrete, rather than continuous.

EXAMPLE 4.1

Problem

A candy dish contains 30 jelly beans and 20 gumdrops. Ten candies are picked at random. What is the probability that 5 of the 10 are gumdrops? The two groups are jelly beans and gumdrops. Since the probability question asks for the probability of picking gumdrops, the group of interest (first group A in the formula) is gumdrops. The size of the group of interest (first group) is 30. The size of the second group is 20. The size of the sample is 10 (jelly beans or gumdrops). Let X = the number of gumdrops in the sample of 10. X takes on the values x = 0, 1, 2, ..., 10. a. What is the probability statement written mathematically? b. What is the hypergeometric probability density function written out to solve this problem? c. What is the answer to the question "What is the probability of drawing 5 gumdrops in 10 picks from the dish?"

Solution 4.1

a. $P(x = 5)$

b. $P(x = 5) = \frac{\binom{30}{5}\binom{20}{5}}{\binom{50}{10}}$

c. $P(x = 5) = 0.215$

TRY IT 4.1

A bag contains letter tiles. Forty-four of the tiles are vowels, and 56 are consonants. Seven tiles are picked at random. You want to know the probability that four of the seven tiles are vowels. What is the group of interest, the size of the group of interest, and the size of the sample?

4.2 BINOMIAL DISTRIBUTION

A more valuable probability density function with many applications is the binomial distribution. This distribution will compute probabilities for any binomial process. A binomial process, often called a Bernoulli process after the first person to fully develop its properties, is any case where there are only two possible outcomes in any one trial, called successes and failures. It gets its name from the binary number system where all numbers are reduced to either 1's or 0's, which is the basis for computer technology and CD music recordings.

Binomial Formula

$$b(x) = \binom{n}{x} p^x q^{n-x}$$

where b(x) is the probability of X successes in n trials when the probability of a success in ANY ONE TRIAL is p. And of course q=(1-p) and is the probability of a failure in any one trial.

We can see now why the combinatorial formula is also called the binomial coefficient because it reappears here again in the binomial probability function. For the binomial formula to work, the probability of a success in any one trial must be the same from trial to trial, or in other words, the outcomes of each trial must be independent. Flipping a coin is a binomial process because the probability of getting a head in one flip does not depend upon what has happened in PREVIOUS flips. (At this time it should be noted that using p for the parameter of the binomial distribution is a violation of the rule that population parameters are designated with Greek letters. In many textbooks θ (pronounced theta) is used instead of p and this is how it should be.

Just like a set of data, a probability density function has a mean and a standard deviation that describes the data set. For the binomial distribution these are given by the formulas:

$$\mu = np$$

$$\sigma = \sqrt{npq}$$

Notice that p is the only parameter in these equations. The binomial distribution is thus seen as coming from the one-parameter family of probability distributions. In short, we know all there is to know about the binomial once we know p, the probability of a success in any one trial.

In probability theory, under certain circumstances, one probability distribution can be used to approximate another. We say that one is the limiting distribution of the other. If a small number is to be drawn from a large population, even if there is no replacement, we can still use the binomial even thought this is not a binomial process. If there is no replacement it violates the independence rule of the binomial. Nevertheless, we can use the binomial to approximate a probability that is really a hypergeometric distribution if we are drawing fewer than 10 percent of the population, i.e. n is less than 10 percent of N in the formula for the hypergeometric function. The rationale for this argument is that when drawing a small percentage of the population we do not alter the probability of a success from draw to draw in any meaningful way. Imagine drawing from not one deck of 52 cards but from 6 decks of cards. The probability of say drawing an ace does not change the conditional probability of what happens on a second draw in the same way it would if there were only 4 aces rather than the 24 aces now to draw from. This ability to use one probability distribution to estimate others will become very valuable to us later.

There are three characteristics of a binomial experiment.

1. There are a fixed number of trials. Think of trials as repetitions of an experiment. The letter n denotes the number of trials.
2. The random variable, xx, number of successes, is discrete.
3. There are only two possible outcomes, called "success" and "failure," for each trial. The letter p denotes the probability of a success on any one trial, and q denotes the probability of a failure on any one trial. $p + q = 1$.
4. The n trials are independent and are repeated using identical conditions. Think of this as drawing WITH replacement. Because the n trials are independent, the outcome of one trial does not help in predicting the outcome of another trial. Another way of saying this is that for each individual trial, the probability, p, of a success and probability, q, of a failure remain the same. For example, randomly guessing at a true-false statistics question has only two outcomes. If a success is guessing correctly, then a failure is guessing incorrectly. Suppose Joe always guesses correctly on any statistics true-false question with a probability $p = 0.6$. Then, $q = 0.4$. This means that for every true-false statistics question Joe answers, his probability of success ($p = 0.6$) and his probability of failure ($q = 0.4$) remain the same.

The outcomes of a binomial experiment fit a binomial probability distribution. The random variable X = the number of successes obtained in the n independent trials.

The mean, μ, and variance, σ^2, for the binomial probability distribution are $\mu = np$ and $\sigma^2 = npq$. The standard deviation, σ, is then $\sigma = \sqrt{npq}$.

Any experiment that has characteristics three and four and where $n = 1$ is called a **Bernoulli Trial** (named after Jacob Bernoulli who, in the late 1600s, studied them extensively). A binomial experiment takes place when the number of successes is counted in one or more Bernoulli Trials.

EXAMPLE 4.2

Suppose you play a game that you can only either win or lose. The probability that you win any game is 55%, and the probability that you lose is 45%. Each game you play is independent. If you play the game 20 times, write the function that describes the probability that you win 15 of the 20 times. Here, if you define X as the number of wins, then X takes on the values 0, 1, 2, 3, ..., 20. The probability of a success is $p = 0.55$. The probability of a failure is $q = 0.45$. The number of trials is $n = 20$. The probability question can be stated mathematically as $P(x = 15)$.

TRY IT 4.2

A trainer is teaching a dolphin to do tricks. The probability that the dolphin successfully performs the trick is 35%, and the probability that the dolphin does not successfully perform the trick is 65%. Out of 20 attempts, you want to find the probability that the dolphin succeeds 12 times. Find the P(X=12) using the binomial Pdf.

EXAMPLE 4.3

A fair coin is flipped 15 times. Each flip is independent. What is the probability of getting more than ten heads? Let X = the number of heads in 15 flips of the fair coin. X takes on the values 0, 1, 2, 3, ..., 15. Since the coin is fair, $p = 0.5$ and $q = 0.5$. The number of trials is $n = 15$. State the probability question mathematically.

Solution 4.3

$P(x > 10)$

EXAMPLE 4.4

Approximately 70% of statistics students do their homework in time for it to be collected and graded. Each student does homework independently. In a statistics class of 50 students, what is the probability that at least 40 will do their homework on time? Students are selected randomly.

a. This is a binomial problem because there is only a success or a __________, there are a fixed number of trials, and the probability of a success is 0.70 for each trial.

Solution 4.4

a. failure
b. If we are interested in the number of students who do their homework on time, then how do we define X?

Solution 4.4

b. X = the number of statistics students who do their homework on time
c. What values does x take on?

Solution 4.4

c. 0, 1, 2, ..., 50
d. What is a "failure," in words?

Solution 4.4

d. Failure is defined as a student who does not complete his or her homework on time.
   The probability of a success is $p = 0.70$. The number of trials is $n = 50$.

e. If $p + q = 1$, then what is q?

Solution 4.4

e. q=0.30
f. The words "at least" translate as what kind of inequality for the probability question P(*x* ____ 40).

Solution 4.4

f. greater than or equal to (≥)

The probability question is $P(x \geq 40)$.

TRY IT 4.4

Sixty-five percent of people pass the state driver's exam on the first try. A group of 50 individuals who have taken the driver's exam is randomly selected. Give two reasons why this is a binomial problem.

TRY IT 4.4

During the 2013 regular NBA season, DeAndre Jordan of the Los Angeles Clippers had the highest field goal completion rate in the league. DeAndre scored with 61.3% of his shots. Suppose you choose a random sample of 80 shots made by DeAndre during the 2013 season. Let X = the number of shots that scored points.

a. What is the probability distribution for X?
b. Using the formulas, calculate the (i) mean and (ii) standard deviation of X.
c. Find the probability that DeAndre scored with 60 of these shots.
d. Find the probability that DeAndre scored with more than 50 of these shots.

4.3 GEOMETRIC DISTRIBUTION

The geometric probability density function builds upon what we have learned from the binomial distribution. In this case the experiment continues until either a success or a failure occurs rather than for a set number of trials. There are three main characteristics of a geometric experiment.

1. There are one or more Bernoulli trials with all failures except the last one, which is a success. In other words, you keep repeating what you are doing until the first success. Then you stop. For example, you throw a dart at a bullseye until you hit the bullseye. The first time you hit the bullseye is a "success" so you stop throwing the dart. It might take six tries until you hit the bullseye. You can think of the trials as failure, failure, failure, failure, failure, success, STOP.
2. In theory, the number of trials could go on forever.
3. The probability, p, of a success and the probability, q, of a failure is the same for each trial. $p + q = 1$ and $q = 1 - p$. For example, the probability of rolling a three when you throw one fair die is $\frac{1}{6}$. This is true no matter how

many times you roll the die. Suppose you want to know the probability of getting the first three on the fifth roll. On rolls one through four, you do not get a face with a three. The probability for each of the rolls is $q = \frac{5}{6}$, the probability of a failure. The probability of getting a three on the fifth roll is $\left(\frac{5}{6}\right)\left(\frac{5}{6}\right)\left(\frac{5}{6}\right)\left(\frac{5}{6}\right)\left(\frac{1}{6}\right) = 0.0804$

4. X = the number of independent trials until the first success.

EXAMPLE 4.5

You play a game of chance that you can either win or lose (there are no other possibilities) **until** you lose. Your probability of losing is $p = 0.57$. What is the probability that it takes five games until you lose? Let X = the number of games you play until you lose (includes the losing game). Then X takes on the values 1, 2, 3, ... (could go on indefinitely). The probability question is $P(x = 5)$.

TRY IT 4.5

You throw darts at a board until you hit the center area. Your probability of hitting the center area is $p = 0.17$. You want to find the probability that it takes eight throws until you hit the center. What values does X take on?

EXAMPLE 4.6

A safety engineer feels that 35% of all industrial accidents in her plant are caused by failure of employees to follow instructions. She decides to look at the accident reports (selected randomly and replaced in the pile after reading) **until** she finds one that shows an accident caused by failure of employees to follow instructions. On average, how many reports would the safety engineer **expect** to look at until she finds a report showing an accident caused by employee failure to follow instructions? What is the probability that the safety engineer will have to examine at least three reports until she finds a report showing an accident caused by employee failure to follow instructions?

Let X = the number of accidents the safety engineer must examine **until** she finds a report showing an accident caused by employee failure to follow instructions. X takes on the values 1, 2, 3, ... The first question asks you to find the **expected value** or the mean. The second question asks you to find $P(x \geq 3)$. ("At least" translates to a "greater than or equal to" symbol).

TRY IT 4.6

An instructor feels that 15% of students get below a C on their final exam. She decides to look at final exams (selected randomly and replaced in the pile after reading) until she finds one that shows a grade below a C. We want to know the probability that the instructor will have to examine at least ten exams until she finds one with a grade below a C. What is the probability question stated mathematically?

EXAMPLE 4.7

Suppose that you are looking for a student at your college who lives within five miles of you. You know that 55% of the 25,000 students do live within five miles of you. You randomly contact students from the college until one says he or she lives within five miles of you. What is the probability that you need to contact four people?

This is a geometric problem because you may have a number of failures before you have the one success you desire. Also, the probability of a success stays approximately the same each time you ask a student if he or she lives within five miles of you. There is no definite number of trials (number of times you ask a student).

a. Let *X* = the number of ____________ you must ask ____________ one says yes.

Solution 4.7

a. Let *X* = the number of students you must ask until one says yes.
b. What values does *X* take on?

Solution 4.7

b. 1, 2, 3, ..., (total number of students)
c. What are p and q?

Solution 4.7

c. $p = 0.55$; $q = 0.45$
d. The probability question is *P*(_______).

Solution 4.7

d. $P(x = 4)$

Notation for the Geometric: G = Geometric Probability Distribution Function

$X \sim G(p)$

Read this as "X is a random variable with a geometric distribution." The parameter is *p*; *p* = the probability of a success for each trial.

The Geometric Pdf tells us the probability that the first occurrence of success requires x number of independent trials, each with success probability p. If the probability of success on each trial is *p*, then the probability that the *x*th trial (out of x trials) is the first success is:

$$P(X = x) = (1 - p)^{x - 1}p$$

for $x = 1, 2, 3, \ldots$.

The expected value of X, the mean of this distribution, is 1/p. This tells us how many trials we have to expect until we get the first success including in the count the trial that results in success. The above form of the Geometric distribution is used for modeling the number of trials until the first success. The number of trials includes the one that is a success: x = all trials including the one that is a success. This can be seen in the form of the formula. If X = number of trials including the success, then we must multiply the probability of failure, (1-p), times the number of failures, that is X-1.

By contrast, the following form of the geometric distribution is used for modeling number of failures until the first success:

$$P(X = x) = (1 - p)^x p$$

for $x = 0, 1, 2, 3, ...$

In this case the trial that is a success is not counted as a trial in the formula: x = number of failures. The expected value, mean, of this distribution is $\mu = \frac{(1 - p)}{p}$. This tells us how many failures to expect before we have a success. In either case, the sequence of probabilities is a geometric sequence.

EXAMPLE 4.8

Assume that the probability of a defective computer component is 0.02. Components are randomly selected. Find the probability that the first defect is caused by the seventh component tested. How many components do you expect to test until one is found to be defective?

Let X = the number of computer components tested until the first defect is found.

X takes on the values 1, 2, 3, ... where $p = 0.02$. X ~ G(0.02)

Find $P(x = 7)$. Answer: $P(x = 7) = (1 - 0.02)^{7-1} \times 0.02 = 0.0177$.

The probability that the seventh component is the first defect is 0.0177.

The graph of X ~ G(0.02) is:

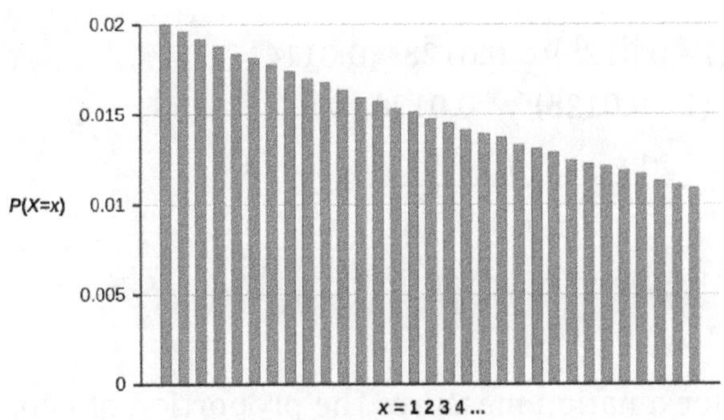

Figure 4.2

The y-axis contains the probability of x, where X = the number of computer components tested. Notice that the probabilities decline by a common increment. This increment is the same ratio between each number and is called a geometric progression and thus the name for this probability density function.

The number of components that you would expect to test until you find the first defective component is the mean, $\mu = 50$.

The formula for the mean for the random variable defined as number of failures until first success is $\mu = \frac{1}{p} = \frac{1}{0.02} = 50$

See Example 4.9 for an example where the geometric random variable is defined as number of trials until first success. The expected value of this formula for the geometric will be different from this version of the distribution.

The formula for the variance is $\sigma^2 = \left(\frac{1}{p}\right)\left(\frac{1}{p} - 1\right) = \left(\frac{1}{0.02}\right)\left(\frac{1}{0.02} - 1\right) = 2{,}450$

The standard deviation is $\sigma = \sqrt{\left(\frac{1}{p}\right)\left(\frac{1}{p} - 1\right)} = \sqrt{\left(\frac{1}{0.02}\right)\left(\frac{1}{0.02} - 1\right)} = 49.5$

EXAMPLE 4.9

The lifetime risk of developing pancreatic cancer is about one in 78 (1.28%). Let X = the number of people you ask before one says he or she has pancreatic cancer. The random variable X in this case includes only the number of trials that were failures and does not count the trial that was a success in finding a person who had the disease. The appropriate formula for this random variable is the second one presented above. Then X is a discrete random variable with a geometric distribution: $X \sim G\left(\frac{1}{78}\right)$ or $X \sim G(0.0128)$.

a. What is the probability of that you ask 9 people before one says he or she has pancreatic cancer? This is asking, what is the probability that you ask 9 people unsuccessfully and the tenth person is a success?
b. What is the probability that you must ask 20 people?
c. Find the (i) mean and (ii) standard deviation of X.

Solution 4.9

a. $P(x = 9) = (1 - 0.0128)^9 * 0.0128 = 0.0114$
b. $P(x = 20) = (1 - 0.0128)^{19} * 0.0128 = 0.01$
c. i. $\text{Mean} = \mu = \frac{(1-p)}{p} = \frac{(1-0.0128)}{0.0128} = 77.12$

   ii. $\text{Standard Deviation} = \sigma = \sqrt{\frac{1-p}{p^2}} = \sqrt{\frac{1-0.0128}{0.0128^2}} \approx 77.62$

TRY IT 4.9

The literacy rate for a nation measures the proportion of people age 15 and over who can read and write. The literacy rate for women in The United Colonies of

Independence is 12%. Let X = the number of women you ask until one says that she is literate.

a. What is the probability distribution of X?
b. What is the probability that you ask five women before one says she is literate?
c. What is the probability that you must ask ten women?

EXAMPLE 4.10

A baseball player has a batting average of 0.320. This is the general probability that he gets a hit each time he is at bat.

What is the probability that he gets his first hit in the third trip to bat?

In this case the sequence is failure, failure success.

Solution 4.10

$P(x=3) = (1-0.32)^{3-1} \times .32 = 0.1480$

In this case the sequence is failure, failure success.

How many trips to bat do you expect the hitter to need before getting a hit?

Solution 4.10

$$\mu = \frac{1}{p} = \frac{1}{0.320} = 3.125 \approx 3$$

This is simply the expected value of successes and therefore the mean of the distribution.

EXAMPLE 4.11

There is an 80% chance that a Dalmatian dog has 13 black spots. You go to a dog show and count the spots on Dalmatians. What is the probability that you will review the spots on 3 dogs before you find one that has 13 black spots?

Solution 4.11

$P(x=3) = (1 - 0.80)^3 \times 0.80 = 0.0064$

4.4 POISSON DISTRIBUTION

Another useful probability distribution is the Poisson distribution, or waiting time distribution. This distribution is used to determine how many checkout clerks are needed to keep the waiting time in line to specified levels, how may telephone lines are needed to keep the system from overloading, and many other practical applications. A modification of the Poisson, the Pascal, invented nearly four centuries ago, is used today by telecommunications companies worldwide for load factors, satellite hookup levels and Internet capacity problems. The distribution gets its name from Simeon Poisson who presented it in 1837 as an extension of the binomial distribution which we will see can be estimated with the Poisson.

There are two main characteristics of a Poisson experiment.

1. The Poisson probability distribution gives the probability of a number of events occurring in a fixed interval of time or space if these events happen with a known average rate.
2. The events are independently of the time since the last event. For example, a book editor might be interested in the number of words spelled incorrectly in a particular book. It might be that, on the average, there are five words spelled incorrectly in 100 pages. The interval is the 100 pages and it is assumed that there is no relationship between when misspellings occur.
3. The random variable X = the number of occurrences in the interval of interest.

EXAMPLE 4.12

A bank expects to receive six bad checks per day, on average. What is the probability of the bank getting fewer than five bad checks on any given day? Of interest is the number of checks the bank receives in one day, so the time interval of interest is one day. Let X = the number of bad checks the bank receives in one day. If the bank expects to receive six bad checks per day then the average is six checks per day. Write a mathematical statement for the probability question.

Solution 4.12

$P(x < 5)$

EXAMPLE 4.13

You notice that a news reporter says "uh," on average, two times per broadcast. What is the probability that the news reporter says "uh" more than two times per broadcast.

This is a Poisson problem because you are interested in knowing the number of times the news reporter says "uh" during a broadcast.

a. What is the interval of interest?

Solution 4.13

a. one broadcast measured in minutes
b. What is the average number of times the news reporter says "uh" during one broadcast?

Solution 4.13

b. 2

c. Let X = ____________. What values does X take on?

Solution 4.13

c. Let X = the number of times the news reporter says "uh" during one broadcast. $x = 0, 1, 2, 3, \ldots$

d. The probability question is $P($_______$)$.

Solution 4.13

d. $P(x > 2)$

Notation for the Poisson: P = Poisson Probability Distribution Function

$X \sim P(\mu)$

Read this as "X is a random variable with a Poisson distribution." The parameter is μ (or λ); μ (or λ) = the mean for the interval of interest. The mean is the number of occurrences that occur on average during the interval period.

The formula for computing probabilities that are from a Poisson process is:

$$P(x) = \frac{\mu^{x} e^{-\mu}}{x!}$$

where $P(X)$ is the probability of X successes, μ is the expected number of successes based upon historical data, e is the natural logarithm approximately equal to 2.718, and X is the number of successes per unit, usually per unit of time.

In order to use the Poisson distribution, certain assumptions must hold. These are: the probability of a success, μ, is unchanged within the interval, there cannot be simultaneous successes within the interval, and finally, that the probability of a success among intervals is independent, the same assumption of the binomial distribution.

In a way, the Poisson distribution can be thought of as a clever way to convert a continuous random variable, usually time, into a discrete random variable by breaking up time into discrete independent intervals. This way of thinking about the Poisson helps us understand why it can be used to estimate the probability for the discrete random variable from the binomial distribution. The Poisson is asking for the probability of a number of successes during a period of time while the binomial is asking for the probability of a certain number of successes for a given number of trials.

EXAMPLE 4.14

Leah's answering machine receives about six telephone calls between 8 a.m. and 10 a.m. What is the probability that Leah receives more than one call in the next 15 minutes?

Let X = the number of calls Leah receives in 15 minutes. (The interval of interest is 15 minutes or $\frac{1}{4}$ hour.)

$x = 0, 1, 2, 3, ...$

If Leah receives, on the average, six telephone calls in two hours, and there are eight 15 minute intervals in two hours, then Leah receives

$\left(\frac{1}{8}\right)(6) = 0.75$ calls in 15 minutes, on average. So, $\mu = 0.75$ for this problem.

$X \sim P(0.75)$

Find $P(x > 1)$. $P(x > 1) = 0.1734$

Probability that Leah receives more than one telephone call in the next 15 minutes is about 0.1734.

The graph of $X \sim P(0.75)$ is:

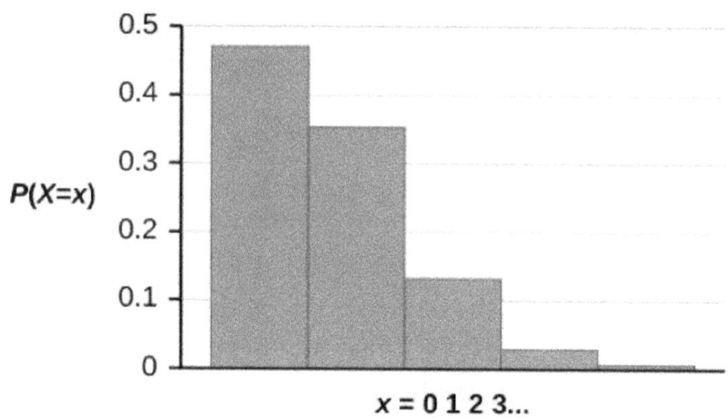

Figure 4.3

The y-axis contains the probability of x where X = the number of calls in 15 minutes.

EXAMPLE 4.15

According to a survey a university professor gets, on average, 7 emails per day. Let X = the number of emails a professor receives per day. The discrete random variable X takes on the values $x = 0, 1, 2$ .... The random variable X has a Poisson distribution: $X \sim P(7)$. The mean is 7 emails.

a. What is the probability that an email user receives exactly 2 emails per day?
b. What is the probability that an email user receives at most 2 emails per day?
c. What is the standard deviation?

Solution 4.5

a. $P\left(x = 2\right) = \frac{\mu^x e^{-\mu}}{x!} = \frac{7^2 e^{-7}}{2!} = 0.022$

b. $P(x \leq 2) = \frac{7^0 e^{-7}}{0!} + \frac{7^1 e^{-7}}{1!} + \frac{7^2 e^{-7}}{2!} = 0.029$

c. Standard Deviation $= \sigma = \sqrt{\mu} = \sqrt{7} \approx 2.65$

EXAMPLE 4.16

Text message users receive or send an average of 41.5 text messages per day.

a. How many text messages does a text message user receive or send per hour?
b. What is the probability that a text message user receives or sends two messages per hour?
c. What is the probability that a text message user receives or sends more than two messages per hour?

Solution 4.16

a. Let X = the number of texts that a user sends or receives in one hour. The average number of texts received per hour is $\frac{41.5}{24} \approx 1.7292$.

b. $P(x = 2) = \frac{\mu^x e^{-\mu}}{x!} = \frac{1.729^2 e^{-1.729}}{2!} = 0.265$

c. $P(x > 2) = 1 - P(x \leq 2) = 1 - \left[\frac{7^0 e^{-7}}{0!} + \frac{7^1 e^{-7}}{1!} + \frac{7^2 e^{-7}}{2!}\right] = 0.250$

EXAMPLE 4.17

On May 13, 2013, starting at 4:30 PM, the probability of low seismic activity for the next 48 hours in Alaska was reported as about 1.02%. Use this information for the next 200 days to find the probability that there will be low seismic activity in ten of the next 200 days. Use both the binomial and Poisson distributions to calculate the probabilities. Are they close?

Solution 4.17

Let X = the number of days with low seismic activity.

Using the binomial distribution:

- $P(x = 10) = \frac{200!}{10!(200 - 10)!} \times .0102^{10} = 0.000039$

Using the Poisson distribution:

- Calculate $\mu = np = 200(0.0102) \approx 2.04$
- $P(x = 10) = \frac{\mu^x e^{-\mu}}{x!} = \frac{2.04^{10} e^{-2.04}}{10!} = 0.000045$

We expect the approximation to be good because n is large (greater than 20) and p is small (less than 0.05). The results are close—both probabilities reported are almost 0.

Estimating the Binomial Distribution with the Poisson Distribution

We found before that the binomial distribution provided an approximation for the hypergeometric distribution. Now we find that the Poisson distribution can provide an approximation for the binomial. We say that the binomial distribution approaches the Poisson. The binomial distribution approaches the Poisson distribution is as *n* gets larger and *p* is small such that *np* becomes a constant value. There are several rules of thumb for when one can say they will use a Poisson to estimate a binomial. One suggests that *np*, the mean of the binomial, should be less than 25. Another author suggests that it should be less than 7. And another, noting that the mean and variance of the Poisson are both the same, suggests that *np* and *npq*, the mean and variance of the binomial, should be greater than 5. There is no one broadly accepted rule of thumb for when one can use the Poisson to estimate the binomial.

As we move through these probability distributions we are getting to more sophisticated distributions that, in a sense, contain the less sophisticated distributions within them. This proposition has been proven by mathematicians. This gets us to the highest level of sophistication in the next probability distribution which can be used as an approximation to all of those that we have discussed so far. This is the normal distribution.

EXAMPLE 4.18

A survey of 500 seniors in the Price Business School yields the following information. 75% go straight to work after graduation. 15% go on to work on their MBA. 9% stay to get a minor in another program. 1% go on to get a Master's in Finance.

What is the probability that more than 2 seniors go to graduate school for their Master's in finance?

Solution 4.18

This is clearly a binomial probability distribution problem. The choices are binary when we define the results as "Graduate School in Finance" versus "all other options." The random variable is discrete, and the events are, we could assume, independent. Solving as a binomial problem, we have:

Binomial Solution

$$n * p = 500 * 0.01 = 5 = \mu$$

$$P(0) = \frac{500!}{0!(500-0)!}0.01^0(1-0.01)^{500-0} = 0.00657$$

$$P(1) = \frac{500!}{1!(500-1)!}0.01^1(1-0.01)^{500-1} = 0.03318$$

$$P(2) = \frac{500!}{2!(500-2)!}0.01^2(1-0.01)^{500-2} = 0.08363$$

Adding all 3 together = 0.12339

$$1 - 0.12339 = 0.87661$$

Poisson approximation

$$n * p = 500 * 0.01 = 5 = \mu$$

$$n * p * (1 - p) = 500 * 0.01 * \left(0.99\right) \approx 5 = \sigma^2 = \mu$$

$$P(X) = \frac{e^{-np}(np)^x}{x!} = \left\{P(0) = \frac{e^{-5} * 5^0}{0!}\right\} + \left\{P(1) = \frac{e^{-5} * 5^1}{1!}\right\} + \left\{P(2) = \frac{e^{-5} * 5^2}{2!}\right\}$$

$$0.0067 + 0.0337 + 0.0842 = 0.1247$$

$$1 - 0.1247 = 0.8753$$

An approximation that is off by 1 one thousandth is certainly an acceptable approximation.

Terminolgy

Bernoulli Trials an experiment with the following characteristics:

1. There are only two possible outcomes called "success" and "failure" for each trial.
2. The probability p of a success is the same for any trial (so the probability $q = 1 - p$ of a failure is the same for any trial).

Binomial Experiment a statistical experiment that satisfies the following three conditions:

1. There are a fixed number of trials, n.
2. There are only two possible outcomes, called "success" and, "failure," for each trial. The letter p denotes the probability of a success on one trial, and q denotes the probability of a failure on one trial.
3. The n trials are independent and are repeated using identical conditions.

Binomial Probability Distribution a discrete random variable (RV) that arises from Bernoulli trials; there are a fixed number, n, of independent trials. "Independent" means that the result of any trial (for example, trial one) does not affect the results of the following trials, and all trials are conducted under the same conditions. Under these circumstances the binomial RV X is defined as the number of successes in n trials. The mean is $\mu = np$ and the standard deviation is $\sigma = \sqrt{npq}$. The probability of exactly x successes in n trials is $P(X = x) = \binom{n}{x} p^x q^{n-x}$.

Geometric Distribution a discrete random variable (RV) that arises from the Bernoulli trials; the trials are repeated until the first success. The geometric variable X is defined as the number of trials until the first success. The mean is $\mu = \frac{1}{p}$ and the standard deviation is $\sigma = \sqrt{\frac{1}{p}\left(\frac{1}{p} - 1\right)}$. The probability of exactly x failures before the first success is given by the formula: $P(X = x) = p(1 - p)^{x-1}$ where one wants to know probabilityforthenumberoftrialsuntilthefirstsuccess:thexthtrailisthefirstsuccess. An alternative formulation of the geometric distribution asks the question: what is the probability of x failures until the first success? In this formulation the trial that resulted in the first success is not counted. The formula for this presentation of the geometric is: $P(X=x)=p(1-p)^x$

The expected value in this form of the geometric distribution is $\mu = \frac{1-p}{p}$

The easiest way to keep these two forms of the geometric distribution straight is to remember that p is the probability of success and (1-p) is the probability of failure. In the formula the exponents simply count the number of successes and number of failures of the desired outcome of the experiment. Of course the sum of these two numbers must add to the number of trials in the experiment.

Geometric Experiment a statistical experiment with the following properties:

1. There are one or more Bernoulli trials with all failures except the last one, which is a success.
2. In theory, the number of trials could go on forever. There must be at least one trial.
3. The probability, p, of a success and the probability, q, of a failure do not change from trial to trial.

Hypergeometric Experiment a statistical experiment with the following properties:

1. You take samples from two groups.
2. You are concerned with a group of interest, called the first group.
3. You sample without replacement from the combined groups.
4. Each pick is not independent, since sampling is without replacement.

Hypergeometric Probability a discrete random variable (RV) that is characterized by:

1. A fixed number of trials.
2. The probability of success is not the same from trial to trial.

We sample from two groups of items when we are interested in only one group. X is defined as the number of successes out of the total number of items chosen.

Poisson Probability Distribution a discrete random variable (RV) that counts the number of times a certain event will occur in a specific interval; characteristics of the variable:

- The probability that the event occurs in a given interval is the same for all intervals.
- The events occur with a known mean and independently of the time since the last event.

The distribution is defined by the mean μ of the event in the interval. The mean is μ = *np*. The standard deviation is $\sigma = \sqrt{\mu}$. The probability of having exactly *x* successes in *r* trials is $P(x) = \frac{\mu^x e^{-\mu}}{x!}$. The Poisson distribution is often used to approximate the binomial distribution, when *n* is "large" and *p* is "small" (a general rule is that *np* should be greater than or equal to 25 and *p* should be less than or equal to 0.01).

Probability Distribution Function (PDF) a mathematical description of a discrete random variable (*RV*), given either in the form of an equation (formula) or in the form of a table listing all the possible outcomes of an experiment and the probability associated with each outcome.

Random Variable (RV) a characteristic of interest in a population being studied; common notation for variables are upper case Latin letters *X*, *Y*, *Z*,...; common notation for a specific value from the domain (set of all possible values of a variable) are lower case Latin letters *x*, *y*, and *z*. For example, if *X* is the number of children in a family, then *x* represents a specific integer 0, 1, 2, 3,.... Variables in statistics differ from variables in intermediate algebra in the two following ways.

- The domain of the random variable (RV) is not necessarily a numerical set; the domain may be expressed in words; for example, if X = hair color then the domain is {black, blond, gray, green, orange}.
- We can tell what specific value *x* the random variable *X* takes only after performing the experiment.

CHAPTER REVIEW

4.0 Introduction

The characteristics of a probability distribution or density function (PDF) are as follows:

1. Each probability is between zero and one, inclusive (*inclusive* means to include zero and one).
2. The sum of the probabilities is one.

4.1 Hypergeometric Distribution

The combinatorial formula can provide the number of unique subsets of size x that can be created from n unique objects to help us calculate probabilities. The combinatorial formula is $\binom{n}{x} = {}_nC_x = \frac{n!}{x!(n-x)!}$

A **hypergeometric experiment** is a statistical experiment with the following properties:

1. You take samples from two groups.
2. You are concerned with a group of interest, called the first group.
3. You sample without replacement from the combined groups.
4. Each pick is not independent, since sampling is without replacement.

The outcomes of a hypergeometric experiment fit a hypergeometric probability distribution. The random variable X = the number of items from the group of interest. $h(x) = \frac{\binom{A}{x}\binom{N-A}{n-x}}{\binom{N}{n}}$.

4.2 Binomial Distribution

A statistical experiment can be classified as a binomial experiment if the following conditions are met:

1. There are a fixed number of trials, *n*.
2. There are only two possible outcomes, called "success" and, "failure" for each trial. The letter *p* denotes the probability of a success on one trial and *q* denotes the probability of a failure on one trial.
3. The *n* trials are independent and are repeated using identical conditions.

The outcomes of a binomial experiment fit a binomial probability distribution. The random variable X = the number of successes obtained in the *n* independent trials. The mean of X can be calculated using the formula $\mu = np$, and the standard deviation is given by the formula $\sigma = \sqrt{npq}$.

The formula for the Binomial probability density function is

$$P(x) = \frac{n!}{x!(n-x)!} \cdot p^x q^{(n-x)}$$

4.3 Geometric Distribution

There are three characteristics of a geometric experiment:

1. There are one or more Bernoulli trials with all failures except the last one, which is a success.
2. In theory, the number of trials could go on forever. There must be at least one trial.
3. The probability, p, of a success and the probability, *q*, of a failure are the same for each trial.

In a geometric experiment, define the discrete random variable *X* as the number of independent trials until the first success. We say that X has a geometric distribution and write $X \sim G(p)$ where *p* is the probability of success in a single trial.

The mean of the geometric distribution $X \sim G(p)$ is $\mu = 1/p1/p$ where x = number of trials until first success for the formula $P(X=x)=(1-p)^{x-1}pP(X=x)=(1-p)x-1p$ where the number of trials is up and including the first success.

An alternative formulation of the geometric distribution asks the question: what is the probability of *x* failures until the first success? In this formulation the trial that resulted in the first success is not counted. The formula for this presentation of the geometric is:

$$P(X = x) = p(1 - p)^x$$

The expected value in this form of the geometric distribution is

$$\mu = \frac{1 - p}{p}$$

The easiest way to keep these two forms of the geometric distribution straight is to remember that *p* is the probability of success and (1-p) is the probability of failure. In the formula the exponents simply count the number of successes and number of failures of the desired outcome of the experiment. Of course the sum of these two numbers must add to the number of trials in the experiment.

4.4 Poisson Distribution

A **Poisson probability distribution** of a discrete random variable gives the probability of a number of events occurring in a fixed interval of time or space, if these events happen at a known average rate and independently of the time since the last event. The Poisson distribution may be used to approximate the binomial, if the probability of success is "small" (less than or equal to 0.01) and the number of trials is "large" (greater than or equal to 25). Other rules of thumb are also suggested by different authors, but all recognize that the Poisson distribution is the limiting distribution of the binomial as *n* increases and *p* approaches zero.

The formula for computing probabilities that are from a Poisson process is:

$$P(x) = \frac{\mu^x e^{-\mu}}{x!}$$

where P(X) is the probability of successes, μ (pronounced mu) is the expected number of successes, e is the natural logarithm approximately equal to 2.718, and X is the number of successes per unit, usually per unit of time.

FORMULA REVIEW

4.1 Hypergeometric Distribution

$$h(x) = \frac{\binom{A}{x}\binom{N-A}{n-x}}{\binom{N}{n}}$$

4.2 Binomial Distribution

$X \sim B(n, p)$ means that the discrete random variable X has a binomial probability distribution with n trials and probability of success p.

X = the number of successes in n independent trials

n = the number of independent trials

X takes on the values $x = 0, 1, 2, 3, \ldots, n$

p = the probability of a success for any trial

q = the probability of a failure for any trial

$p + q = 1$

$q = 1 - p$

The mean of X is $\mu = np$. The standard deviation of X is $\sigma = \sqrt{npq}$.

$$P(x) = \frac{n!}{x!(n-x)!} \cdot p^x q^{(n-x)}$$

where P(X) is the probability of X successes in n trials when the probability of a success in ANY ONE TRIAL is p.

4.3 Geometric Distribution

$P(X = x) = p(1 - p)^{x - 1}$

$X \sim G(p)$ means that the discrete random variable X has a geometric probability distribution with probability of success in a single trial p.

X = the number of independent trials until the first success

X takes on the values $x = 1, 2, 3, \ldots$

p = the probability of a success for any trial

q = the probability of a failure for any trial $p + q = 1$

$q = 1 - p$

The mean is $\mu = \frac{1}{p}$.

The standard deviation is $\sigma = \sqrt{\frac{1-p}{p^2}} = \sqrt{\frac{1}{p}\left(\frac{1}{p} - 1\right)}$.

4.4 Poisson Distribution

$X \sim P(\mu)$ means that X has a Poisson probability distribution where X = the number of occurrences in the interval of interest.

X takes on the values $x = 0, 1, 2, 3, \ldots$

The mean μ or λ is typically given.

The variance is $\sigma^2 = \mu$, and the standard deviation is $\sigma = \sqrt{\mu}$.

When $P(\mu)$ is used to approximate a binomial distribution, $\mu = np$ where n represents the number of independent trials and p represents the probability of success in a single trial.

$$P(x) = \frac{\mu^{x} e^{-\mu}}{x!}$$

PRACTICE

4.0 Introduction

Use the following information to answer the next five exercises: A company wants to evaluate its attrition rate, in other words, how long new hires stay with the company. Over the years, they have established the following probability distribution.

Let X = the number of years a new hire will stay with the company.

Let $P(x)$ = the probability that a new hire will stay with the company x years.

1.

Complete **Table 4.1** using the data provided.

Table 4.1

x	P(x)
0	0.12
1	0.18
2	0.30
3	0.15
4	
5	0.10
6	0.05

2. P(x = 4) = _______

3. P(x ≥ 5) = _______

4. On average, how long would you expect a new hire to stay with the company?

5. What does the column "P(x)" sum to?

Use the following information to answer the next six exercises: A baker is deciding how many batches of muffins to make to sell in his bakery. He wants to make enough to sell every one and no fewer. Through observation, the baker has established a probability distribution.

Table 4.2

x	P(x)
1	0.15
2	0.35
3	0.40
4	0.10

6. Define the random variable X.
7. What is the probability the baker will sell more than one batch? $P(x > 1)$ = _______
8. What is the probability the baker will sell exactly one batch? $P(x = 1)$ = _______
9. On average, how many batches should the baker make?

Use the following information to answer the next four exercises: Ellen has music practice three days a week. She practices for all of the three days 85% of the time, two days 8% of the time, one day 4% of the time, and no days 3% of the time. One week is selected at random.

10. Define the random variable X.
11. Construct a probability distribution table for the data.
12. We know that for a probability distribution function to be discrete, it must have two characteristics. One is that the sum of the probabilities is one. What is the other characteristic?

Use the following information to answer the next five exercises: Javier volunteers in community events each month. He does not do more than five events in a month. He attends exactly five events 35% of the time, four events 25% of the time, three events 20% of the time, two events 10% of the time, one event 5% of the time, and no events 5% of the time.

13. Define the random variable X.
14. What values does x take on?
15. Construct a PDF table.
16. Find the probability that Javier volunteers for less than three events each month. $P(x < 3)$ = _______
17. Find the probability that Javier volunteers for at least one event each month. $P(x > 0)$ = _______

4.1 Hypergeometric Distribution

Use the following information to answer the next five exercises: Suppose that a group of statistics students is divided into two groups: business majors and non-business majors. There are 16 business majors in the group and seven non-business majors in the group. A random sample of nine students is taken. We are interested in the number of business majors in the sample.

18. In words, define the random variable X.
19. What values does X take on?

4.2 Binomial Distribution

Use the following information to answer the next eight exercises: The Higher Education Research Institute at UCLA collected data from 203,967 incoming first-time, full-time freshmen from 270 four-year colleges and universities in the U.S. 71.3% of those students replied that, yes, they believe that same-sex couples should have the right to legal marital status. Suppose that you randomly pick eight first-time, full-time freshmen from the survey. You are interested in the number that believes that same sex-couples should have the right to legal marital status.

20. In words, define the random variable X.

21. $X \sim$ _____(_____,_____)

22. What values does the random variable X take on?

23. Construct the probability distribution function (PDF).

Table 4.3

x	P(x)

24. On average (μ), how many would you expect to answer yes?

25. What is the standard deviation (σ)?

26. What is the probability that at most five of the freshmen reply "yes"?

27. What is the probability that at least two of the freshmen reply "yes"?

4.3 Geometric Distribution

Use the following information to answer the next six exercises: The Higher Education Research Institute at UCLA collected data from 203,967 incoming first-time, full-time freshmen from 270 four-year colleges and universities in the U.S. 71.3% of those students replied that, yes, they believe that same-sex couples should have the right to legal marital status. Suppose that you randomly select freshman from the study until you find one who replies "yes." You are interested in the number of freshmen you must ask.

28. In words, define the random variable X.

29. $X \sim$ _____(_____,_____)

30. What values does the random variable X take on?

31. Construct the probability distribution function (PDF). Stop at $x = 6$.

Table 4.4

x	P(x)
1	
2	
3	
4	
5	
6	

32. On average (μ), how many freshmen would you expect to have to ask until you found one who replies "yes?"

33. What is the probability that you will need to ask fewer than three freshmen?

4.4 Poisson Distribution

Use the following information to answer the next six exercises: On average, a clothing store gets 120 customers per day.

34. Assume the event occurs independently in any given day. Define the random variable X.

35. What values does X take on?

36. What is the probability of getting 150 customers in one day?

37. What is the probability of getting 35 customers in the first four hours? Assume the store is open 12 hours each day.

38. What is the probability that the store will have more than 12 customers in the first hour?

39. What is the probability that the store will have fewer than 12 customers in the first two hours?

40. Which type of distribution can the Poisson model be used to approximate? When would you do this?

Use the following information to answer the next six exercises: On average, eight teens in the U.S. die from motor vehicle injuries per day. As a result, states across the country are debating raising the driving age.

41. Assume the event occurs independently in any given day. In words, define the random variable X.

42. X ~ _____(_____,_____)

43. What values does X take on?

44. For the given values of the random variable X, fill in the corresponding probabilities.

45. Is it likely that there will be no teens killed from motor vehicle injuries on any given day in the U.S? Justify your answer numerically.

46. Is it likely that there will be more than 20 teens killed from motor vehicle injuries on any given day in the U.S.? Justify your answer numerically.

HOMEWORK

4.1 Hypergeometric Distribution

47. A group of Martial Arts students is planning on participating in an upcoming demonstration. Six are students of Tae Kwon Do; seven are students of Shotokan Karate. Suppose that eight students are randomly picked to be in the first demonstration. We are interested in the number of Shotokan Karate students in that first demonstration.

a. In words, define the random variable X.

b. List the values that X may take on.

c. How many Shotokan Karate students do we expect to be in that first demonstration?

48. In one of its Spring catalogs, L.L. Bean® advertised footwear on 29 of its 192 catalog pages. Suppose we randomly survey 20 pages. We are interested in the number of pages that advertise footwear. Each page may be picked at most once.

a. In words, define the random variable X.

b. List the values that X may take on.

c. How many pages do you expect to advertise footwear on them?

d. Calculate the standard deviation.

49. Suppose that a technology task force is being formed to study technology awareness among instructors. Assume that ten people will be randomly chosen to be on the committee from a group of 28 volunteers, 20 who are technically proficient and eight who are not. We are interested in the number on the committee who are **not** technically proficient.

a. In words, define the random variable X.

b. List the values that X may take on.

c. How many instructors do you expect on the committee who are not technically proficient?

d. Find the probability that at least five on the committee are not technically proficient.

e. Find the probability that at most three on the committee are not technically proficient.

50. Suppose that nine Massachusetts athletes are scheduled to appear at a charity benefit. The nine are randomly chosen from eight volunteers from the Boston Celtics and four volunteers from the New England Patriots. We are interested in the number of Patriots picked.

a. In words, define the random variable X.
b. List the values that X may take on.
c. Are you choosing the nine athletes with or without replacement?

51. A bridge hand is defined as 13 cards selected at random and without replacement from a deck of 52 cards. In a standard deck of cards, there are 13 cards from each suit: hearts, spades, clubs, and diamonds. What is the probability of being dealt a hand that does not contain a heart?

a. What is the group of interest?
b. How many are in the group of interest?
c. How many are in the other group?
d. Let X = _________. What values does X take on?
e. The probability question is P(_______).
f. Find the probability in question.
g. Find the (i) mean and (ii) standard deviation of X.

4.2 Binomial Distribution

52. According to a recent article the average number of babies born with significant hearing loss (deafness) is approximately two per 1,000 babies in a healthy baby nursery. The number climbs to an average of 30 per 1,000 babies in an intensive care nursery.

Suppose that 1,000 babies from healthy baby nurseries were randomly surveyed. Find the probability that exactly two babies were born deaf.

Use the following information to answer the next four exercises. Recently, a nurse commented that when a patient calls the medical advice line claiming to have the flu, the chance that he or she truly has the flu (and not just a nasty cold) is only about 4%. Of the next 25 patients calling in claiming to have the flu, we are interested in how many actually have the flu.

53. Define the random variable and list its possible values.

54. State the distribution of X.

55. Find the probability that at least four of the 25 patients actually have the flu.

56. On average, for every 25 patients calling in, how many do you expect to have the flu?

57. People visiting video rental stores often rent more than one DVD at a time. The probability distribution for DVD rentals per customer at Video To Go is given

Table 4.5. There is five-video limit per customer at this store, so nobody ever rents more than five DVDs.

Table 4.5

x	P(x)
0	0.03
1	0.50
2	0.24
3	
4	0.07
5	0.04

a. Describe the random variable *X* in words.
b. Find the probability that a customer rents three DVDs.
c. Find the probability that a customer rents at least four DVDs.
d. Find the probability that a customer rents at most two DVDs.

58. A school newspaper reporter decides to randomly survey 12 students to see if they will attend Tet (Vietnamese New Year) festivities this year. Based on past years, she knows that 18% of students attend Tet festivities. We are interested in the number of students who will attend the festivities.

a. In words, define the random variable *X*.
b. List the values that *X* may take on.
c. Give the distribution of *X*. *X* ~ _____(_____,_____)
d. How many of the 12 students do we expect to attend the festivities?
e. Find the probability that at most four students will attend.
f. Find the probability that more than two students will attend.

Use the following information to answer the next two exercises: The probability that the San Jose Sharks will win any given game is 0.3694 based on a 13-year win history of 382 wins out of 1,034 games played (as of a certain date). An upcoming monthly schedule contains 12 games.

59. The expected number of wins for that upcoming month is:

a. 1.67
b. 12
c. $\frac{382}{1043}$
d. 4.43

Let *X* = the number of games won in that upcoming month.

60. What is the probability that the San Jose Sharks win six games in that upcoming month?

a. 0.1476
b. 0.2336
c. 0.7664
d. 0.8903

61. What is the probability that the San Jose Sharks win at least five games in that upcoming month

a. 0.3694
b. 0.5266
c. 0.4734
d. 0.2305

62. A student takes a ten-question true-false quiz, but did not study and randomly guesses each answer. Find the probability that the student passes the quiz with a grade of at least 70% of the questions correct.

63. A student takes a 32-question multiple-choice exam, but did not study and randomly guesses each answer. Each question has three possible choices for the answer. Find the probability that the student guesses more than 75% of the questions correctly.

64. Six different colored dice are rolled. Of interest is the number of dice that show a one.

a. In words, define the random variable X.
b. List the values that X may take on.
c. On average, how many dice would you expect to show a one?
d. Find the probability that all six dice show a one.
e. Is it more likely that three or that four dice will show a one? Use numbers to justify your answer numerically.

65. More than 96 percent of the very largest colleges and universities (more than 15,000 total enrollments) have some online offerings. Suppose you randomly pick 13 such institutions. We are interested in the number that offer distance learning courses.

a. In words, define the random variable X.
b. List the values that X may take on.
c. Give the distribution of X. X ~ _____(_____,_____)
d. On average, how many schools would you expect to offer such courses?
e. Find the probability that at most ten offer such courses.
f. Is it more likely that 12 or that 13 will offer such courses? Use numbers to justify your answer numerically and answer in a complete sentence.

66. Suppose that about 85% of graduating students attend their graduation. A group of 22 graduating students is randomly chosen.

a. In words, define the random variable X.
b. List the values that X may take on.
c. Give the distribution of X. X ~ _____(_____,_____)
d. How many are expected to attend their graduation?

e. Find the probability that 17 or 18 attend.
f. Based on numerical values, would you be surprised if all 22 attended graduation? Justify your answer numerically.

67. At The Fencing Center, 60% of the fencers use the foil as their main weapon. We randomly survey 25 fencers at The Fencing Center. We are interested in the number of fencers who do not use the foil as their main weapon.

a. In words, define the random variable X.
b. List the values that X may take on.
c. Give the distribution of X. X ~ _____(_____,_____)
d. How many are expected to not to use the foil as their main weapon?
e. Find the probability that six do not use the foil as their main weapon.
f. Based on numerical values, would you be surprised if all 25 did not use foil as their main weapon? Justify your answer numerically.

68. Approximately 8% of students at a local high school participate in after-school sports all four years of high school. A group of 60 seniors is randomly chosen. Of interest is the number who participated in after-school sports all four years of high school.

a. In words, define the random variable X.
b. List the values that X may take on.
c. Give the distribution of X. X ~ _____(_____,_____)
d. How many seniors are expected to have participated in after-school sports all four years of high school?
e. Based on numerical values, would you be surprised if none of the seniors participated in after-school sports all four years of high school? Justify your answer numerically.
f. Based upon numerical values, is it more likely that four or that five of the seniors participated in after-school sports all four years of high school? Justify your answer numerically.

69. The chance of an IRS audit for a tax return with over $25,000 in income is about 2% per year. We are interested in the expected number of audits a person with that income has in a 20-year period. Assume each year is independent.

a. In words, define the random variable X.
b. List the values that X may take on.
c. Give the distribution of X. X ~ _____(_____,_____)
d. How many audits are expected in a 20-year period?
e. Find the probability that a person is not audited at all.
f. Find the probability that a person is audited more than twice.

70. It has been estimated that only about 30% of California residents have adequate earthquake supplies. Suppose you randomly survey 11 California residents. We are interested in the number who have adequate earthquake supplies.

a. In words, define the random variable X.
b. List the values that X may take on.
c. Give the distribution of X. X ~ _____(_____,_____)
d. What is the probability that at least eight have adequate earthquake supplies?
e. Is it more likely that none or that all of the residents surveyed will have adequate earthquake supplies? Why?

How many residents do you expect will have adequate earthquake supplies?

71. There are two similar games played for Chinese New Year and Vietnamese New Year. In the Chinese version, fair dice with numbers 1, 2, 3, 4, 5, and 6 are used, along with a board with those numbers. In the Vietnamese version, fair dice with pictures of a gourd, fish, rooster, crab, crayfish, and deer are used. The board has those six objects on it, also. We will play with bets being $1. The player places a bet on a number or object. The "house" rolls three dice. If none of the dice show the number or object that was bet, the house keeps the $1 bet. If one of the dice shows the number or object bet (and the other two do not show it), the player gets back his or her $1 bet, plus $1 profit. If two of the dice show the number or object bet (and the third die does not show it), the player gets back his or her $1 bet, plus $2 profit. If all three dice show the number or object bet, the player gets back his or her $1 bet, plus $3 profit. Let X = number of matches and Y = profit per game.

a. In words, define the random variable X.
b. List the values that X may take on.
c. List the values that Y may take on. Then, construct one PDF table that includes both X and Y and their probabilities.
d. Calculate the average expected matches over the long run of playing this game for the player.
e. Calculate the average expected earnings over the long run of playing this game for the player.
f. Determine who has the advantage, the player or the house.

72. According to The World Bank, only 9% of the population of Uganda had access to electricity as of 2009. Suppose we randomly sample 150 people in Uganda. Let X = the number of people who have access to electricity.

a. What is the probability distribution for X?

b. Using the formulas, calculate the mean and standard deviation of X.
c. Find the probability that 15 people in the sample have access to electricity.
d. Find the probability that at most ten people in the sample have access to electricity.
e. Find the probability that more than 25 people in the sample have access to electricity.

73. The literacy rate for a nation measures the proportion of people age 15 and over that can read and write. The literacy rate in Afghanistan is 28.1%. Suppose you choose 15 people in Afghanistan at random. Let X = the number of people who are literate.

a. Sketch a graph of the probability distribution of X.
b. Using the formulas, calculate the (i) mean and (ii) standard deviation of X.
c. Find the probability that more than five people in the sample are literate. Is it is more likely that three people or four people are literate.

4.3 Geometric Distribution

74. A consumer looking to buy a used red Miata car will call dealerships until she finds a dealership that carries the car. She estimates the probability that any independent dealership will have the car will be 28%. We are interested in the number of dealerships she must call.

a. In words, define the random variable X.
b. List the values that X may take on.
c. Give the distribution of X. X ~ _____(_____,_____)
d. On average, how many dealerships would we expect her to have to call until she finds one that has the car?
e. Find the probability that she must call at most four dealerships.
f. Find the probability that she must call three or four dealerships.

75. Suppose that the probability that an adult in America will watch the Super Bowl is 40%. Each person is considered independent. We are interested in the number of adults in America we must survey until we find one who will watch the Super Bowl.

a. In words, define the random variable X.
b. List the values that X may take on.
c. Give the distribution of X. X ~ _____(_____,_____)
d. How many adults in America do you expect to survey until you find one who will watch the Super Bowl?
e. Find the probability that you must ask seven people.

f. Find the probability that you must ask three or four people.

76. It has been estimated that only about 30% of California residents have adequate earthquake supplies. Suppose we are interested in the number of California residents we must survey until we find a resident who does **not** have adequate earthquake supplies.

a. In words, define the random variable X.
b. List the values that X may take on.
c. Give the distribution of X. X ~ _____(_____,_____)
d. What is the probability that we must survey just one or two residents until we find a California resident who does not have adequate earthquake supplies?
e. What is the probability that we must survey at least three California residents until we find a California resident who does not have adequate earthquake supplies?
f. How many California residents do you expect to need to survey until you find a California resident who does not have adequate earthquake supplies?
g. How many California residents do you expect to need to survey until you find a California resident who does have adequate earthquake supplies?

77. In one of its Spring catalogs, L.L. Bean® advertised footwear on 29 of its 192 catalog pages. Suppose we randomly survey 20 pages. We are interested in the number of pages that advertise footwear. Each page may be picked more than once.

a. In words, define the random variable X.
b. List the values that X may take on.
c. Give the distribution of X. X ~ _____(_____,_____)
d. How many pages do you expect to advertise footwear on them?
e. Is it probable that all twenty will advertise footwear on them? Why or why not?
f. What is the probability that fewer than ten will advertise footwear on them?
g. Reminder: A page may be picked more than once. We are interested in the number of pages that we must randomly survey until we find one that has footwear advertised on it. Define the random variable X and give its distribution.
h. What is the probability that you only need to survey at most three pages in order to find one that advertises footwear on it?
i. How many pages do you expect to need to survey in order to find one that advertises footwear?

78. Suppose that you are performing the probability experiment of rolling one fair six-sided die. Let F be the event of rolling a four or a five. You are interested in how many times you need to roll the die in order to obtain the first four or five as the outcome.

- p = probability of success (event F occurs)
- q = probability of failure (event F does not occur)

a. Write the description of the random variable X.
b. What are the values that X can take on?
c. Find the values of p and q.
d. Find the probability that the first occurrence of event F (rolling a four or five) is on the second trial.

79. Ellen has music practice three days a week. She practices for all of the three days 85% of the time, two days 8% of the time, one day 4% of the time, and no days 3% of the time. One week is selected at random. What values does X take on?

80. The World Bank records the prevalence of HIV in countries around the world. According to their data, "Prevalence of HIV refers to the percentage of people ages 15 to 49 who are infected with HIV."[1] In South Africa, the prevalence of HIV is 17.3%. Let X = the number of people you test until you find a person infected with HIV.

a. Sketch a graph of the distribution of the discrete random variable X.
b. What is the probability that you must test 30 people to find one with HIV?
c. What is the probability that you must ask ten people?
d. Find the (i) mean and (ii) standard deviation of the distribution of X.

81. According to a recent Pew Research poll, 75% of millennials (people born between 1981 and 1995) have a profile on a social networking site. Let X = the number of millennials you ask until you find a person without a profile on a social networking site.

a. Describe the distribution of X.
b. Find the (i) mean and (ii) standard deviation of X.
c. What is the probability that you must ask ten people to find one person without a social networking site?
d. What is the probability that you must ask 20 people to find one person without a social networking site?
e. What is the probability that you must ask *at most* five people?

4.4 Poisson Distribution

82. The switchboard in a Minneapolis law office gets an average of 5.5 incoming phone calls during the noon hour on Mondays. Experience shows that the existing staff can handle up to six calls in an hour. Let X = the number of calls received at noon.

a. Find the mean and standard deviation of X.
b. What is the probability that the office receives at most six calls at noon on Monday?
c. Find the probability that the law office receives six calls at noon. What does this mean to the law office staff who get, on average, 5.5 incoming phone calls at noon?
d. What is the probability that the office receives more than eight calls at noon?

83. The maternity ward at Dr. Jose Fabella Memorial Hospital in Manila in the Philippines is one of the busiest in the world with an average of 60 births per day. Let X = the number of births in an hour.

a. Find the mean and standard deviation of X.
b. Sketch a graph of the probability distribution of X.
c. What is the probability that the maternity ward will deliver three babies in one hour?
d. What is the probability that the maternity ward will deliver at most three babies in one hour?
e. What is the probability that the maternity ward will deliver more than five babies in one hour?

84. A manufacturer of Christmas tree light bulbs knows that 3% of its bulbs are defective. Find the probability that a string of 100 lights contains at most four defective bulbs using both the binomial and Poisson distributions.

85. The average number of children a Japanese woman has in her lifetime is 1.37. Suppose that one Japanese woman is randomly chosen.

a. In words, define the random variable X.
b. List the values that X may take on.
c. Find the probability that she has no children.
d. Find the probability that she has fewer children than the Japanese average.
e. Find the probability that she has more children than the Japanese average.

86. The average number of children a Spanish woman has in her lifetime is 1.47. Suppose that one Spanish woman is randomly chosen.

a. In words, define the Random Variable X.
b. List the values that X may take on.

c. Find the probability that she has no children.
d. Find the probability that she has fewer children than the Spanish average.
e. Find the probability that she has more children than the Spanish average .

87. Fertile, female cats produce an average of three litters per year. Suppose that one fertile, female cat is randomly chosen. In one year, find the probability she produces:

a. In words, define the random variable X.
b. List the values that X may take on.
c. Give the distribution of X. X ~ _______
d. Find the probability that she has no litters in one year.
e. Find the probability that she has at least two litters in one year.
f. Find the probability that she has exactly three litters in one year.

88. The chance of having an extra fortune in a fortune cookie is about 3%. Given a bag of 144 fortune cookies, we are interested in the number of cookies with an extra fortune. Two distributions may be used to solve this problem, but only use one distribution to solve the problem.

a. In words, define the random variable X.
b. List the values that X may take on.
c. How many cookies do we expect to have an extra fortune?
d. Find the probability that none of the cookies have an extra fortune.
e. Find the probability that more than three have an extra fortune.
f. As *n* increases, what happens involving the probabilities using the two distributions? Explain in complete sentences.

89. According to the South Carolina Department of Mental Health web site, for every 200 U.S. women, the average number who suffer from anorexia is one. Out of a randomly chosen group of 600 U.S. women determine the following.

a. In words, define the random variable X.
b. List the values that X may take on.
c. Give the distribution of X. X ~ _____(_____,_____)
d. How many are expected to suffer from anorexia?
e. Find the probability that no one suffers from anorexia.
f. Find the probability that more than four suffer from anorexia.

90. The chance of an IRS audit for a tax return with over $25,000 in income is about 2% per year. Suppose that 100 people with tax returns over $25,000 are randomly picked. We are interested in the number of people audited in one year. Use a Poisson distribution to answer the following questions.

a. In words, define the random variable X.
b. List the values that X may take on.
c. How many are expected to be audited?
d. Find the probability that no one was audited.
e. Find the probability that at least three were audited.

91. Approximately 8% of students at a local high school participate in after-school sports all four years of high school. A group of 60 seniors is randomly chosen. Of interest is the number that participated in after-school sports all four years of high school.

a. In words, define the random variable X.
b. List the values that X may take on.
c. How many seniors are expected to have participated in after-school sports all four years of high school?
d. Based on numerical values, would you be surprised if none of the seniors participated in after-school sports all four years of high school? Justify your answer numerically.
e. Based on numerical values, is it more likely that four or that five of the seniors participated in after-school sports all four years of high school? Justify your answer numerically.

92. On average, Pierre, an amateur chef, drops three pieces of egg shell into every two cake batters he makes. Suppose that you buy one of his cakes.

a. In words, define the random variable X.
b. List the values that X may take on.
c. On average, how many pieces of egg shell do you expect to be in the cake?
d. What is the probability that there will not be any pieces of egg shell in the cake?
e. Let's say that you buy one of Pierre's cakes each week for six weeks. What is the probability that there will not be any egg shell in any of the cakes?
f. Based upon the average given for Pierre, is it possible for there to be seven pieces of shell in the cake? Why?

Use the following information to answer the next two exercises: The average number of times per week that Mrs. Plum's cats wake her up at night because they want to play is ten. We are interested in the number of times her cats wake her up each week.

93. In words, the random variable X = _______________

a. the number of times Mrs. Plum's cats wake her up each week.
b. the number of times Mrs. Plum's cats wake her up each hour.

c. the number of times Mrs. Plum's cats wake her up each night.
d. the number of times Mrs. Plum's cats wake her up.

94. Find the probability that her cats will wake her up no more than five times next week.

a. 0.5000 b. 0.9329
c. 0.0378 d. 0.0671

REFERENCES

4.2 Binomial Distribution

"Access to electricity (% of population)," The World Bank, 2013. Available online at http://data.worldbank.org/indicator/EG.ELC.ACCS.ZS?order=wbapi_data_value_2009%20wbapi_data_value%20wbapi_data_value-first&sort=asc (accessed May 15, 2015).

"Distance Education." Wikipedia. Available online at http://en.wikipedia.org/wiki/Distance_education (accessed May 15, 2013).

"NBA Statistics – 2013," ESPN NBA, 2013. Available online at http://espn.go.com/nba/statistics/_/seasontype/2 (accessed May 15, 2013).

Newport, Frank. "Americans Still Enjoy Saving Rather than Spending: Few demographic differences seen in these views other than by income," GALLUP® Economy, 2013. Available online at http://www.gallup.com/poll/162368/americans-enjoy-saving-rather-spending.aspx (accessed May 15, 2013).

Pryor, John H., Linda DeAngelo, Laura Palucki Blake, Sylvia Hurtado, Serge Tran. *The American Freshman: National Norms Fall 2011*. Los Angeles: Cooperative Institutional Research Program at the Higher Education Research Institute at UCLA, 2011. Also available online at http://heri.ucla.edu/PDFs/pubs/TFS/Norms/Monographs/TheAmericanFreshman2011.pdf (accessed May 15, 2013).

"The World FactBook," Central Intelligence Agency. Available online at https://www.cia.gov/library/publications/the-world-factbook/geos/af.html (accessed May 15, 2013).

"What are the key statistics about pancreatic cancer?" American Cancer Society, 2013. Available online at http://www.cancer.org/cancer/pancreaticcancer/detailedguide/pancreatic-cancer-key-statistics (accessed May 15, 2013).

4.3 Geometric Distribution

"Millennials: A Portrait of Generation Next," PewResearchCenter. Available online at http://www.pewsocialtrends.org/files/2010/10/millennials-confident-connected-open-to-change.pdf (accessed May 15, 2013).

"Millennials: Confident. Connected. Open to Change." Executive Summary by PewResearch Social & Demographic Trends, 2013. Available online at http://www.pewsocialtrends.org/2010/02/24/millennials-confident-connected-open-to-change/ (accessed May 15, 2013).

"Prevalence of HIV, total (% of populations ages 15-49)," The World Bank, 2013. Available online at http://data.worldbank.org/indicator/SH.DYN.AIDS.ZS?order=wbapi_data_value_2011+wbapi_data_value+wbapi_data_value-last&sort=desc (accessed May 15, 2013).

Pryor, John H., Linda DeAngelo, Laura Palucki Blake, Sylvia Hurtado, Serge Tran. *The American Freshman: National Norms Fall 2011.* Los Angeles: Cooperative Institutional Research Program at the Higher Education Research Institute at UCLA, 2011. Also available online at http://heri.ucla.edu/PDFs/pubs/TFS/Norms/Monographs/TheAmericanFreshman2011.pdf (accessed May 15, 2013).

"Summary of the National Risk and Vulnerability Assessment 2007/8: A profile of Afghanistan," The European Union and ICON-Institute. Available online at http://ec.europa.eu/europeaid/where/asia/documents/afgh_brochure_summary_en.pdf (accessed May 15, 2013).

"The World FactBook," Central Intelligence Agency. Available online at https://www.cia.gov/library/publications/the-world-factbook/geos/af.html (accessed May 15, 2013).

"UNICEF reports on Female Literacy Centers in Afghanistan established to teach women and girls basic resading [sic] and writing skills," UNICEF Television. Video available online at http://www.unicefusa.org/assets/video/afghan-female-literacy-centers.html (accessed May 15, 2013).

4.4 Poisson Distribution

"ATL Fact Sheet," Department of Aviation at the Hartsfield-Jackson Atlanta International Airport, 2013. Available online at http://www.atl.com/about-atl/atl-factsheet/ (accessed February 6, 2019).

Center for Disease Control and Prevention. "Teen Drivers: Fact Sheet," Injury Prevention & Control: Motor Vehicle Safety, October 2, 2012. Available online at http://www.cdc.gov/Motorvehiclesafety/Teen_Drivers/teendrivers_factsheet.html (accessed May 15, 2013).

"Children and Childrearing," Ministry of Health, Labour, and Welfare. Available online at http://www.mhlw.go.jp/english/policy/children/children-childrearing/index.html (accessed May 15, 2013).

"Eating Disorder Statistics," South Carolina Department of Mental Health, 2006.

Available online at http://www.state.sc.us/dmh/anorexia/statistics.htm (accessed May 15, 2013).

"Giving Birth in Manila: The maternity ward at the Dr Jose Fabella Memorial Hospital in Manila, the busiest in the Philippines, where there is an average of 60 births a day," theguardian, 2013. Available online at http://www.theguardian.com/world/gallery/2011/jun/08/philippines-health#/?picture=375471900&index=2 (accessed May 15, 2013).

"How Americans Use Text Messaging," Pew Internet, 2013. Available online at http://pewinternet.org/Reports/2011/Cell-Phone-Texting-2011/Main-Report.aspx (accessed May 15, 2013).

Lenhart, Amanda. "Teens, Smartphones & Testing: Texting volume is up while the frequency of voice calling is down. About one in four teens say they own smartphones," Pew Internet, 2012. Available online at http://www.pewinternet.org/~/media/Files/Reports/2012/PIP_Teens_Smartphones_and_Texting.pdf (accessed May 15, 2013).

"One born every minute: the maternity unit where mothers are THREE to a bed," MailOnline. Available online at http://www.dailymail.co.uk/news/article-2001422/Busiest-maternity-ward-planet-averages-60-babies-day-mothers-bed.html (accessed May 15, 2013).

Vanderkam, Laura. "Stop Checking Your Email, Now." CNNMoney, 2013. Available online at http://management.fortune.cnn.com/2012/10/08/stop-checking-your-email-now/ (accessed May 15, 2013).

"World Earthquakes: Live Earthquake News and Highlights," World Earthquakes, 2012. http://www.world-earthquakes.com/index.php?option=ethq_prediction (accessed May 15, 2013).

SOLUTIONS

1.

Table 4.6

x	P(x)
0	0.12
1	0.18
2	0.30
3	0.15
4	0.10
5	0.10
6	0.05

3. 0.10 + 0.05 = 0.15

5. 1

7. 0.35 + 0.40 + 0.10 = 0.85

9. 1(0.15) + 2(0.35) + 3(0.40) + 4(0.10) = 0.15 + 0.70 + 1.20 + 0.40 = 2.45

11.

Table 4.7

x	P(x)
0	0.03
1	0.04
2	0.08
3	0.85

13. Let X = the number of events Javier volunteers for each month.

15.

Table 4.8

x	P(x)
0	0.05
1	0.05
2	0.10
3	0.20
4	0.25
5	0.35

17. 1 – 0.05 = 0.95

18. X = the number of business majors in the sample.

19. 2, 3, 4, 5, 6, 7, 8, 9

20. X = the number that reply "yes"

22. 0, 1, 2, 3, 4, 5, 6, 7, 8

24. 5.7

26. 0.4151

28. X = the number of freshmen selected from the study until one replied "yes" that same-sex couples should have the right to legal marital status.

30. 1,2,...

32. 1.4

35. 0, 1, 2, 3, 4, ...

37. 0.0485

39. 0.0214

41. X = the number of U.S. teens who die from motor vehicle injuries per day.

43. 0, 1, 2, 3, 4, ...

45. No

48. a. X = the number of pages that advertise footwear

b. 0, 1, 2, 3, ..., 20

c. 3.03

d. 1.5197

50. a. X = the number of Patriots picked

b. 0, 1, 2, 3, 4

c. Without replacement

53. X = the number of patients calling in claiming to have the flu, who actually have the flu. X = 0, 1, 2, ...25

55. 0.0165

57. a. X = the number of DVDs a Video to Go customer rents

b. 0.12

c. 0.11

d. 0.77

59. d. 4.43

61. c

63.
- X = number of questions answered correctly
- $X \sim B\left(32, \frac{1}{3}\right)$
- We are interested in MORE THAN 75% of 32 questions correct. 75% of 32 is 24. We want to find $P(x > 24)$. The event "more than 24" is the complement of "less than or equal to 24."
- $P(x > 24) = 0$
- The probability of getting more than 75% of the 32 questions correct when randomly guessing is very small and practically zero.

65. a. X = the number of college and universities that offer online offerings.

b. 0, 1, 2, ..., 13

c. $X \sim B(13, 0.96)$

d. 12.48

   e. 0.0135
   f. $P(x = 12) = 0.3186$ $P(x = 13) = 0.5882$ More likely to get 13.

67. a. X = the number of fencers who do **not** use the foil as their main weapon
   b. 0, 1, 2, 3,... 25
   c. $X \sim B(25,0.40)$
   d. 10
   e. 0.0442
   f. The probability that all 25 not use the foil is almost zero. Therefore, it would be very surprising.

69. *a.* X = the number of audits in a 20-year period
   b. 0, 1, 2, ..., 20
   c. $X \sim B(20, 0.02)$
   d. 0.4
   e. 0.6676
   f. 0.0071

71. 1. X = the number of matches
   2. 0, 1, 2, 3
   3. In dollars: −1, 1, 2, 3
   4. $\frac{1}{2}$
   5. The answer is −0.0787. You lose about eight cents, on average, per game.
   6. The house has the advantage.

73. a. $X \sim B(15, 0.281)$

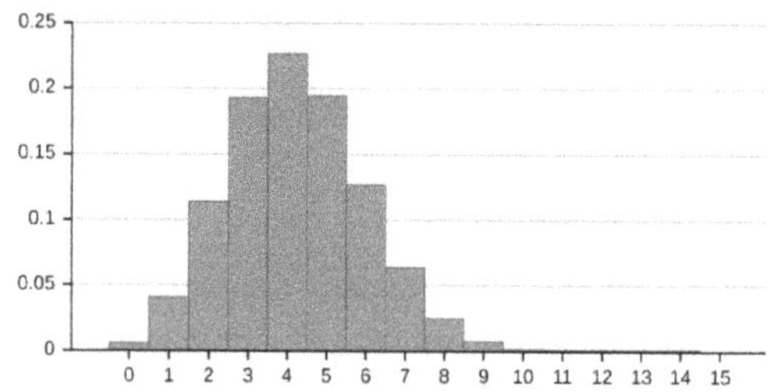

Figure 4.4

   b. i. Mean = $\mu = np = 15(0.281) = 4.215$
      ii. Standard Deviation = $\sigma = \sqrt{npq} = \sqrt{15(0.281)(0.719)} = 1.7409$
   c. $P(x > 5)=1 - 0.7754 = 0.2246$
      $P(x = 3) = 0.1927$
      $P(x = 4) = 0.2259$

   It is more likely that four people are literate that three people are.

75. a. X = the number of adults in America who are surveyed until one says he or she will watch the Super Bowl.

b. $X \sim G(0.40)$

c. 2.5

d. 0.0187

e. 0.2304

77. a. X = the number of pages that advertise footwear

b. X takes on the values 0, 1, 2, ..., 20

$X \sim B(20, \frac{29}{192})$

c. 3.02

d. No

e. 0.9997

f. X = the number of pages we must survey until we find one that advertises footwear. $X \sim G(\frac{29}{192})$

h. 0.3881

i. 6.6207 pages

79. 0, 1, 2, and 3

81. *a.* $X \sim G(0.25)$

b.

i. Mean $= \mu = \frac{1}{p} = \frac{1}{0.25} = 4$

ii. Standard Deviation $= \sigma = \sqrt{\frac{1-p}{p^2}} = \sqrt{\frac{1-0.25}{0.25^2}} \approx 3.4641$

a. $P(x = 10) = 0.0188$

b. $P(x = 20) = 0.0011$

c. $P(x \leq 5) = 0.7627$

82. a. $X \sim P(5.5)$; $\mu = 5.5$ $\sigma = \sqrt{5.5} \approx 2.3452$

b. $P(x \leq 6) \approx 0.6860$

c. There is a 15.7% probability that the law staff will receive more calls than they can handle.

d. $P(x > 8) = 1 - P(x \leq 8) \approx 1 - 0.8944 = 0.1056$

84. Let X = the number of defective bulbs in a string.

- $\mu = np = 100(0.03) = 3$
- $X \sim P(3)$
- $P(x \leq 4) \approx 0.8153$

Using the binomial distribution:

- $X \sim B(100, 0.03)$

- $P(x \leq 4) = 0.8179$

The Poisson approximation is very good—the difference between the probabilities is only 0.0026.

86. a. X = the number of children for a Spanish woman

b. 0, 1, 2, 3,...

c. 0.2299

d. 0.5679

e. 0.4321

88. a. X = the number of fortune cookies that have an extra fortune

b. 0, 1, 2, 3,... 144

c. 4.32

d. 0.0124 or 0.0133

e. 0.6300 or 0.6264

f. As n gets larger, the probabilities get closer together.

90. a. X = the number of people audited in one year

b. 0, 1, 2, ..., 100

c. 2

d. 0.1353

e. 0.3233

92. a. X = the number of shell pieces in one cake

b. 0, 1, 2, 3,...

c. 1.5

d. 0.2231

e. 0.0001

f. Yes

94. d

5 Chapter

CONTINUOUS RANDOM VARIABLES AND ITS APPLICATIONS

Figure 5.1 The heights of these radish plants are continuous random variables. (Credit: Rev Stan)

INTRODUCTION: AN OVERVIEW

Continuous random variables has numerous applications. Examples include baseball batting averages, IQ scores, the duration of long-distance telephone calls, the quantity of cash an individual possesses, the lifespan of a computer chip, investment return rates, and SAT scores. The domain of reliability relies on numerous continuous random variables, similar to all sectors of risk analysis.

NOTE

The values of discrete and continuous random variables can be ambiguous. For example, if X is equal to the number of miles (to the nearest mile) you drive to work, then *X* is a discrete random variable. You count the miles. If *X* is the distance you drive to work, then you measure values of *X* and *X* is a continuous random variable. For a second example, if *X* is equal to the number of books in a backpack, then *X* is a discrete random variable. If *X* is the weight of a book, then *X* is a continuous random variable because weights are measured. How the random variable is defined is very important.

5.1 PROPERTIES OF CONTINUOUS PROBABILITY DENSITY FUNCTIONS

The graph of a continuous probability distribution is a curve. Probability is represented by area under the curve. We have already met this concept when we developed relative frequencies with histograms in Chapter 2. The relative area for a range of values was the probability of drawing at random an observation in that group. Again with the Poisson distribution in Chapter 4, the graph in Example 4.14 used boxes to represent the probability of specific values of the random variable. In this case, we were being a bit casual because the random variables of a Poisson distribution are discrete, whole numbers, and a box has width. Notice that the horizontal axis, the random variable x, purposefully did not mark the points along the axis. The probability of a specific value of a continuous random variable will be zero because the area under a point is zero. Probability is area.

The curve is called the probability density function (abbreviated as pdf). We use the symbol f(x) to represent the curve. f(x) is the function that corresponds to the graph; we use the density function f(x) to draw the graph of the probability distribution.

Area under the curve is given by a different function called the cumulative distribution function (abbreviated as cdf). The cumulative distribution function is used to evaluate probability as area. Mathematically, the cumulative probability density function is the integral of the pdf, and the probability between two values of a continuous random variable will be the integral of the pdf between these two values: the area under the curve between these values. Remember that the area under the pdf for all possible values of the random variable is one, certainty. Probability thus can be seen as the relative percent of certainty between the two values of interest.

- The outcomes are measured, not counted.
- The entire area under the curve and above the x-axis is equal to one.
- Probability is found for intervals of *x* values rather than for individual *x* values.
- $P(c < x < d)$ is the probability that the random variable *X* is in the interval between the values *c* and *d*. $P(c < x < d)$ is the area under the curve, above the *x*-axis, to the right of *c* and the left of *d*.

- $P(x = c) = 0$ The probability that x takes on any single individual value is zero. The area below the curve, above the x-axis, and between $x = c$ and $x = c$ has no width, and therefore no area (area = 0). Since the probability is equal to the area, the probability is also zero.
- $P(c < x < d)$ is the same as $P(c \leq x \leq d)$ because probability is equal to area.

We will find the area that represents probability by using geometry, formulas, technology, or probability tables. In general, integral calculus is needed to find the area under the curve for many probability density functions. When we use formulas to find the area in this textbook, the formulas were found by using the techniques of integral calculus.

There are many continuous probability distributions. When using a continuous probability distribution to model probability, the distribution used is selected to model and fit the particular situation in the best way.

In this chapter and the next, we will study the uniform distribution, the exponential distribution, and the normal distribution. The following graphs illustrate these distributions.

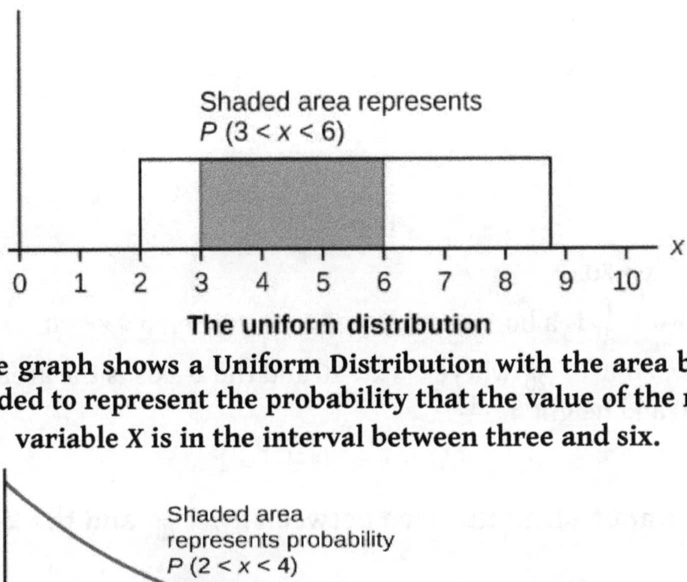

Figure 5.2 The graph shows a Uniform Distribution with the area between $x = 3$ and $x = 6$ shatded to represent the probability that the value of the random variable X is in the interval between three and six.

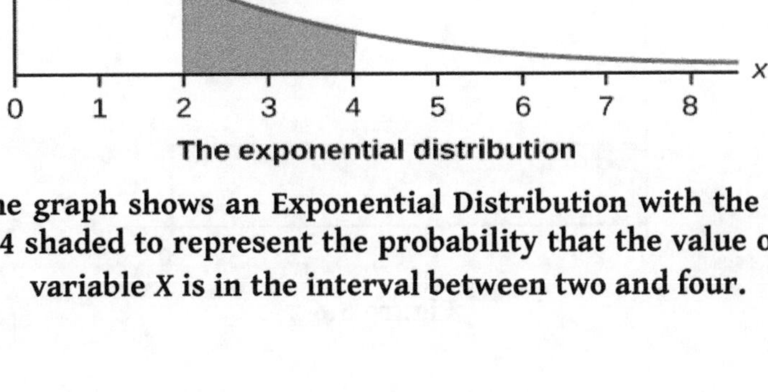

Figure 5.3 The graph shows an Exponential Distribution with the area between $x = 2$ and $x = 4$ shaded to represent the probability that the value of the random variable X is in the interval between two and four.

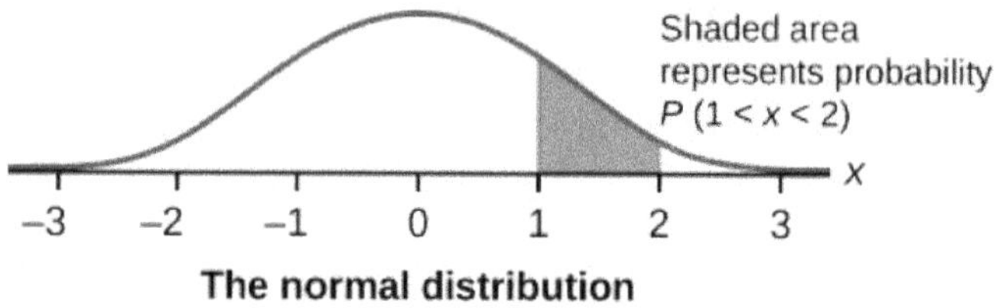

Figure 5.4 The graph shows the Standard Normal Distribution with the area between x = 1 and x = 2 shaded to represent the probability that the value of the random variable X is in the interval between one and two.

For continuous probability distributions, PROBABILITY = AREA.

Example 5.1

Consider the function $f(x) = \frac{1}{20}$ for $0 \le x \le 20$. x = a real number. The graph of $f(x) = \frac{1}{20}$ is a horizontal line. However, since $0 \le x \le 20$, $f(x)$ is restricted to the portion between $x = 0$ and $x = 20$, inclusive.

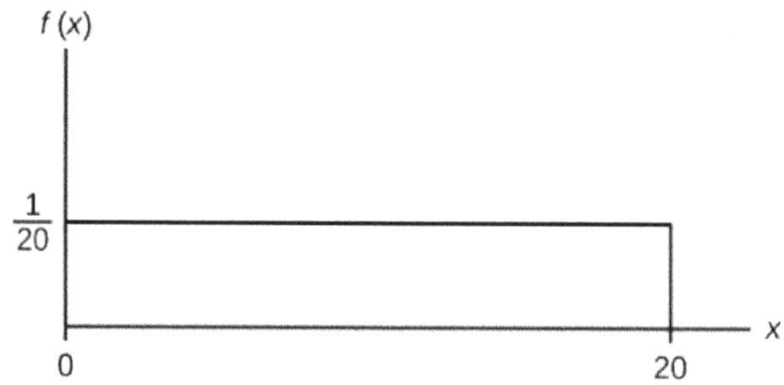

Figure 5.5

$f(x) = \frac{1}{20}$ for $0 \le x \le 20$.

The graph of $f(x) = \frac{1}{20}$ is a horizontal line segment when $0 \le x \le 20$.

The area between $f(x) = \frac{1}{20}$ where $0 \le x \le 20$ and the x-axis is the area of a rectangle with base = 20 and height $= \frac{1}{20}$.

$$\text{AREA} = 20\left(\frac{1}{20}\right) = 1$$

Suppose we want to find the area between $f(x) = \frac{1}{20}$ and the x-axis where $0 < x < 2$.

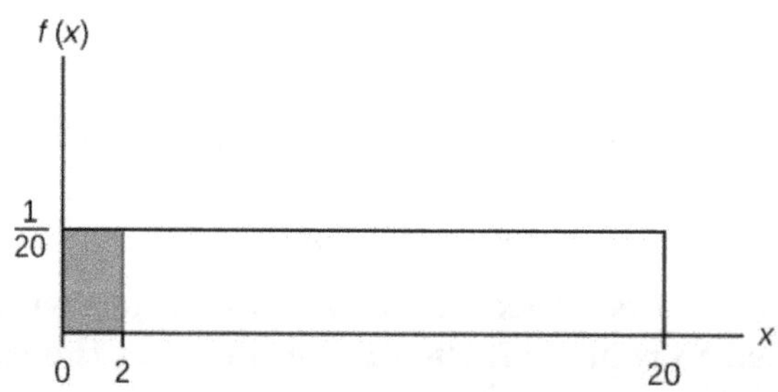

Figure 5.6

$$\text{AREA} = (2-0)\left(\frac{1}{20}\right) = 0.1$$

(2–0)=2=base of a rectangle(2–0)=2=base of a rectangle

Reminder

area of a rectangle = (base)(height).

The area corresponds to a probability. The probability that x is between zero and two is 0.1, which can be written mathematically as $P(0 < x < 2) = P(x < 2) = 0.1$.

Suppose we want to find the area between $f(x) = \frac{1}{20}$ and the x-axis where $4 < x < 15$.

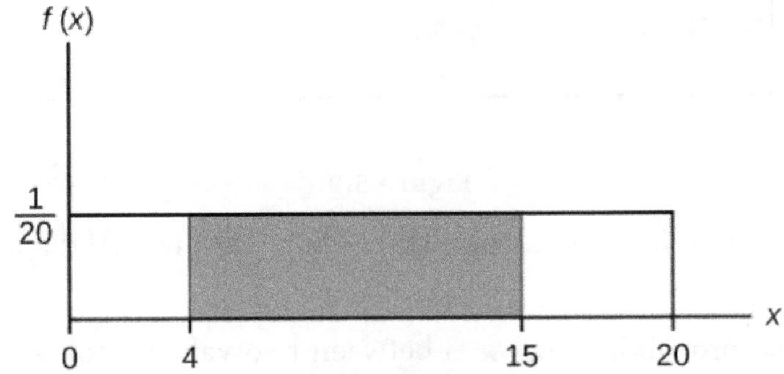

Figure 5.7

$$\text{AREA} = (15-4)\left(\frac{1}{20}\right) = 0.55$$

(15 – 4) = 11 = the base of a rectangle(15 – 4) = 11 = the base of a rectangle

The area corresponds to the probability $P(4 < x < 15) = 0.55$.

Suppose we want to find $P(x = 15)$. On an x-y graph, $x = 15$ is a vertical line. A vertical line has no width (or zero width). Therefore, $P(x = 15)$ = (base)(height $= (0)\left(\frac{1}{20}\right) = 0$

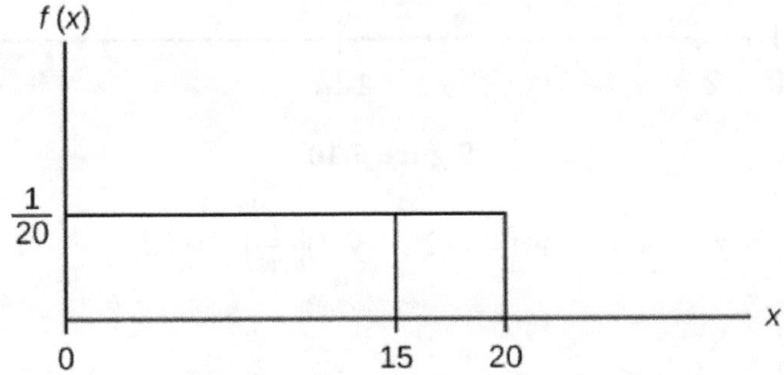

Figure 5.8

$P(X \leq x)$, which can also be written as $P(X < x)$ for continuous distributions, is called the cumulative distribution function or CDF. Notice the "less than or equal to" symbol. We can also use the CDF to calculate $P(X > x)$. The CDF gives "area to the left" and $P(X > x)$ gives "area to the right." We calculate $P(X > x)$ for continuous distributions as follows: $P(X > x) = 1 - P(X < x)$.

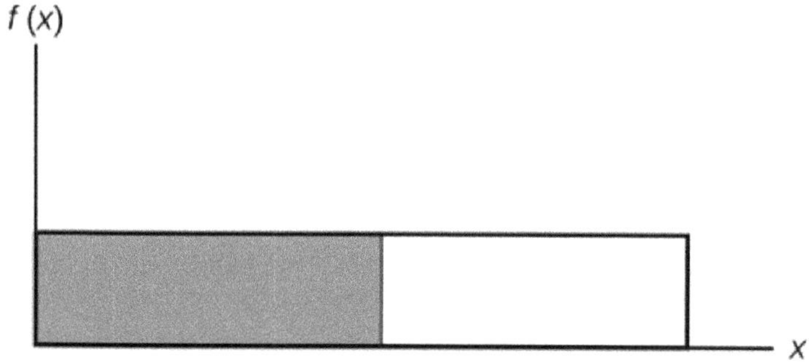

Figure 5.9

Label the graph with $f(x)$ and x. Scale the x and y axes with the maximum x and y values. $f f(x) = \frac{1}{20}$, $0 \leq x \leq 20$.

To calculate the probability that x is between two values, look at the following graph. Shade the region between $x = 2.3$ and $x = 12.7$. Then calculate the shaded area of a rectangle.

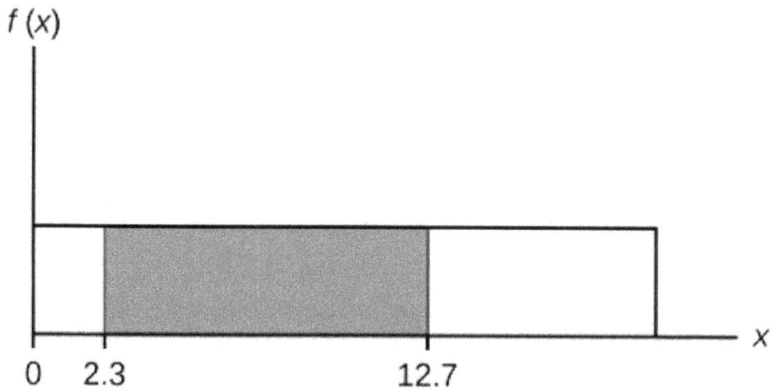

Figure 5.10

$$P(2.3 < x < 12.7) = (\text{base})(\text{height}) = (12.7 - 2.3)\left(\frac{1}{20}\right) = 0.52$$

TRY IT 5.1

Consider the function $f(x) = \frac{1}{8}$ for $0 \leq x \leq 8$. Draw the graph of $f(x)$ and find $P(2.5 < x < 7.5)$.

5.2 THE UNIFORM DISTRIBUTION

The uniform distribution is a continuous probability distribution and is concerned with events that are equally likely to occur. When working out problems that have a uniform distribution, be careful to note if the data is inclusive or exclusive of endpoints.

The mathematical statement of the uniform distribution is

$$f(x) = \frac{1}{b-a} \text{ for } a \le x \le b$$

where a = the lowest value of x and b = the highest value of x.

Formulas for the theoretical mean and standard deviation are

$$\mu = \frac{a+b}{2} \text{ and } \sigma = \sqrt{\frac{(b-a)^2}{12}}$$

TRY IT

The data that follow record the total weight, to the nearest pound, of fish caught by passengers on 35 different charter fishing boats on one summer day. The sample mean = 7.9 and the sample standard deviation = 4.33. The data follow a uniform distribution where all values between and including zero and 14 are equally likely. State the values of a and b. Write the distribution in proper notation, and calculate the theoretical mean and standard deviation.

Table 5.1

1	12	4	10	4	14	11
7	11	4	13	2	4	6
3	10	0	12	6	9	10
5	13	4	10	14	12	11
6	10	11	0	11	13	2

EXAMPLE 5.2

The amount of time, in minutes, that a person must wait for a bus is uniformly distributed between zero and 15 minutes, inclusive.

a. What is the probability that a person waits fewer than 12.5 minutes?

Solution 5.2

a. Let X = the number of minutes a person must wait for a bus. a = 0 and b = 15. $X \sim U(0, 15)$.Write the probability density function. $f(x) = \frac{1}{15-0} = \frac{1}{15}$ for $0 \le x \le 15$.

Find $P(x < 12.5)$. Draw a graph.

$$P(x < k) = (\text{base})(\text{height}) = (12.5 - 0)\left(\frac{1}{15}\right) = 0.8333$$

The probability a person waits less than 12.5 minutes is 0.8333.

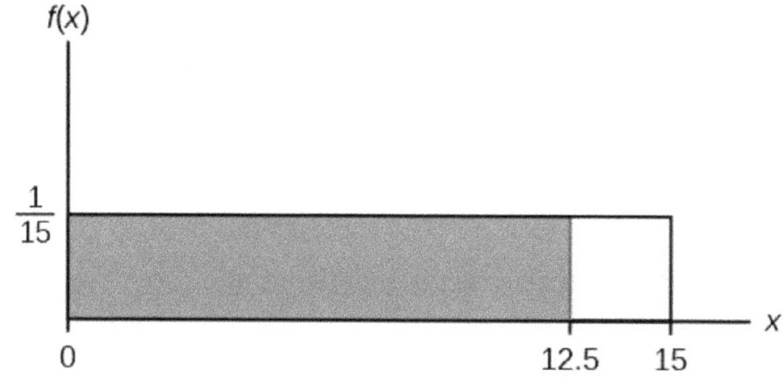

Figure 5.11

b. On the average, how long must a person wait? Find the mean, μ, and the standard deviation, σ.

Solution 5.2

b $\mu = \frac{a+b}{2} = \frac{15+0}{2} = 7.5$. On the average, a person must wait 7.5 minutes.

$\sigma = \sqrt{\frac{(b-a)^2}{12}} = \sqrt{\frac{(15-0)^2}{12}} = 4.3$. The Standard deviation is 4.3 minutes.

c. Ninety percent of the time, the time a person must wait falls below what value?

NOTE

This asks for the 90th percentile.

Solution 5.2

c. Find the 90th percentile. Draw a graph. Let k = the 90th percentile.

$$P(x < k) = (\text{base})(\text{height}) = (k - 0)(\frac{1}{15})$$

$$0.90 = (k)\left(\frac{1}{15}\right)$$

$k = (0.90)(15) = 13.5$

The 90th percentile is 13.5 minutes. Ninety percent of the time, a person must wait at most 13.5 minutes.

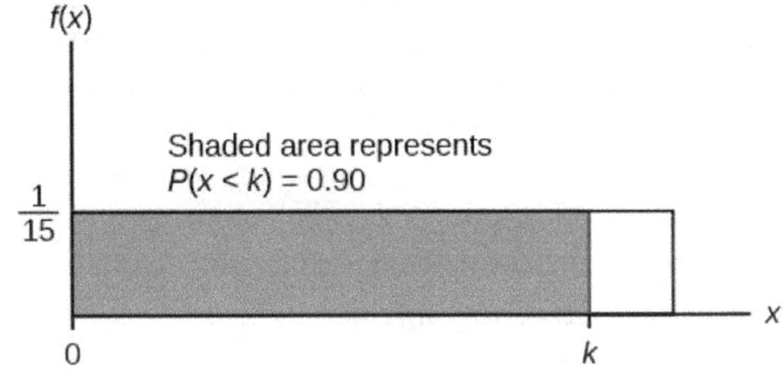

Figure 5.12

TRY IT 5.2

The total duration of baseball games in the major league in the 2011 season is uniformly distributed between 447 hours and 521 hours inclusive.

a. Find *a* and *b* and describe what they represent.
b. Write the distribution.
c. Find the mean and the standard deviation.
d. What is the probability that the duration of games for a team for the 2011 season is between 480 and 500 hours?

5.3 THE EXPONENTIAL DISTRIBUTION

The exponential distribution is often concerned with the amount of time until some specific event occurs. For example, the amount of time (beginning now) until an earthquake occurs has an exponential distribution. Other examples include the length of time, in minutes, of long distance business telephone calls, and the amount of time, in months, a car battery lasts. It can be shown, too, that the value of the change that you have in your pocket or purse approximately follows an exponential distribution.

Values for an exponential random variable occur in the following way. There are fewer large values and more small values. For example, marketing studies have shown that the amount of money customers spend in one trip to the supermarket follows an exponential distribution. There are more people who spend small amounts of money and fewer people who spend large amounts of money.

Exponential distributions are commonly used in calculations of product reliability, or the length of time a product lasts.

The random variable for the exponential distribution is continuous and often measures a passage of time, although it can be used in other applications. Typical questions may be, "what is the probability that some event will occur within the next x hours or days, or what is the probability that some event will occur between

x_1 hours and x_2 hours, or what is the probability that the event will take more than x_1 hours to perform?" In short, the random variable X equals (*a*) the time between events or (*b*) the passage of time to complete an action, e.g. wait on a customer. The probability density function is given by:

$$f(x) = \frac{1}{\mu}e^{-\frac{1}{\mu}x}$$

where μ is the historical average waiting time.

and has a mean and standard deviation of 1/μ.

An alternative form of the exponential distribution formula recognizes what is often called the decay factor. The decay factor simply measures how rapidly the probability of an event declines as the random variable X increases. When the notation using the decay parameter *m* is used, the probability density function is presented as:

$$f(x) \quad = \quad me^{-mx}$$

where $m = \frac{1}{\mu}$

In order to calculate probabilities for specific probability density functions, the cumulative density function is used. The cumulative density function (cdf) is simply the integral of the pdf and is:

$$F\left(x\right) = \int_0^{\infty}\left[\frac{1}{\mu}e^{-\frac{x}{\mu}}\right] = 1 - e^{-\frac{x}{\mu}}$$

EXAMPLE 5.3

Let X = amount of time (in minutes) a postal clerk spends with a customer. The time is known from historical data to have an average amount of time equal to four minutes.

It is given that μ = 4 minutes, that is, the average time the clerk spends with a customer is 4 minutes. Remember that we are still doing probability and thus we have to be told the population parameters such as the mean. To do any calculations, we need to know the mean of the distribution: the historical time to provide a service, for example. Knowing the historical mean allows the calculation of the decay parameter, *m*.

$m = \frac{1}{\mu}$. Therefore, $m = \frac{1}{4} = 0.25$.

When the notation used the decay parameter, *m*, the probability density function is presented as $f(x) = me^{-mx}$, which is simply the original formula with *m* substituted for $\frac{1}{\mu}$, or $f(x) = \frac{1}{\mu}e^{-\frac{1}{\mu}x}$.

To calculate probabilities for an exponential probability density function, we need to use the cumulative density function. As shown below, the curve for the cumulative density function is:

$f(x) = 0.25e^{-0.25x}$ where *x* is at least zero and $m = 0.25$.

For example, $f(5) = 0.25e^{(-0.25)(5)} = 0.072$. In other words, the function has a value of .072 when $x = 5$.

The graph is as follows:

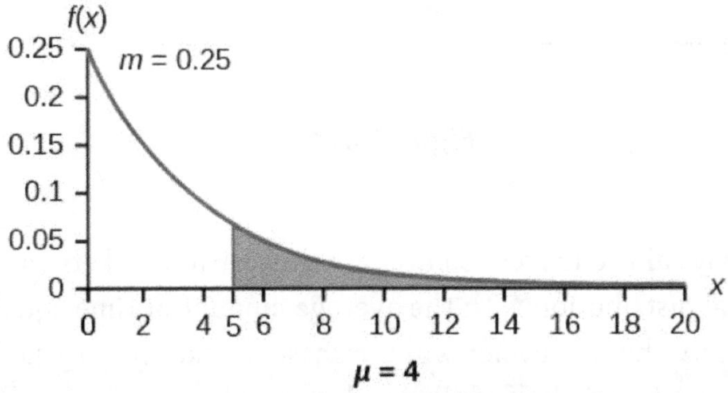

Figure 5.13

Notice the graph is a declining curve. When $x = 0$,
$f(x) = 0.25e^{(-0.25)(0)} = (0.25)(1) = 0.25 = m$. The maximum value on the *y*-axis is always *m*, one divided by the mean.

TRY IT 5.3

The amount of time spouses shop for anniversary cards can be modeled by an exponential distribution with the average amount of time equal to eight minutes. Write the distribution, state the probability density function, and graph the distribution.

EXAMPLE 5.4

a. Using the information in Example 5.3, find the probability that a clerk spends four to five minutes with a randomly selected customer.

Solution 5.4

a. Find $P(4 < x < 5)$.

The **cumulative distribution function (CDF)** gives the area to the left.

$P(x < x) = 1 - e^{-mx}$

$P(x < 5) = 1 - e^{(-0.25)(5)} = 0.7135$ and $P(x < 4) = 1 - e^{(-0.25)(4)} = 0.6321$

$P(4 < x < 5) = 0.7135 - 0.6321 = 0.0814$

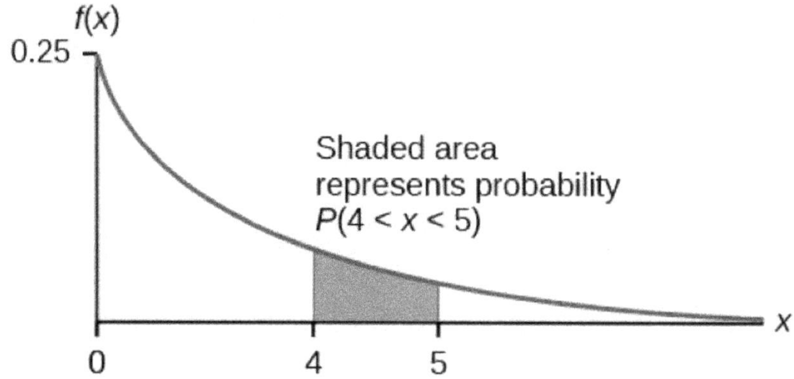

Figure 5.14

TRY IT 5.4

The number of days ahead travelers purchase their airline tickets can be modeled by an exponential distribution with the average amount of time equal to 15 days. Find the probability that a traveler will purchase a ticket fewer than ten days in advance. How many days do half of all travelers wait?

EXAMPLE 5.5

On the average, a certain computer part lasts ten years. The length of time the computer part lasts is exponentially distributed.

a. What is the probability that a computer part lasts more than 7 years?

Solution 5.5

a. Let x = the amount of time (in years) a computer part lasts.

$$\mu = 10 \text{ so } m = \frac{1}{\mu} = \frac{1}{10} = 0.1$$

Find $P(x > 7)$. Draw the graph.

$P(x > 7) = 1 - P(x < 7)$.

Since $P(X < x) = 1 - e^{-mx}$ then $P(X > x) = 1 - (1 - e^{-mx}) = e^{-mx}$

$P(x > 7) = e^{(-0.1)(7)} = 0.4966$. The probability that a computer part lasts more than seven years is 0.4966.

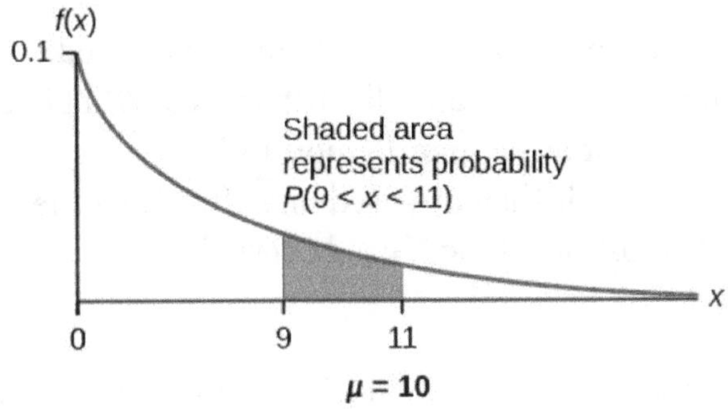

Figure 5.15

b. On the average, how long would five computer parts last if they are used one after another?

Solution 5.5

b. On the average, one computer part lasts ten years. Therefore, five computer parts, if they are used one right after the other would last, on the average, (5)(10) = 50 years.

d. What is the probability that a computer part lasts between nine and 11 years?

Solution 5.5

d. Find $P(9 < x < 11)$. Draw the graph.

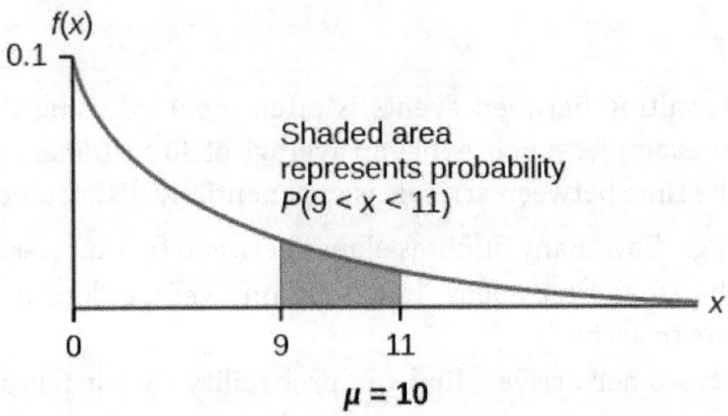

Figure 5.16

Problem

d. What is the probability that a computer part lasts between nine and 11 years?
$P(9 < x < 11) = P(x < 11) - P(x < 9) = (1 - e^{(-0.1)(11)}) - (1 - e^{(-0.1)(9)}) = 0.6671 - 0.5934 = 0.0737$. The probability that a computer part lasts between nine and 11 years is 0.0737.

TRY IT 5.5

On average, a pair of running shoes can last 18 months if used every day. The length of time running shoes last is exponentially distributed. What is the probability that a pair of running shoes last more than 15 months? On average, how long would six pairs of running shoes last if they are used one after the other? Eighty percent of running shoes last at most how long if used every day?

EXAMPLE 5.6

Suppose that the length of a phone call, in minutes, is an exponential random variable with decay parameter 112112. The decay p[parameter is another way to view 1/λ. If another person arrives at a public telephone just before you, find the probability that you will have to wait more than five minutes. Let X = the length of a phone call, in minutes.

What is m, μ, and σ? The probability that you must wait more than five minutes is _______ .

Solution 5.6

- $m = \frac{1}{12}$
- $\mu = 12$
- $\sigma = 12$

$P(x > 5) = 0.6592$

EXAMPLE 5.7

The time spent waiting between events is often modeled using the exponential distribution. For example, suppose that an average of 30 customers per hour arrive at a store and the time between arrivals is exponentially distributed.

a. On average, how many minutes elapse between two successive arrivals?
b. When the store first opens, how long on average does it take for three customers to arrive?
c. After a customer arrives, find the probability that it takes less than one minute for the next customer to arrive.
d. After a customer arrives, find the probability that it takes more than five minutes for the next customer to arrive.
e. Is an exponential distribution reasonable for this situation?

Solution 5.7

a. Since we expect 30 customers to arrive per hour (60 minutes), we expect on average one customer to arrive every two minutes on average.

b. Since one customer arrives every two minutes on average, it will take six minutes on average for three customers to arrive.

c. Let X = the time between arrivals, in minutes. By part a, $\mu = 2$, so $m = \frac{1}{2} = 0.5$.
The cumulative distribution function is $P(X < x) = 1 - e^{(-0.5)(x)}$

Therefore $P(X < 1) = 1 - e^{(-0.5)(1)} = 0.3935$.

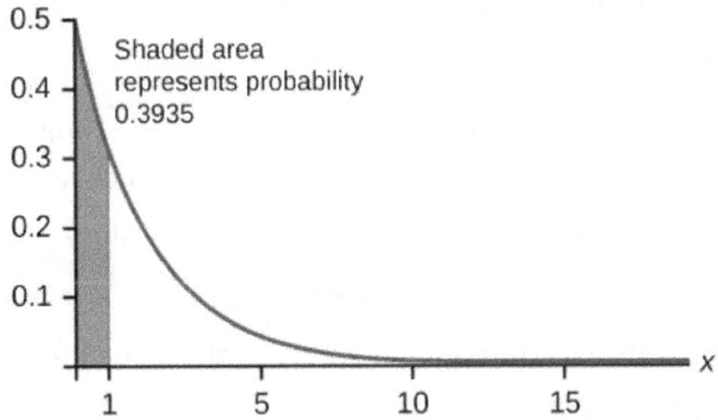

Figure 5.17

d. $P(X > 5) = 1 - P(X < 5) = 1 - (1 - e^{(-0.5)(5)}) = e^{-2.5} \approx 0.0821$.

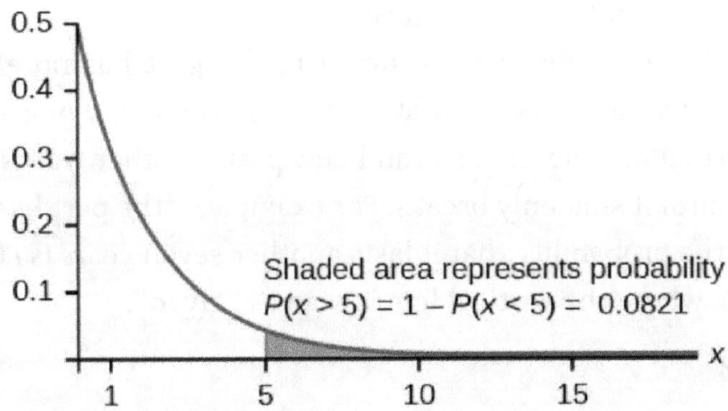

Figure 5.18

e. This model assumes that a single customer arrives at a time, which may not be reasonable since people might shop in groups, leading to several customers arriving at the same time. It also assumes that the flow of customers does not change throughout the day, which is not valid if some times of the day are busier than others.

Memorylessness of the Exponential Distribution

Recall that the amount of time between customers for the postal clerk discussed earlier is exponentially distributed with a mean of two minutes. Suppose that five minutes have elapsed since the last customer arrived. Since an unusually long amount of time has now elapsed, it would seem to be more likely for a customer to arrive within the next minute. With the exponential distribution, this is not the case–the additional time spent waiting for the next customer does not depend on how much time has already elapsed since the last customer. This is referred to as the memoryless property. The exponential and geometric probability density functions are the only probability functions that have the memoryless property. Specifically, the memoryless property says that

$P(X > r + t \mid X > r) = P(X > t)$ for all $r \geq 0$ and $t \geq 0$

For example, if five minutes have elapsed since the last customer arrived, then the probability that more than one minute will elapse before the next customer arrives is computed by using $r = 5$ and $t = 1$ in the foregoing equation.

$P(X > 5 + 1 \mid X > 5) = P(X > 1) = e^{(-0.5)(1)} = 0.6065.$

This is the same probability as that of waiting more than one minute for a customer to arrive after the previous arrival.

The exponential distribution is often used to model the longevity of an electrical or mechanical device. In Example 5.5, the lifetime of a certain computer part has the exponential distribution with a mean of ten years. The memoryless property says that knowledge of what has occurred in the past has no effect on future probabilities. In this case it means that an old part is not any more likely to break down at any particular time than a brand new part. In other words, the part stays as good as new until it suddenly breaks. For example, if the part has already lasted ten years, then the probability that it lasts another seven years is $P(X > 17|X > 10) = P(X > 7) = 0.4966$, where the vertical line is read as "given".

EXAMPLE 5.8

Refer back to the postal clerk again where the time a postal clerk spends with his or her customer has an exponential distribution with a mean of four minutes. Suppose a customer has spent four minutes with a postal clerk. What is the probability that he or she will spend at least an additional three minutes with the postal clerk?

The decay parameter of X is $m = \frac{1}{4} = 0.25$, so $X \sim Exp(0.25)$.

The cumulative distribution function is $P(X < x) = 1 - e^{-0.25x}$.

We want to find $P(X > 7|X > 4)$. The memoryless property says that $P(X > 7|X > 4) = P(X > 3)$, so we just need to find the probability that a customer spends more than three minutes with a postal clerk.

This is $P(X > 3) = 1 - P(X < 3) = 1 - (1 - e^{-0.25 \cdot 3}) = e^{-0.75} \approx 0.4724$.

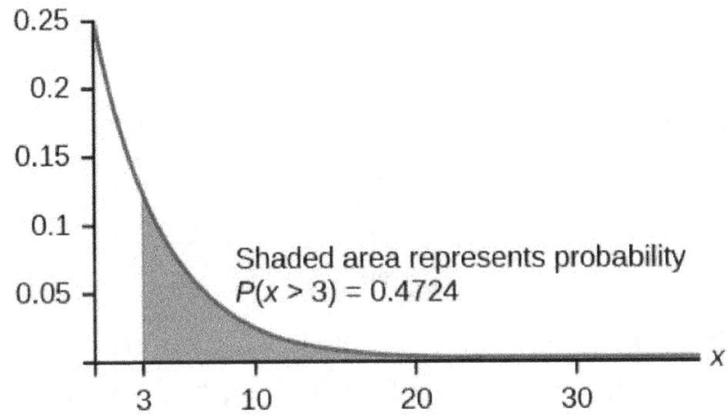

Figure 5.19

Relationship between the Poisson and the Exponential Distribution

There is an interesting relationship between the exponential distribution and the Poisson distribution. Suppose that the time that elapses between two successive events follows the exponential distribution with a mean of μ units of time. Also assume that these times are independent, meaning that the time between events is not affected by the times between previous events. If these assumptions hold, then the number of events per unit time follows a Poisson distribution with mean μ. Recall that if X has the Poisson distribution with mean μ, then $P(X = x) = \frac{\mu^x e^{-\mu}}{x!}$. The formula for the exponential distribution: $P(X = x) = me^{-mx} = \frac{1}{\mu}e^{-\frac{1}{\mu}x}$ Where m = the rate parameter, or μ = average time between occurrences.

We see that the exponential is the cousin of the Poisson distribution and they are linked through this formula. There are important differences that make each distribution relevant for different types of probability problems.

First, the Poisson has a discrete random variable, x, where time; a continuous variable is artificially broken into discrete pieces. We saw that the number of occurrences of an event in a given time interval, x, follows the Poisson distribution.

For example, the **number** of times the telephone rings per hour. By contrast, the time **between** occurrences follows the exponential distribution. For example. The telephone just rang, how long will it be until it rings again? We are measuring length of time of the interval, a continuous random variable, exponential, not events during an interval, Poisson.

The Exponential Distribution v. the Poisson Distribution

A visual way to show both the similarities and differences between these two distributions is with a time line.

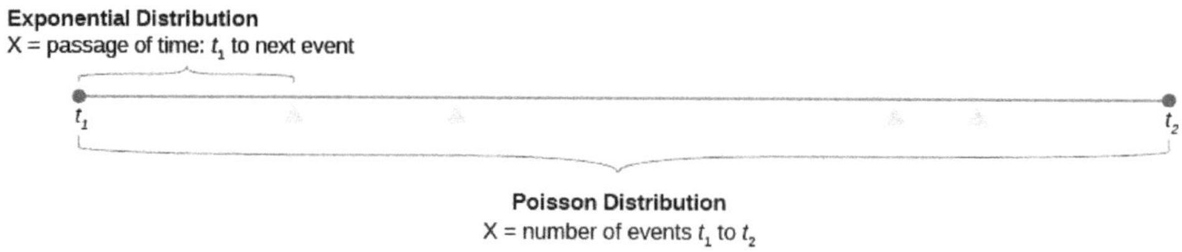

Figure 5.20

The random variable for the Poisson distribution is discrete and thus counts events during a given time period, t_1 to t_2 on Figure 5.20, and calculates the probability of that number occurring. The number of events, four in the graph, is measured in counting numbers; therefore, the random variable of the Poisson is a discrete random variable.

The exponential probability distribution calculates probabilities of the passage of time, a continuous random variable. In Figure 5.20 this is shown as the bracket from t_1 to the next occurrence of the event marked with a triangle.

Classic Poisson distribution questions are "how many people will arrive at my checkout window in the next hour?".

Classic exponential distribution questions are "how long it will be until the next person arrives," or a variant, "how long will the person remain here once they have arrived?".

Again, the formula for the exponential distribution is:

$$f(x) = me^{-mx} \text{ or } f(x) = \frac{1}{\mu}e^{-\frac{1}{\mu}x}$$

We see immediately the similarity between the exponential formula and the Poisson formula.

$$P(x) = \frac{\mu^x e^{-\mu}}{x!}$$

Both probability density functions are based upon the relationship between time and exponential growth or decay. The "e" in the formula is a constant with the approximate value of 2.71828 and is the base of the natural logarithmic exponential growth formula. When people say that something has grown exponentially this is what they are talking about.

An example of the exponential and the Poisson will make clear the differences been the two. It will also show the interesting applications they have.

Poisson Distribution Suppose that historically 10 customers arrive at the checkout lines each hour. Remember that this is still probability so we have to be told these historical values. We see this is a Poisson probability problem.

We can put this information into the Poisson probability density function and get a general formula that will calculate the probability of any specific number of customers arriving in the next hour.

The formula is for any value of the random variable we chose, and so the x is put into the formula. This is the formula:

$$f(x) = \frac{10^x e^{-10}}{x!}$$

As an example, the probability of 15 people arriving at the checkout counter in the next hour would be

$$P(x = 15) = \frac{10^{15} e^{-10}}{15!} = 0.0611$$

Here we have inserted x = 15 and calculated the probability that in the next hour 15 people will arrive is .061.

Exponential Distribution

If we keep the same historical facts that 10 customers arrive each hour, but we now are interested in the service time a person spends at the counter, then we would use the exponential distribution. The exponential probability function for any value of x, the random variable, for this particular checkout counter historical data is:

$$f(x) = \frac{1}{.1}e^{\frac{-x}{.1}} = 10e^{-10x}$$

To calculate μ, the historical average service time, we simply divide the number of people that arrive per hour, 10 , into the time period, one hour, and have μ = 0.1. Historically, people spend 0.1 of an hour at the checkout counter, or 6 minutes. This explains the .1 in the formula.

There is a natural confusion with μ in both the Poisson and exponential formulas. They have different meanings, although they have the same symbol. The mean of the exponential is one divided by the mean of the Poisson. If you are given the historical number of arrivals you have the mean of the Poisson. If you are given an historical length of time between events you have the mean of an exponential.

Continuing with our example at the checkout clerk; if we wanted to know the probability that a person would spend 9 minutes or less checking out, then we use this formula. First, we convert to the same time units which are parts of one hour. Nine minutes is 0.15 of one hour. Next we note that we are asking for a range of values. This is always the case for a continuous random variable. We write the probability question as:

$$p(x \leq 9) = 1 - 10e^{-10x}$$

We can now put the numbers into the formula and we have our result.

$$p(x = .15) = 1 - 10e^{-10(.15)} = 0.7769$$

The probability that a customer will spend 9 minutes or less checking out is 0.7769.

We see that we have a high probability of getting out in less than nine minutes and a tiny probability of having 15 customers arriving in the next hour.

Key Terms

Conditional Probability the likelihood that an event will occur given that another event has already occurred.

decay parameter The decay parameter describes the rate at which probabilities decay to zero for increasing values of x. It is the value *m* in the probability density function $f(x) = me^{(-mx)}$ of an exponential random variable. It is also equal to $m = \frac{1}{\mu}$, where μ is the mean of the random variable.

Exponential Distribution a continuous random variable (RV) that appears when we are interested in the intervals of time between some random events, for example, the length of time between emergency arrivals at a hospital. The mean is $\mu = \frac{1}{m}$ and the standard deviation is $\sigma = \frac{1}{m}$. The probability density function is $f(x)=me^{-mx}$ or $f(x) = \frac{1}{\mu}e^{-\frac{1}{\mu}x}, x \geq 0$ and the cumulative distribution function is $P(X \leq x) = 1 - e^{-mx}$ or $P(X \leq x) = 1 - e^{-\frac{1}{\mu}x}$

memoryless property For an exponential random variable X, the memoryless property is the statement that knowledge of what has occurred in the past has no effect on future probabilities. This means that the probability that X exceeds *x* + *t*, given that it has exceeded *x*, is the same as the probability that X would exceed *t* if we had no knowledge about it. In symbols we say that $P(X > x + t|X > x) = P(X > t)$.

Poisson distribution If there is a known average of μ events occurring per unit time, and these events are independent of each other, then the number of events X occurring in one unit of time has the Poisson distribution. The probability of x events occurring in one unit time is equal to $P(X = x) = \frac{\mu^x e^{-\mu}}{x!}$.

Uniform Distribution a continuous random variable (RV) that has equally likely outcomes over the domain, $a < x < b$; it is often referred as the rectangular distribution because the graph of the pdf has the form of a rectangle. The mean is $\mu = \frac{a+b}{2}$ and the standard deviation is $\sigma = \sqrt{\frac{(b-a)^2}{12}}$. The probability density function is $f(x) = \frac{1}{b-a}$ for $a < x < b$ or $a \le x \le b$. The cumulative distribution is $P(X \le x) = \frac{x-a}{b-a}$.

CHAPTER REVIEW

5.1 Properties of Continuous Probability Density Functions

The probability density function (pdf) is used to describe probabilities for continuous random variables. The area under the density curve between two points corresponds to the probability that the variable falls between those two values. In other words, the area under the density curve between points a and b is equal to $P(a < x < b)$. The cumulative distribution function (cdf) gives the probability as an area. If X is a continuous random variable, the probability density function (pdf), $f(x)$, is used to draw the graph of the probability distribution. The total area under the graph of $f(x)$ is one. The area under the graph of $f(x)$ and between values a and b gives the probability $P(a < x < b)$.

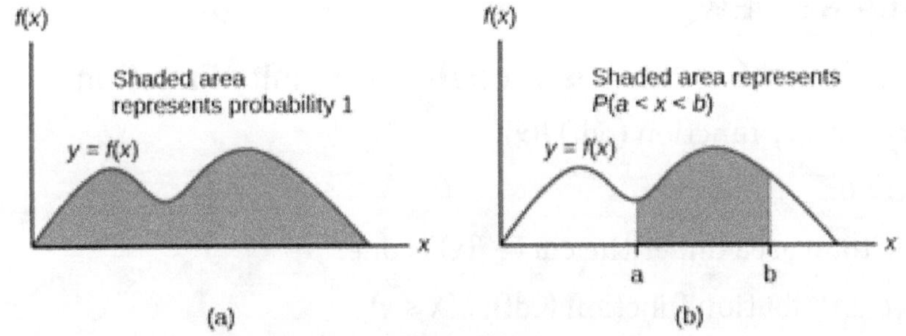

Figure 5.21

The cumulative distribution function (cdf) of X is defined by $P(X \le x)$. It is a function of x that gives the probability that the random variable is less than or equal to x.

5.2 The Uniform Distribution

If X has a uniform distribution where $a < x < b$ or $a \le x \le b$, then X takes on values between a and b (may include a and b). All values x are equally likely. We write $X \sim U(a, b)$. The mean of X is $\mu = \frac{a+b}{2}$. The standard deviation of X is $\sigma = \sqrt{\frac{(b-a)^2}{12}}$. The probability density function of X is $f(x) = \frac{1}{b-a}$ for $a \le x \le b$. The cumulative distribution function of X is $P(X \le x) = \frac{x-a}{b-a}$. X is continuous.

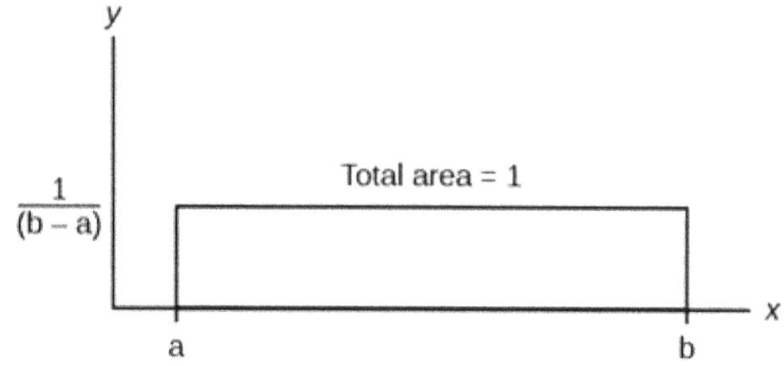

Figure 5.22

The probability $P(c < X < d)$ may be found by computing the area under $f(x)$, between c and d. Since the corresponding area is a rectangle, the area may be found simply by multiplying the width and the height.

5.3 The Exponential Distribution

If X has an **exponential distribution** with mean μ, then the decay parameter is $m = \frac{1}{\mu}$. The probability density function of X is $f(x) = me^{-mx}$ (or equivalently $f(x) = \frac{1}{\mu}e^{-x/\mu}$. The cumulative distribution function of X is $P(X \le x) = 1 - e^{-mx}$.

FORMULA REVIEW

5.1 Properties of Continuous Probability Density Functions

Probability density function (pdf) $f(x)$:

- $f(x) \ge 0$
- The total area under the curve $f(x)$ is one.

Cumulative distribution function (cdf): $P(X \le x)$

5.2 The Uniform Distribution

X = a real number between a and b (in some instances, X can take on the values a and b). a = smallest X; b = largest X

$X \sim U(a, b)$

The mean is $\mu = \frac{a+b}{2}$

The standard deviation is $\sigma = \sqrt{\frac{(b-a)^2}{12}}$

Probability density function: $f(x) = \frac{1}{b-a}$ for $a \le X \le b$

Area to the Left of *x*: $P(X < x) = (x - a)\left(\frac{1}{b-a}\right)$

Area to the Right of *x*: $P(X > x) = (b - x)\left(\frac{1}{b-a}\right)$

Area Between *c* and *d*: $P(c < x < d)$ = (base)(height) = $(d - c)\left(\frac{1}{b-a}\right)$

- pdf: $f(x) = \frac{1}{b-a}$ for $a \le x \le b$
- cdf: $P(X \le x) = \frac{x-a}{b-a}$
- mean $\mu = \frac{a+b}{2}$
- standard deviation $\sigma = \sqrt{\frac{(b-a)^2}{12}}$
- $P(c < X < d) = (d - c)\left(\frac{1}{b-a}\right)$

5.3 The Exponential Distribution

pdf: $f(x) = me^{(-mx)}$ where $x \ge 0$ and $m > 0$

- cdf: $P(X \le x) = 1 - e^{(-mx)}$
- mean $\mu = \frac{1}{m}$
- standard deviation $\sigma = \mu$
- Additionally
    - $P(X > x) = e^{(-mx)}$
    - $P(a < X < b) = e^{(-ma)} - e^{(-mb)}$
- Poisson probability: $P(X = x) = \frac{\mu^x e^{-\mu}}{x!}$ with mean and variance of μ

PRACTICE

5.1 Properties of Continuous Probability Density Functions

1. Which type of distribution does the graph illustrate?

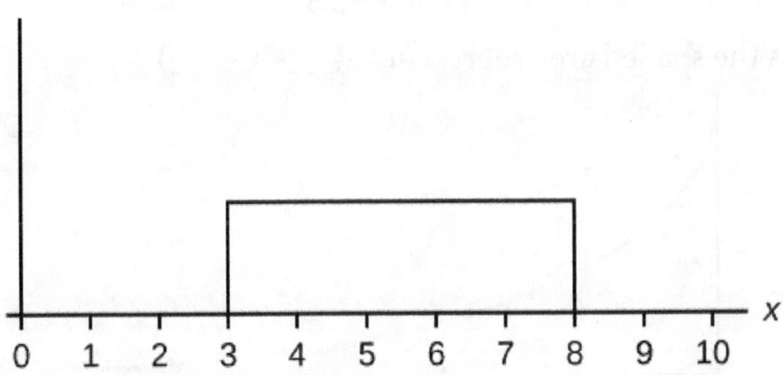

Figure 5.23

2. Which type of distribution does the graph illustrate?

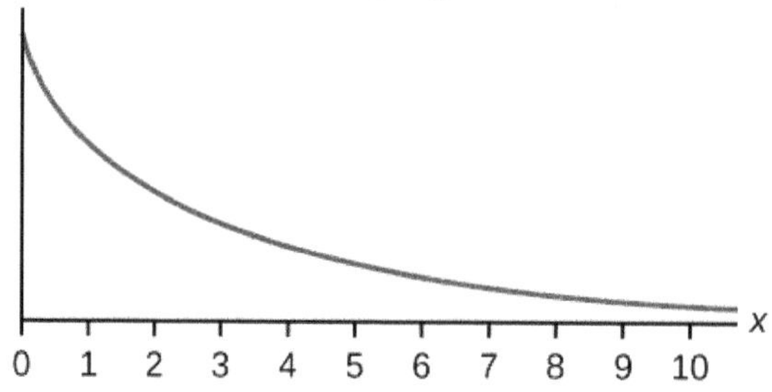

Figure 5.24

3. Which type of distribution does the graph illustrate?

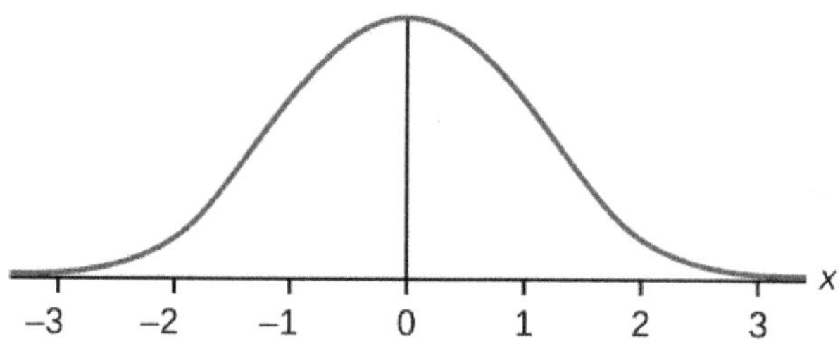

Figure 5.25

4. What does the shaded area represent? $P(\_\_\_ < x < \_\_\_)$

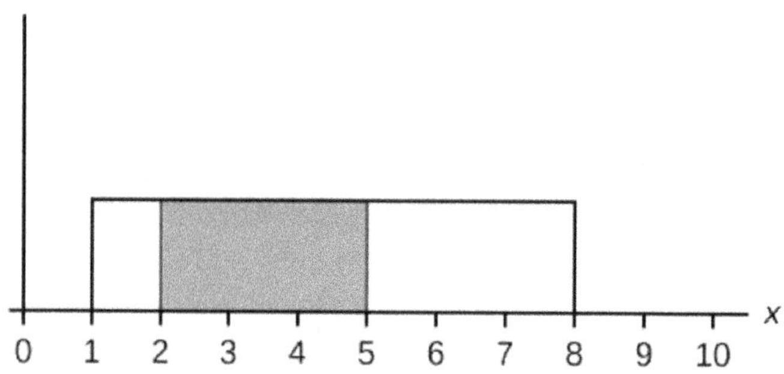

Figure 5.26

5. What does the shaded area represent? $P(\_\_\_ < x < \_\_\_)$

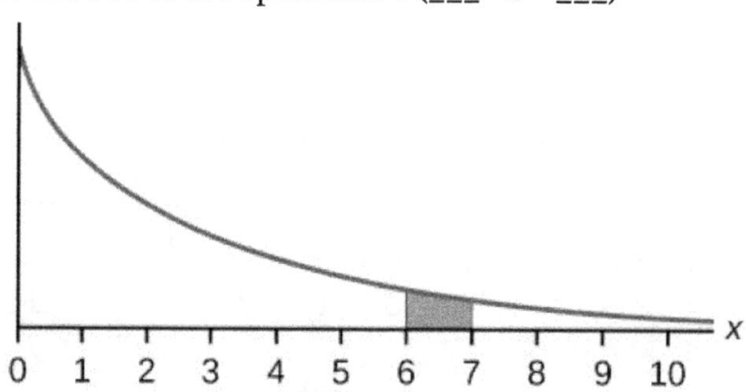

Figure 5.27

6. For a continuous probability distribution, $0 \le x \le 15$. What is $P(x > 15)$?
7. What is the area under $f(x)$ if the function is a continuous probability density function?
8. For a continuous probability distribution, $0 \le x \le 10$. What is $P(x = 7)$?
9. A continuous probability function is restricted to the portion between $x = 0$ and 7. What is $P(x = 10)$?
10. $f(x)$ for a continuous probability function is $\frac{1}{5}$, and the function is restricted to $0 \le x \le 5$. What is $P(x < 0)$?
11. $f(x)$, a continuous probability function, is equal to $\frac{1}{12}$, and the function is restricted to $0 \le x \le 12$. What is $P(0 < x < 12)$?
12. Find the probability that x falls in the shaded area.

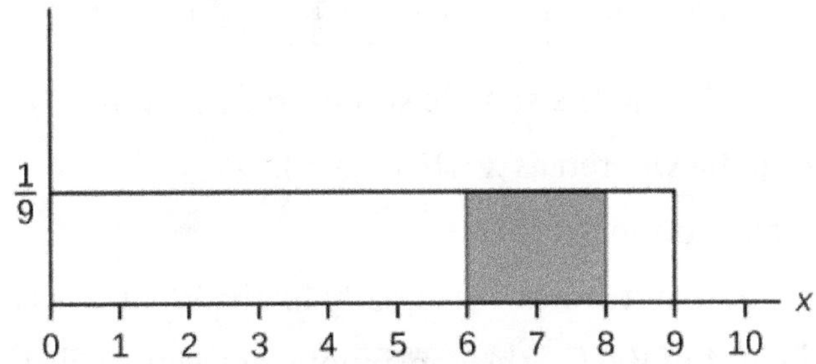

Figure 5.28

13. Find the probability that x falls in the shaded area.

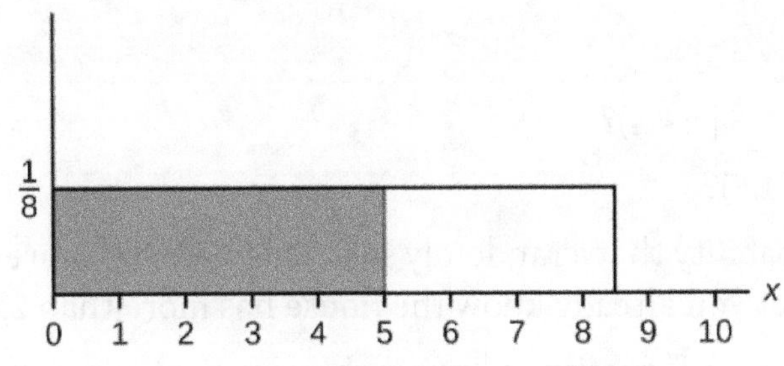

Figure 5.29

14. Find the probability that x falls in the shaded area.

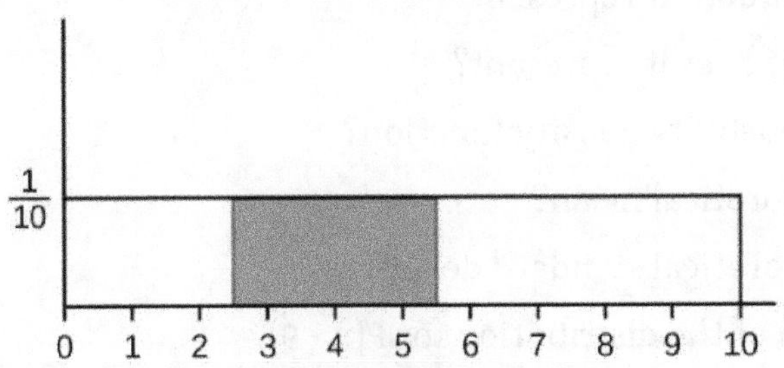

Figure 5.30

15. $f(x)$, a continuous probability function, is equal to $\frac{1}{3}$ and the function is restricted to $1 \le x \le 4$. Describe $P\left(x > \frac{3}{2}\right)$.

5.2 The Uniform Distribution

Use the following information to answer the next ten questions. The data that follow are the square footage (in 1,000 feet squared) of 28 homes.

Table 5.2

1.5	2.4	3.6	2.6	1.6	2.4	2.0
3.5	2.5	1.8	2.4	2.5	3.5	4.0
2.6	1.6	2.2	1.8	3.8	2.5	1.5
2.8	1.8	4.5	1.9	1.9	3.1	1.6

The sample mean = 2.50 and the sample standard deviation = 0.8302.

The distribution can be written as $X \sim U(1.5, 4.5)$.

16. What type of distribution is this?
17. In this distribution, outcomes are equally likely. What does this mean?
18. What is the height of $f(x)$ for the continuous probability distribution?
19. What are the constraints for the values of x?
20. Graph $P(2 < x < 3)$.
21. What is $P(2 < x < 3)$?
22. What is $P(x < 3.5 \mid x < 4)$?
23. What is $P(x = 1.5)$?
24. Find the probability that a randomly selected home has more than 3,000 square feet given that you already know the house has more than 2,000 square feet.

    Use the following information to answer the next eight exercises. A distribution is given as $X \sim U(0, 12)$.

25. What is a? What does it represent?
26. What is b? What does it represent?
27. What is the probability density function?
28. What is the theoretical mean?
29. What is the theoretical standard deviation?
30. Draw the graph of the distribution for $P(x > 9)$.
31. Find $P(x > 9)$.

Use the following information to answer the next eleven exercises. The age of cars in the staff parking lot of a suburban college is uniformly distributed from six months (0.5 years) to 9.5 years.

32. What is being measured here?

33. In words, define the random variable X.

34. Are the data discrete or continuous?

35. The interval of values for x is ______.

36. The distribution for X is ______.

37. Write the probability density function.

38. Graph the probability distribution.

a. Sketch the graph of the probability distribution.

Figure 5.31

b. Identify the following values:

i. Lowest value for $\overline{x}$: _______

ii. Highest value for $\overline{x}$: _______

iii. Height of the rectangle: _______

iv. Label for x-axis (words): _______

v. Label for y-axis (words): _______

39. Find the average age of the cars in the lot.

40. Find the probability that a randomly chosen car in the lot was less than four years old.

a. Sketch the graph, and shade the area of interest.

Figure 5.32

b. Find the probability. $P(x < 4) =$ _______

41. Considering only the cars less than 7.5 years old, find the probability that a randomly chosen car in the lot was less than four years old.

a. Sketch the graph, shade the area of interest.

Figure 5.33

b. Find the probability. $P(x < 4 \mid x < 7.5)$ = _______

42. What has changed in the previous two problems that made the solutions different?

43. Find the third quartile of ages of cars in the lot. This means you will have to find the value such that 3434, or 75%, of the cars are at most (less than or equal to) that age.

a. Sketch the graph, and shade the area of interest.

Figure 5.34

b. Find the value k such that $P(x < k) = 0.75$.

c. The third quartile is _______

5.3 The Exponential Distribution

Use the following information to answer the next ten exercises. A customer service representative must spend different amounts of time with each customer to resolve various concerns. The amount of time spent with each customer can be modeled by the following distribution: $X \sim Exp(0.2)$

44. What type of distribution is this?

45. Are outcomes equally likely in this distribution? Why or why not?

46. What is *m*? What does it represent?

47. What is the mean?

48. What is the standard deviation?

49. State the probability density function.

50. Graph the distribution.

51. Find $P(2 < x < 10)$.

52. Find $P(x > 6)$.

53. Find the 70th percentile.

Use the following information to answer the next seven exercises. A distribution is given as $X \sim Exp(0.75)$.

54. What is *m*?

55. What is the probability density function?

56. What is the cumulative distribution function?

57. Draw the distribution.

58. Find $P(x < 4)$.

59. Find the 30th percentile.

60. Find the median.

61. Which is larger, the mean or the median?

Use the following information to answer the next 16 exercises. Carbon-14 is a radioactive element with a half-life of about 5,730 years. Carbon-14 is said to decay exponentially. The decay rate is 0.000121. We start with one gram of carbon-14. We are interested in the time (years) it takes to decay carbon-14.

62. What is being measured here?

63. Are the data discrete or continuous?

64. In words, define the random variable X.

65. What is the decay rate (*m*)?

66. The distribution for *X* is ______.

67. Find the amount (percent of one gram) of carbon-14 lasting less than 5,730 years. This means, find $P(x < 5,730)$.

a. Sketch the graph, and shade the area of interest.

Figure 5.35

b. Find the probability. $P(x < 5{,}730) =$ __________

68. Find the percentage of carbon-14 lasting longer than 10,000 years.

a. Sketch the graph, and shade the area of interest.

Figure 5.36

b. Find the probability. $P(x > 10{,}000) =$ ________

69. Thirty percent (30%) of carbon-14 will decay within how many years?

a. Sketch the graph, and shade the area of interest.

Figure 5.37

b. Find the value k such that $P(x < k) = 0.30$.

HOMEWORK

5.1 Properties of Continuous Probability Density Functions

For each probability and percentile problem, draw the picture.

70. Consider the following experiment. You are one of 100 people enlisted to take part in a study to determine the percent of nurses in America with an R.N. (registered nurse) degree. You ask nurses if they have an R.N. degree. The nurses answer "yes" or "no." You then calculate the percentage of nurses with an R.N. degree. You give that percentage to your supervisor.

a. What part of the experiment will yield discrete data?
b. What part of the experiment will yield continuous data?

71. When age is rounded to the nearest year, do the data stay continuous, or do they become discrete? Why?

5.2 The Uniform Distribution

For each probability and percentile problem, draw the picture.

72. Births are approximately uniformly distributed between the 52 weeks of the year. They can be said to follow a uniform distribution from one to 53 (spread of 52 weeks).

a. Graph the probability distribution.
b. $f(x)$ = _________
c. μ = _________
d. σ = _________
e. Find the probability that a person is born at the exact moment week 19 starts. That is, find $P(x = 19)$ = _________
f. $P(2 < x < 31)$ = _________
g. Find the probability that a person is born after week 40.
h. $P(12 < x \mid\mid x < 28)$ = _________

73. A random number generator picks a number from one to nine in a uniform manner.

a. Graph the probability distribution.
b. $f(x)$ = _________
c. μ = _________
d. σ = _________
e. $P(3.5 < x < 7.25)$ = _________
f. $P(x > 5.67)$
g. $P(x > 5 \mid\mid x > 3)$ = _________

74. According to a study by Dr. John McDougall of his live-in weight loss program at St. Helena Hospital, the people who follow his program lose between six and 15 pounds a month until they approach trim body weight. Let's suppose that the weight loss is uniformly distributed. We are interested in the weight loss of a randomly selected individual following the program for one month.

a. Define the random variable. X = _________
b. Graph the probability distribution.
c. $f(x)$ = _________
d. μ = _________
e. σ = _________
f. Find the probability that the individual lost more than ten pounds in a month.
g. Suppose it is known that the individual lost more than ten pounds in a month. Find the probability that he lost less than 12 pounds in the month.
h. $P(7 < x < 13 || x > 9)$ = __________. State this in a probability question, similarly to parts g and h, draw the picture, and find the probability.

75. A subway train on the Red Line arrives every eight minutes during rush hour. We are interested in the length of time a commuter must wait for a train to arrive. The time follows a uniform distribution.

a. Define the random variable. X = _______
b. Graph the probability distribution.
c. $f(x)$ = _______
d. μ = _______
e. σ = _______
f. Find the probability that the commuter waits less than one minute.
g. Find the probability that the commuter waits between three and four minutes.

76. The age of a first grader on September 1 at Garden Elementary School is uniformly distributed from 5.8 to 6.8 years. We randomly select one first grader from the class.

a. Define the random variable. X = _________
b. Graph the probability distribution.
c. $f(x)$ = _________
d. μ = _________
e. σ = _________
f. Find the probability that she is over 6.5 years old.
g. Find the probability that she is between four and six years old.

Use the following information to answer the next three exercises. The Sky Train from the terminal to the rental–car and long–term parking center is supposed to arrive every eight minutes. The waiting times for the train are known to follow a uniform distribution.

77. What is the average waiting time (in minutes)?

a. zero
b. two
c. three
d. four

78. The probability of waiting more than seven minutes given a person has waited more than four minutes is?

a. 0.125
b. 0.25
c. 0.5
d. 0.75

79. The time (in minutes) until the next bus departs a major bus depot follows a distribution with $f(x) = 120120$ where x goes from 25 to 45 minutes.

a. Define the random variable. X = ________
b. Graph the probability distribution.
c. The distribution is ______________ (name of distribution). It is ______________ (discrete or continuous).
d. μ = ________
e. σ = ________
f. Find the probability that the time is at most 30 minutes. Sketch and label a graph of the distribution. Shade the area of interest. Write the answer in a probability statement.
g. Find the probability that the time is between 30 and 40 minutes. Sketch and label a graph of the distribution. Shade the area of interest. Write the answer in a probability statement.
h. $P(25 < x < 55)$ = __________. State this in a probability statement, similarly to parts g and h, draw the picture, and find the probability.

80. Suppose that the value of a stock varies each day from \$16 to \$25 with a uniform distribution.

a. Find the probability that the value of the stock is more than \$19.
b. Find the probability that the value of the stock is between \$19 and \$22.
c. Given that the stock is greater than \$18, find the probability that the stock is more than \$21.

81. A fireworks show is designed so that the time between fireworks is between one and five seconds, and follows a uniform distribution.

a. Find the average time between fireworks.
b. Find probability that the time between fireworks is greater than four seconds.

82. The number of miles driven by a truck driver falls between 300 and 700, and follows a uniform distribution.

a. Find the probability that the truck driver goes more than 650 miles in a day.
b. Find the probability that the truck drivers goes between 400 and 650 miles in a day.

5.3 The Exponential Distribution

83. Suppose that the length of long distance phone calls, measured in minutes, is known to have an exponential distribution with the average length of a call equal to eight minutes.

a. Define the random variable. X = ________________.
b. Is X continuous or discrete?
c. μ = ________
d. σ = ________
e. Draw a graph of the probability distribution. Label the axes.
f. Find the probability that a phone call lasts less than nine minutes.
g. Find the probability that a phone call lasts more than nine minutes.
h. Find the probability that a phone call lasts between seven and nine minutes.
i. If 25 phone calls are made one after another, on average, what would you expect the total to be? Why?

84. Suppose that the useful life of a particular car battery, measured in months, decays with parameter 0.025. We are interested in the life of the battery.

a. Define the random variable. X = __________________________________.
b. Is X continuous or discrete?
c. On average, how long would you expect one car battery to last?
d. On average, how long would you expect nine car batteries to last, if they are used one after another?
e. Find the probability that a car battery lasts more than 36 months.
f. Seventy percent of the batteries last at least how long?

85. The percent of persons (ages five and older) in each state who speak a language at home other than English is approximately exponentially distributed with a mean of 9.848. Suppose we randomly pick a state.

a. Define the random variable. X = __________________________________.
b. Is X continuous or discrete?

c. μ = ________
d. σ = ________
e. Draw a graph of the probability distribution. Label the axes.
f. Find the probability that the percent is less than 12.
g. Find the probability that the percent is between eight and 14.
h. The percent of all individuals living in the United States who speak a language at home other than English is 13.8.
   i. Why is this number different from 9.848%?
   ii. What would make this number higher than 9.848%?

86. The time (in years) after reaching age 60 that it takes an individual to retire is approximately exponentially distributed with a mean of about five years. Suppose we randomly pick one retired individual. We are interested in the time after age 60 to retirement.

a. Define the random variable. X = ____________________________________.
b. Is X continuous or discrete?
c. μ = ________
d. σ = ________
e. Draw a graph of the probability distribution. Label the axes.
f. Find the probability that the person retired after age 70.
g. Do more people retire before age 65 or after age 65?
h. In a room of 1,000 people over age 80, how many do you expect will NOT have retired yet?

87. The cost of all maintenance for a car during its first year is approximately exponentially distributed with a mean of $150.

a. Define the random variable. X = ____________________________________.
b. μ = ________
c. σ = ________
d. Draw a graph of the probability distribution. Label the axes.
e. Find the probability that a car required over $300 for maintenance during its first year.

Use the following information to answer the next three exercises. The average lifetime of a certain new cell phone is three years. The manufacturer will replace any cell phone failing within two years of the date of purchase. The lifetime of these cell phones is known to follow an exponential distribution.

88. The decay rate is:

a. 0.3333
b. 0.5000
c. 2
d. 3

89. What is the probability that a phone will fail within two years of the date of purchase?

a. 0.8647 b. 0.4866

c. 0.2212 d. 0.9997

90. What is the median lifetime of these phones (in years)?

a. 0.1941 b. 1.3863

c. 2.0794 d. 5.5452

91. At a 911 call center, calls come in at an average rate of one call every two minutes. Assume that the time that elapses from one call to the next has the exponential distribution.

a. On average, how much time occurs between five consecutive calls?
b. Find the probability that after a call is received, it takes more than three minutes for the next call to occur.
c. Ninety-percent of all calls occur within how many minutes of the previous call?
d. Suppose that two minutes have elapsed since the last call. Find the probability that the next call will occur within the next minute.
e. Find the probability that less than 20 calls occur within an hour.

92. In major league baseball, a no-hitter is a game in which a pitcher, or pitchers, doesn't give up any hits throughout the game. No-hitters occur at a rate of about three per season. Assume that the duration of time between no-hitters is exponential.

a. What is the probability that an entire season elapses with a single no-hitter?
b. If an entire season elapses without any no-hitters, what is the probability that there are no no-hitters in the following season?
c. What is the probability that there are more than 3 no-hitters in a single season?

93. During the years 1998–2012, a total of 29 earthquakes of magnitude greater than 6.5 have occurred in Papua New Guinea. Assume that the time spent waiting between earthquakes is exponential.

a. What is the probability that the next earthquake occurs within the next three months?
b. Given that six months has passed without an earthquake in Papua New Guinea, what is the probability that the next three months will be free of earthquakes?
c. What is the probability of zero earthquakes occurring in 2014?
d. What is the probability that at least two earthquakes will occur in 2014?

94. According to the American Red Cross, about one out of nine people in the U.S. have Type B blood. Suppose the blood types of people arriving at a blood drive are independent. In this case, the number of Type B blood types that arrive roughly follows the Poisson distribution.

a. If 100 people arrive, how many on average would be expected to have Type B blood?
b. What is the probability that over 10 people out of these 100 have type B blood?
c. What is the probability that more than 20 people arrive before a person with type B blood is found?

95. A web site experiences traffic during normal working hours at a rate of 12 visits per hour. Assume that the duration between visits has the exponential distribution.

a. Find the probability that the duration between two successive visits to the web site is more than ten minutes.
b. The top 25% of durations between visits are at least how long?
c. Suppose that 20 minutes have passed since the last visit to the web site. What is the probability that the next visit will occur within the next 5 minutes?
d. Find the probability that less than 7 visits occur within a one-hour period.

96. At an urgent care facility, patients arrive at an average rate of one patient every seven minutes. Assume that the duration between arrivals is exponentially distributed.

a. Find the probability that the time between two successive visits to the urgent care facility is less than 2 minutes.
b. Find the probability that the time between two successive visits to the urgent care facility is more than 15 minutes.
c. If 10 minutes have passed since the last arrival, what is the probability that the next person will arrive within the next five minutes?
d. Find the probability that more than eight patients arrive during a half-hour period.

REFERENCES

5.2 The Uniform Distribution

McDougall, John A. The McDougall Program for Maximum Weight Loss. Plume, 1995.

5.3 The Exponential Distribution

Data from the United States Census Bureau.

Data from World Earthquakes, 2013. Available online at http://www.world-earthquakes.com/ (accessed June 11, 2013).

"No-hitter." Baseball-Reference.com, 2013. Available online at http://www.baseball-reference.com/bullpen/No-hitter (accessed June 11, 2013).

Zhou, Rick. "Exponential Distribution lecture slides." Available online at www.public.iastate.edu/~riczw/stat330s11/lecture/lec13.pdf (accessed June 11, 2013).

SOLUTIONS

1. Uniform Distribution

3. Normal Distribution

5. $P(6 < x < 7)$

7. one

9. zero

11. one

13. 0.625

15. The probability is equal to the area from $x = \frac{3}{2}$ to $x = 4$ above the x-axis and up to $f(x) = \frac{1}{3}$.

17. It means that the value of x is just as likely to be any number between 1.5 and 4.5.

19. $1.5 \le x \le 4.5$

21. 0.3333

23. zero

24. 0.6

26. b is 12, and it represents the highest value of x.

28. six

30.

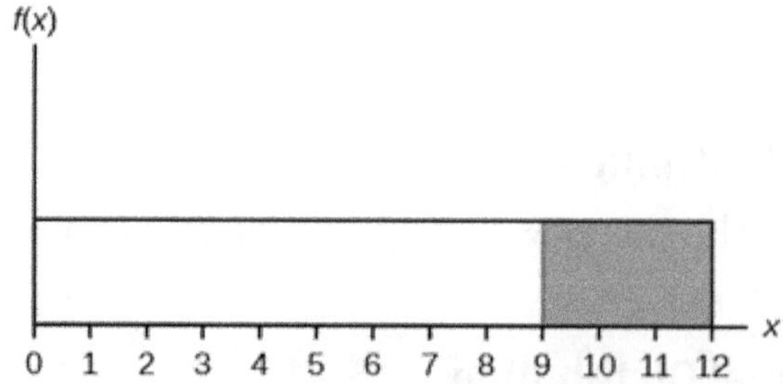

Figure 5.38

33. X = The age (in years) of cars in the staff parking lot

35. 0.5 to 9.5

37. $f(x) = \frac{1}{9}$ where x is between 0.5 and 9.5, inclusive.

39. $\mu = 5$

41.

a. Check student's solution.

b. $\frac{3.5}{7}$

43.

a. Check student's solution.

b. $k = 7.25$

c. 7.25

45. No, outcomes are not equally likely. In this distribution, more people require a little bit of time, and fewer people require a lot of time, so it is more likely that someone will require less time.

47. five

49. $f(x) = 0.2e^{-0.2x}$

51. 0.5350

53. 6.02

55. $f(x) = 0.75e^{-0.75x}$

57.

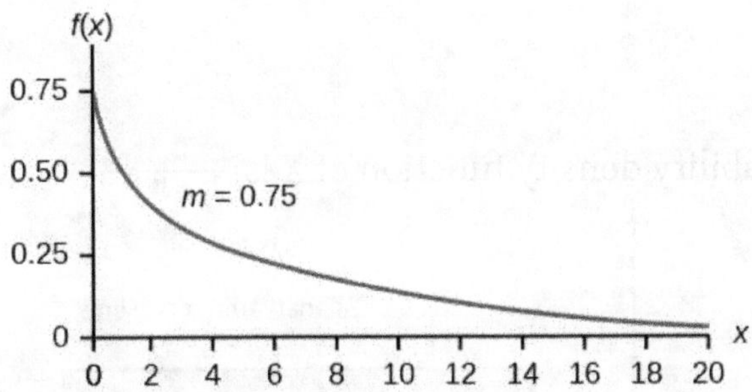

Figure 5.39

59. 0.4756

61. The mean is larger. The mean is $\frac{1}{m} = \frac{1}{0.75} \approx 1.33$, which is greater than 0.9242.

63. continuous

65. $m = 0.000121$

67. a. Check student's solution

b. $P(x < 5{,}730) = 0.5001$

69. a. Check student's solution.

b. $k = 2947.73$

71. Age is a measurement, regardless of the accuracy used.

73. a. Check student's solution.

b. $f(x) = \frac{1}{8}$ where $1 \le x \le 9$

c. five

d. 2.3

e. $\frac{15}{32}$

f. $\frac{333}{800}$

g. $\frac{2}{3}$

75. a. X represents the length of time a commuter must wait for a train to arrive on the Red Line.

b. Graph the probability distribution.

c. $f(x) = \frac{1}{8}$ where $0 \le x \le 8$

d. four

e. 2.31

f. $\frac{1}{8}$

g. $\frac{1}{8}$

77. d

78. b

80. a. The probability density function of X is $\frac{1}{25 - 16} = \frac{1}{9}$.

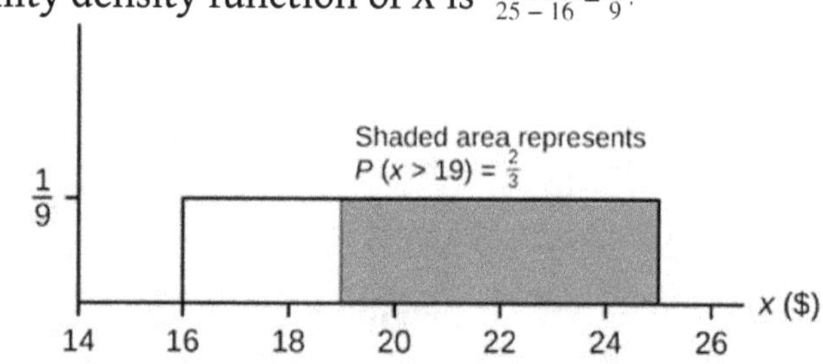

Figure 5.40

b. $P(19 < X < 22) = (22 - 19)\left(\frac{1}{9}\right) = \frac{3}{9} = \frac{1}{3}$.

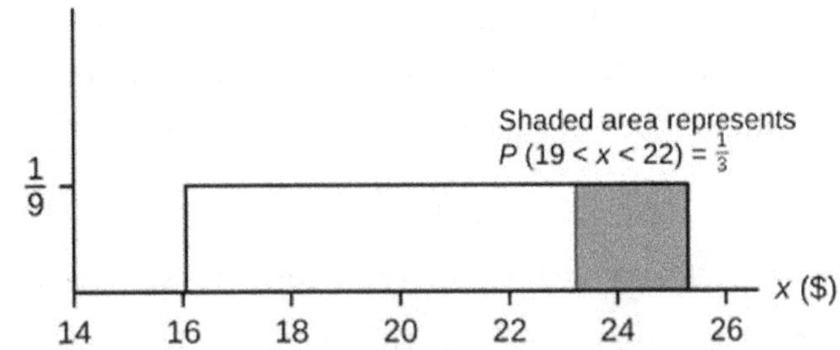

Figure 5.41

c. This is a conditional probability question. P(x > 21 || x > 18). You can do this two ways:

- Draw the graph where a is now 18 and b is still 25. The height is $\frac{1}{(25-18)} = \frac{1}{7}$

  So, $P(x > 21 \mid x > 18) = (25 - 21)\left(\frac{1}{7}\right) = 4/7.$

- Use the formula: $P(x > 21 \mid x > 18) = \frac{P(x > 21 \cap x > 18)}{P(x > 18)} = \frac{P(x > 21)}{P(x > 18)} = \frac{(25-21)}{(25-18)} = \frac{4}{7}.$

82.

a. $P(X > 650) = \frac{700-650}{700-300} = \frac{50}{400} = \frac{1}{8} = 0.125.$

b. $P(400 < X < 650) = \frac{650-400}{700-300} = \frac{250}{400} = 0.625$

84.

a. X = the useful life of a particular car battery, measured in months.
b. X is continuous.
c. 40 months
d. 360 months
e. 0.4066
f. 14.27

86.

a. X = the time (in years) after reaching age 60 that it takes an individual to retire
b. X is continuous.
c. five
d. five
e. Check student's solution.
f. 0.1353
g. before
h. 18.3

88. a

90. c

92. Let X = the number of no-hitters throughout a season. Since the duration of time between no-hitters is exponential. the number of no-hitters per season is Poisson with mean $\lambda = 3$. Therefore, $(X = 0) = \frac{3^0 e^{-3}}{0!} = e^{-3} \approx 0.0498$

NOTE

You could let T = duration of time between no-hitters. Since the time is exponential and there are 3 no-hitters per season, then the time between no-hitters is $\frac{1}{3}$ season. For the exponential, $\mu = \frac{1}{3}$ Therefore, $m = \frac{1}{\mu} = 3$ and $T \sim Exp(3)$.

a. The desired probability is $P(T > 1) = 1 - P(T < 1) = 1 - (1 - e^{-3}) = e^{-3} \approx 0.0498$.

b. Let T = duration of time between no-hitters. We find $P(T > 2|T > 1)$, and by the **memoryless property** this is simply $P(T > 1)$, which we found to be 0.0498 in part a.

c. Let X = the number of no-hitters is a season. Assume that X is Poisson with mean $\lambda = 3$. Then $P(X > 3) = 1 - P(X \leq 3) = 0.3528$.

94. a. $\frac{100}{9} = 11.11$

b. $P(X > 10) = 1 - P(X \leq 10) = 1 - \text{Poissoncdf}(11.11, 10) \approx 0.5532$.

c. The number of people with Type B blood encountered roughly follows the Poisson distribution, so the number of people X who arrive between successive Type B arrivals is roughly exponential with mean $\mu = 9$ and $m = \frac{1}{9}$. The cumulative distribution function of X is $P(X < x) = 1 - e^{-\frac{x}{9}}$. Thus, $P(X > 20) = 1 - P(X \leq 20) = 1 - \left(1 - e^{-\frac{20}{9}}\right) \approx 0.1084.$

NOTE

We could also deduce that each person arriving has a 8/9 chance of not having Type B blood. So the probability that none of the first 20 people arrive have Type B blood is $\left(\frac{8}{9}\right)^{20} \approx 0.0948$ (The geometric distribution is more appropriate than the exponential because the number of people between Type B people is discrete instead of continuous.)

96. Let T = duration (in minutes) between successive visits. Since patients arrive at a rate of one patient everv seven minutes, $\mu = 7$ and the decay constant is $m = \frac{1}{7}$. The cdf is $P(T < t) = 1 - e^{\frac{t}{7}}$

a. $P(T < 2) = 1 - 1 - e^{-\frac{2}{7}} \approx 0.2485.$

b. $P(T > 15) = 1 - P(T < 15) = 1 - \left(1 - e^{-\frac{15}{7}}\right) \approx e^{-\frac{15}{7}} \approx 0.1173.$

c. $P(T > 15|T > 10) = P(T > 5) = 1 - \left(1 - e^{-\frac{5}{7}}\right) = e^{-\frac{5}{7}} \approx 0.4895$.

Let X = # of patients arriving during a half-hour period. Then X has the Poisson distribution with a mean of $\frac{30}{7}$, $X \sim \text{Poisson}\left(\frac{30}{7}\right)$. Find $P(X > 8) = 1 - P(X \leq 8)$ ≈ 0.0311.

6
Chapter

THE NORMAL DISTRIBUTION AND ITS APPLICATIONS

Figure 6.1 If you ask enough people about their shoe size, you will find that your graphed data is shaped like a bell curve and can be described as normally distributed. (credit: Ömer Ünlü)

INTRODUCTION: AN OVERVIEW

The normal probability density function, a continuous distribution, is the most significant of all distributions. It is extensively utilized and much more extensively misused. The graph exhibits a bell-shaped configuration. The bell curve is prevalent throughout nearly all fields of study. These encompass psychology, business, economics, the sciences, nursing, and mathematics. Certain lecturers may employ the normal distribution to ascertain your grade. The majority of IQ scores exhibit a normal distribution. Real estate values frequently conform to a normal distribution. The normal distribution is highly significant, although it is not universally applicable to all real-world scenarios. It is important to note that we are still discussing the distribution of population data. This discourse pertains to probability, specifically with population data that may exhibit a normal distribution. If such a distribution is confirmed, we can ascertain the probabilities of certain events, analogous to our approach with population data that may follow binomial or Poisson distributions. This warning is provided because the subsequent chapter will demonstrate that the normal distribution represents a concept distinct from raw data and serves as the cornerstone of inferential statistics. The normal distribution has two parameters (two numerical descriptive measures): the mean (μ) and the standard deviation (σ). If X is a quantity to be measured that has a normal distribution with mean (μ) and standard deviation (σ), we designate this by writing the following formula of the normal probability density function:

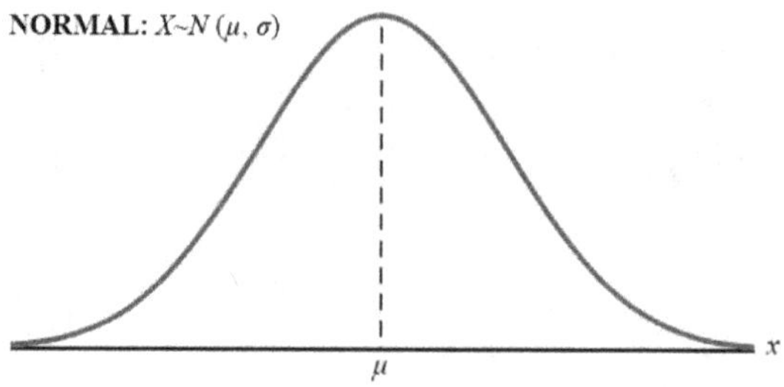

Figure 6.2

The probability density function is a rather complicated function. Do not memorize it. It is not necessary.

$$f(x) = \frac{1}{\sigma \cdot \sqrt{2 \cdot \pi}} \cdot e^{-\frac{1}{2} \cdot \left(\frac{x - \mu}{\sigma}\right)^2}$$

The curve is symmetric about a vertical line drawn through the mean, μ. The mean is the same as the median, which is the same as the mode, because the graph is symmetric about μ. As the notation indicates, the normal distribution depends only on the mean and the standard deviation. Note that this is unlike several probability density functions we have already studied, such as the Poisson, where the mean is equal to µμ and the standard deviation simply the square root of the mean, or the binomial, where p is used to determine both the mean and standard deviation. Since the area under the curve must equal one, a change in the standard deviation, σ, causes a change in the shape of the normal curve; the curve becomes fatter and wider or skinnier and taller depending on σ. A change in μ causes the graph to shift to the left or right. This means there are an infinite number of normal probability distributions. One of special interest is called the **standard normal distribution.**

6.1 THE STANDARD NORMAL DISTRIBUTION

The standard normal distribution is a normal distribution of standardized values called z-scores. A z-score is measured in units of the standard deviation.

The mean for the standard normal distribution is zero, and the standard deviation is one. What this does is dramatically simplify the mathematical calculation of probabilities. Take a moment and substitute zero and one in the appropriate places in the above formula and you can see that the equation collapses into one that can be much more easily solved using integral calculus. The transformation $z = \frac{x - \mu}{\sigma}$ produces the distribution $Z \sim N(0, 1)$. The value x in the given equation comes from a known normal distribution with known mean μ and known standard deviation σ. The z-score tells how many standard deviations a particular x is away from the mean.

Z-Scores

If X is a normally distributed random variable and $X \sim N(\mu, \sigma)$, then the z-score for a particular x is:

$$z = \frac{x - \mu}{\sigma}$$

The z-score tells you how many standard deviations the value x is above (to the right of) or below (to the left of) the mean, μ. Values of x that are larger than the mean have positive z-scores, and values of x that are smaller than the mean have negative z-scores. If x equals the mean, then x has a z-score of zero.

EXAMPLE 6.1

Suppose $X \sim N(5, 6)$. This says that X is a normally distributed random variable with mean $\mu = 5$ and standard deviation $\sigma = 6$. Suppose $x = 17$. Then:

$$z = \frac{x - \mu}{\sigma} = \frac{17 - 5}{6} = 2$$

This means that $x = 17$ is two standard deviations (2σ) above or to the right of the mean $\mu = 5$.

Now suppose $x = 1$. Then: $z = \frac{x - \mu}{\sigma} = \frac{1 - 5}{6} = -0.67$ (rounded to two decimal places)

This means that $x = 1$ is 0.67 standard deviations (-0.67σ) below or to the left of the mean $\mu = 5$.

The Empirical Rule If X is a random variable and has a normal distribution with mean μ and standard deviation σ, then the Empirical Rule states the following:

- About 68% of the x values lie between -1σ and $+1\sigma$ of the mean μ (within one standard deviation of the mean).
- About 95% of the x values lie between -2σ and $+2\sigma$ of the mean μ (within two standard deviations of the mean).

- About 99.7% of the x values lie between -3σ and $+3\sigma$ of the mean μ (within three standard deviations of the mean). Notice that almost all the x values lie within three standard deviations of the mean.
- The z-scores for $+1\sigma$ and -1σ are +1 and −1, respectively.
- The z-scores for $+2\sigma$ and -2σ are +2 and −2, respectively.
- The z-scores for $+3\sigma$ and -3σ are +3 and −3 respectively.

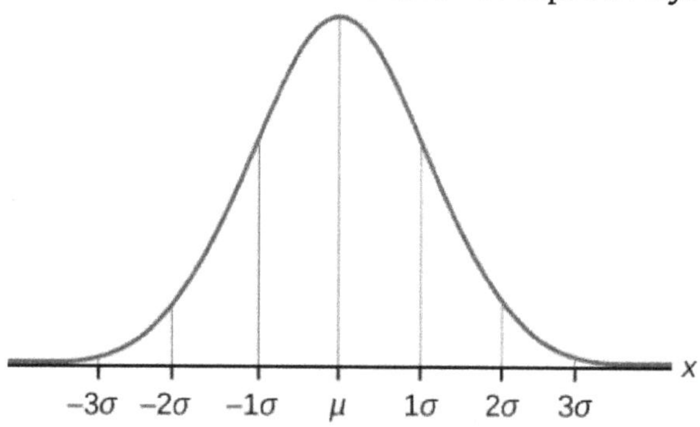

Figure 6.3

EXAMPLE 6.2

Suppose x has a normal distribution with mean 50 and standard deviation 6.

- About 68% of the x values lie within one standard deviation of the mean. Therefore, about 68% of the x values lie between $-1\sigma = (-1)(6) = -6$ and $1\sigma = (1)(6) = 6$ of the mean 50. The values $50 - 6 = 44$ and $50 + 6 = 56$ are within one standard deviation from the mean 50. The z-scores are −1 and +1 for 44 and 56, respectively.
- About 95% of the x values lie within two standard deviations of the mean. Therefore, about 95% of the x values lie between $-2\sigma = (-2)(6) = -12$ and $2\sigma = (2)(6) = 12$. The values $50 - 12 = 38$ and $50 + 12 = 62$ are within two standard deviations from the mean 50. The z-scores are −2 and +2 for 38 and 62, respectively.
- About 99.7% of the x values lie within three standard deviations of the mean. Therefore, about 99.7% of the x values lie between $-3\sigma = (-3)(6) = -18$ and $3\sigma = (3)(6) = 18$ of the mean 50. The values $50 - 18 = 32$ and $50 + 18 = 68$ are within three standard deviations from the mean 50. The z-scores are −3 and +3 for 32 and 68, respectively.

6.2 USING THE NORMAL DISTRIBUTION

The shaded area in the following graph indicates the area to the right of x. This area is represented by the probability $P(X > x)$. Normal tables provide the probability between the mean, zero for the standard normal distribution, and a specific value such as x_1. This is the unshaded part of the graph from the mean to x_1.

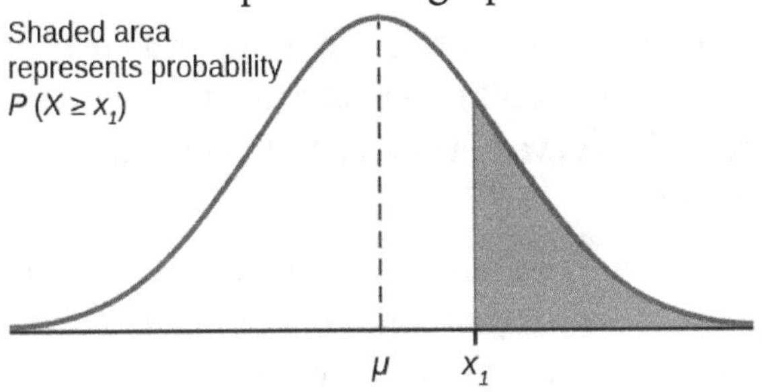

Figure 6.4

Because the normal distribution is symmetrical , if x1x1 were the same distance to the left of the mean the area, probability, in the left tail, would be the same as the shaded area in the right tail. Also, bear in mind that because of the symmetry of this distribution, one-half of the probability is to the right of the mean and one-half is to the left of the mean.

Calculations of Probabilities

To find the probability for probability density functions with a continuous random variable we need to calculate the area under the function across the values of X we are interested in. For the normal distribution this seems a difficult task given the complexity of the formula. There is, however, a simply way to get what we want. Here again is the formula for the normal distribution:

$$f(x) = \frac{1}{\sigma \cdot \sqrt{2 \cdot \pi}} \cdot e^{-\frac{1}{2} \cdot \left(\frac{x-\mu}{\sigma}\right)^2}$$

Looking at the formula for the normal distribution it is not clear just how we are going to solve for the probability doing it the same way we did it with the previous probability functions. There we put the data into the formula and did the math.

To solve this puzzle we start knowing that the area under a probability density function is the probability.

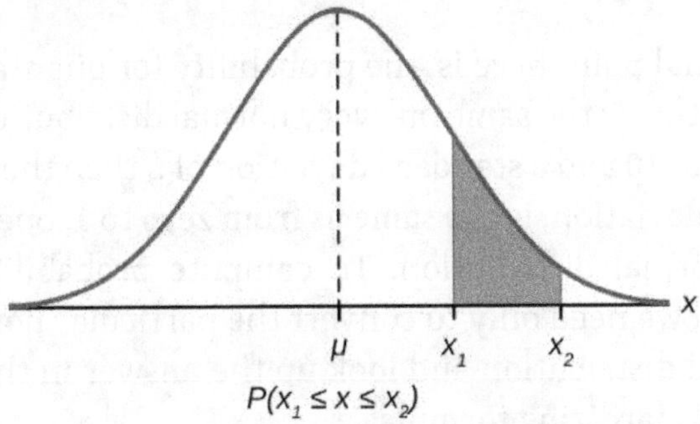

Figure 6.5

This shows that the area between X_1 and X_2 is the probability as stated in the formula: P $(X_1 \le x \le X_2)$

The mathematical tool needed to find the area under a curve is integral calculus. The integral of the normal probability density function between the two points x_1 and x_2 is the area under the curve between these two points and is the probability between these two points.

Doing these integrals is no fun and can be very time consuming. But now, remembering that there are an infinite number of normal distributions out there, we can consider the one with a mean of zero and a standard deviation of 1. This particular normal distribution is given the name Standard Normal Distribution. Putting these values into the formula it reduces to a very simple equation. We can now quite easily calculate all probabilities for any value of x, for this particular normal distribution, that has a mean of zero and a standard deviation of 1. These have been produced and are available here in the appendix to the text or everywhere on the web. They are presented in various ways. The table in this text is the most common presentation and is set up with probabilities for one-half the distribution beginning with zero, the mean, and moving outward. The shaded area in the graph at the top of the table in Statistical Tables represents the probability from zero to the specific Z value noted on the horizontal axis, Z.

The only problem is that even with this table, it would be a ridiculous coincidence that our data had a mean of zero and a standard deviation of one. The solution is to convert the distribution we have with its mean and standard deviation to this new Standard Normal Distribution. The Standard Normal has a random variable called Z.

Using the standard normal table, typically called the normal table, to find the probability of one standard deviation, go to the Z column, reading down to 1.0 and then read at column 0. That number, 0.3413 is the probability from zero to 1 standard deviation. At the top of the table is the shaded area in the distribution which is the probability for one standard deviation. The table has solved our integral calculus problem. But only if our data has a mean of zero and a standard deviation of 1.

However, the essential point here is, the probability for one standard deviation on one normal distribution is the same on every normal distribution. If the population data set has a mean of 10 and a standard deviation of 5 then the probability from 10 to 15, one standard deviation, is the same as from zero to 1, one standard deviation on the standard normal distribution. To compute probabilities, areas, for any normal distribution, we need only to convert the particular normal distribution to the standard normal distribution and look up the answer in the tables. As review, here again is the standardizing formula:

$$Z = \frac{x - \mu}{\sigma}$$

where Z is the value on the standard normal distribution, X is the value from a normal distribution one wishes to convert to the standard normal, μ and σ are, respectively, the mean and standard deviation of that population. Note that the equation uses μ and σ which denotes population parameters. This is still dealing with probability so we always are dealing with the population, with known parameter values and a known distribution. It is also important to note that because the normal distribution is symmetrical it does not matter if the z-score is positive or negative when calculating a probability. One standard deviation to the left (negative Z-score) covers the same area as one standard deviation to the right (positive Z-score). This fact is why the Standard Normal tables do not provide areas for the left side of the distribution. Because of this symmetry, the Z-score formula is sometimes written as:

$$Z = \frac{|x - \mu|}{\sigma}$$

Where the vertical lines in the equation means the absolute value of the number.

What the standardizing formula is really doing is computing the number of standard deviations X is from the mean of its own distribution. The standardizing formula and the concept of counting standard deviations from the mean is the secret of all that we will do in this statistics class. The reason this is true is that **all** of statistics boils down to variation, and the counting of standard deviations is a measure of variation.

This formula, in many disguises, will reappear over and over throughout this course.

EXAMPLE 6.3

The final exam scores in a statistics class were normally distributed with a mean of 63 and a standard deviation of five.

a. Find the probability that a randomly selected student scored more than 65 on the exam.
b. Find the probability that a randomly selected student scored less than 85.

Solution 6.3

a. Let X = a score on the final exam. X ~ $N(63, 5)$, where $\mu = 63$ and $\sigma = 5$.

Draw a graph.

Then, find $P(x > 65)$.

$P(x > 65) = 0.3446$

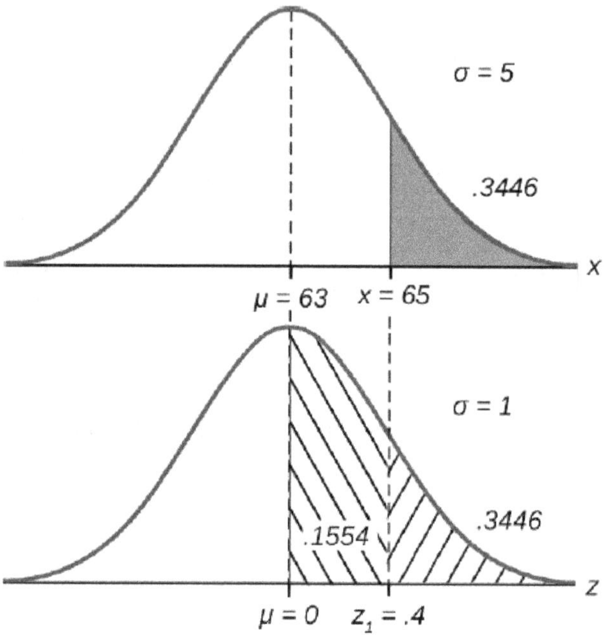

Figure 6.6

$$Z_1 = \frac{x_1 - \mu}{\sigma} = \frac{65 - 63}{5} = 0.4$$

$P(x \geq x1) = P(Z \geq Z1) = 0.3446$

The probability that any student selected at random scores more than 65 is 0.3446. Here is how we found this answer.

The normal table provides probabilities from zero to the value Z_1. For this problem the question can be written as: $P(X \geq 65) = P(Z \geq Z_1)$, which is the area in the tail. To find this area the formula would be $0.5 - P(X \leq 65)$. One half of the probability is above the mean value because this is a symmetrical distribution. The graph shows how to find the area in the tail by subtracting that portion from the mean, zero, to the Z_1 value. The final answer is: $P(X \geq 63) = P(Z \geq 0.4) = 0.3446$

$$z = \frac{65 - 63}{5} = 0.4$$

Area to the left of Z1 to the mean of zero is 0.1554

$P(x > 65) = P(z > 0.4) = 0.5 - 0.1554 = 0.3446$

Solution 6.3

b. $Z = \frac{x - \mu}{\sigma} = \frac{85 - 63}{5} = 4.4$ which is larger than the maximum value on the Standard Normal Table. Therefore, the probability that one student scores less than 85 is approximately one or 100%.

A score of 85 is 4.4 standard deviations from the mean of 63 which is beyond the range of the standard normal table. Therefore, the probability that one student scores less than 85 is approximately one (or 100%).

TRY IT 6.3

The golf scores for a school team were normally distributed with a mean of 68 and a standard deviation of three.

Find the probability that a randomly selected golfer scored less than 65.

EXAMPLE 6.4

A personal computer is used for office work at home, research, communication, personal finances, education, entertainment, social networking, and a myriad of other things. Suppose that the average number of hours a household personal computer is used for entertainment is two hours per day. Assume the times for entertainment are normally distributed and the standard deviation for the times is half an hour.

a. Find the probability that a household personal computer is used for entertainment between 1.8 and 2.75 hours per day.

Solution 6.4

a. Let X = the amount of time (in hours) a household personal computer is used for entertainment. $X \sim N(2, 0.5)$ where $\mu = 2$ and $\sigma = 0.5$.

Find $P(1.8 < x < 2.75)$.

The probability for which you are looking is the area between $x = 1.8$ and $x = 2.75$. $P(1.8 < x < 2.75) = 0.5886$

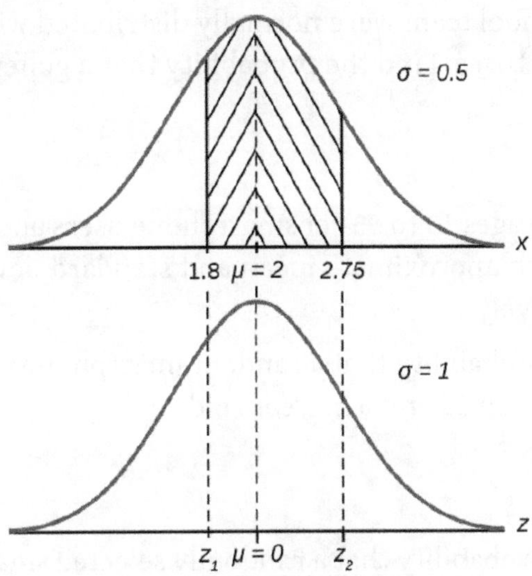

Figure 6.7

$P(1.8 \le x \le 2.75) = P(Zi \le Z \le Z2)$

The probability that a household personal computer is used between 1.8 and 2.75 hours per day for entertainment is 0.5886.

b. Find the maximum number of hours per day that the bottom quartile of households uses a personal computer for entertainment.

Solution 6.4

b. To find the maximum number of hours per day that the bottom quartile of households uses a personal computer for entertainment, find the 25th percentile, k, where $P(x < k) = 0.25$.

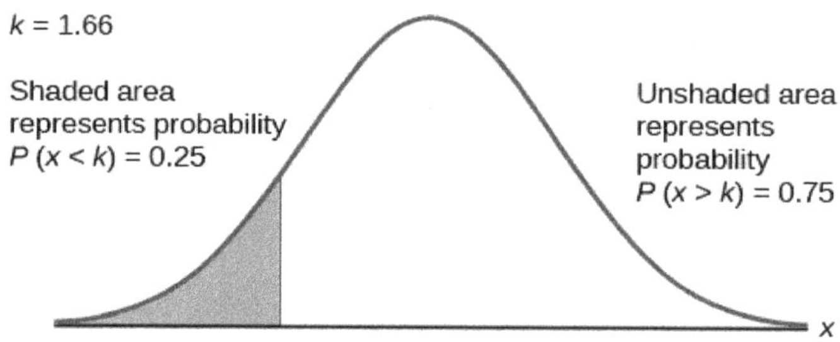

Figure 6.8

$f(Z) = 0.5 - 0.25 = 0.25$, *therefore* $Z \approx -0.675$ *(or just 0.67 using the table* $z = \frac{x-\mu}{\sigma} = \frac{x-2}{0.5} = -0.675$, *therefore* $x = -0.675 * 0.5 + 2 = 1.66$ *hours.*

The maximum number of hours per day that the bottom quartile of households uses a personal computer for entertainment is 1.66 hours.

TRY IT 6.4

The golf scores for a school team were normally distributed with a mean of 68 and a standard deviation of three. Find the probability that a golfer scored between 66 and 70.

EXAMPLE 6.5

In the United States the ages 13 to 55+ of smartphone users approximately follow a normal distribution with approximate mean and standard deviation of 36.9 years and 13.9 years, respectively.

a. Determine the probability that a random smartphone user in the age range 13 to 55+ is between 23 and 64.7 years old.

Solution 6.5

a. 0.8186
b. Determine the probability that a randomly selected smartphone user in the age range 13 to 55+ is at most 50.8 years old.

Solution 6.5

b. 0.8413

EXAMPLE 6.6

A citrus farmer who grows mandarin oranges finds that the diameters of mandarin oranges harvested on his farm follow a normal distribution with a mean diameter of 5.85 cm and a standard deviation of 0.24 cm.

a. Find the probability that a randomly selected mandarin orange from this farm has a diameter larger than 6.0 cm. Sketch the graph.

Solution 6.6

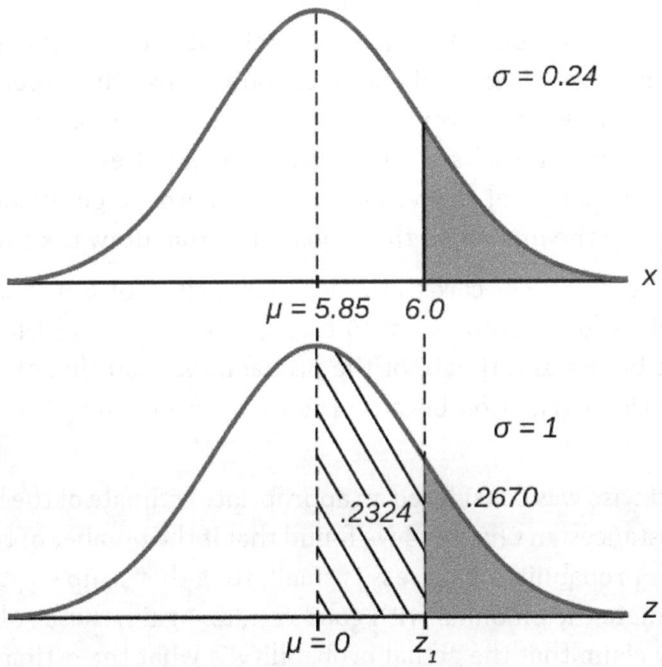

Figure 6.9

$$Z_1 = \frac{6 - 5.85}{.24} = .625$$

$P(x \geq 6) = P(z \geq 0.625) = 0.2670$

b. The middle 20% of mandarin oranges from this farm have diameters between ______ and _____.

Solution 6.6

$$f(Z) = \frac{0.20}{2} = 0.10 \text{ , therefore } Z \approx \pm 0.25$$

$$Z = \frac{x - \mu}{\sigma} = \frac{x - 5.85}{0.24} = \pm 0.25 \rightarrow \pm 0.25 * 0.24 + 5.85 = (5.79, 5.91)$$

6.3 ESTIMATING THE BINOMIAL WITH THE NORMAL DISTRIBUTION

We found earlier that various probability density functions are the limiting distributions of others; thus, we can estimate one with another under certain circumstances. We will find here that the normal distribution can be used to estimate a binomial process. The Poisson was used to estimate the binomial previously, and the binomial was used to estimate the hypergeometric distribution.

In the case of the relationship between the hypergeometric distribution and the binomial, we had to recognize that a binomial process assumes that the probability of a success remains constant from trial to trial: a head on the last flip cannot have an effect on the probability of a head on the next flip. In the hypergeometric distribution this is the essence of the question because the experiment assumes that any "draw" is without replacement. If one draws without replacement, then all subsequent "draws" are conditional probabilities. We found that if the hypergeometric experiment draws only a small percentage of the total objects, then we can ignore the impact on the probability from draw to draw.

Imagine that there are 312 cards in a deck comprised of 6 normal decks. If the experiment called for drawing only 10 cards, less than 5% of the total, than we will accept the binomial estimate of the probability, even though this is actually a hypergeometric distribution because the cards are presumably drawn without replacement.

The Poisson likewise was considered an appropriate estimate of the binomial under certain circumstances. In Chapter 4 we found that if the number of trials of interest is large and the probability of success is small, such that $\mu=np < 7$, the Poisson can be used to estimate the binomial with good results. Again, these rules of thumb do not in any way claim that the actual probability is what the estimate determines, only that the difference is in the third or fourth decimal and is thus *de minimus*.

Here, again, we find that the normal distribution makes particularly accurate estimates of a binomial process under certain circumstances. Figure 6.10 is a frequency distribution of a binomial process for the experiment of flipping three coins where the random variable is the number of heads. The sample space is listed below the distribution. The experiment assumed that the probability of a success is 0.5; the probability of a failure, a tail, is thus also 0.5. In observing Figure 6.10 we are struck by the fact that the distribution is symmetrical. The root of this result is that the probabilities of success and failure are the same, 0.5. If the probability of success were smaller than 0.5, the distribution becomes skewed right. Indeed, as the probability of success diminishes, the degree of skewness increases. If the probability of success increases from 0.5, then the skewness increases in the lower tail, resulting in a left-skewed distribution.

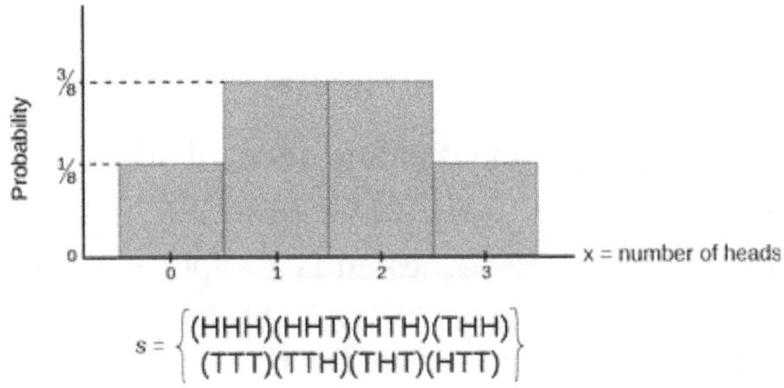

Figure 6.10

The reason the skewness of the binomial distribution is important is because if it is to be estimated with a normal distribution, then we need to recognize that the normal distribution is symmetrical. The closer the underlying binomial distribution is to being symmetrical, the better the estimate that is produced by the normal distribution. Figure 6.11 shows a symmetrical normal distribution transposed on a graph of a binomial distribution where p = 0.2 and n = 5. The discrepancy between the estimated probability using a normal distribution and the probability of the original binomial distribution is apparent. The criteria for using a normal distribution to estimate a binomial thus addresses this problem by requiring BOTH np AND $n(1 - p)$ are greater than five. Again, this is a rule of thumb, but is effective and results in acceptable estimates of the binomial probability.

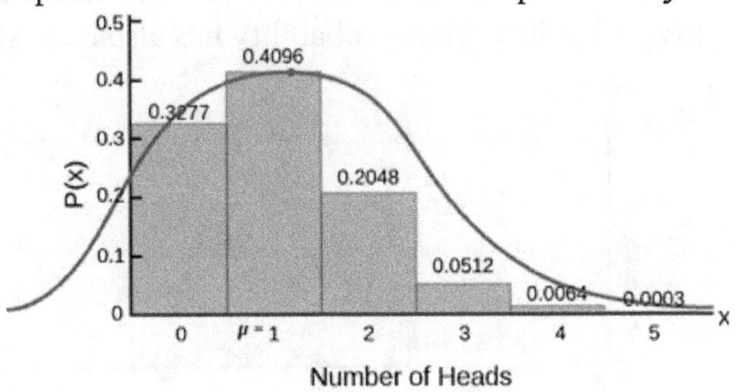

Figure 6.11

EXAMPLE 6.7

Imagine that it is known that only 10% of Australian Shepherd puppies are born with what is called "perfect symmetry" in their three colors, black, white, and copper. Perfect symmetry is defined as equal coverage on all parts of the dog when looked at in the face and measuring left and right down the centerline. A kennel would have a good reputation for breeding Australian Shepherds if they had a high percentage of dogs that met this criterion. During the past 5 years and out of the 100 dogs born to Dundee Kennels, 16 were born with this coloring characteristic.

What is the probability that, in 100 births, more than 16 would have this characteristic?

Solution 6.7

If we assume that one dog's coloring is independent of other dogs' coloring, a bit of a brave assumption, this becomes a classic binomial probability problem.

The statement of the probability requested is 1 - $[p(X = 0) + p(X = 1) + p(X = 2) + ... + p(X = 16)]$. This requires us to calculate 17 binomial formulas and add them together and then subtract from one to get the right hand part of the distribution. Alternatively, we can use the normal distribution to get an acceptable answer and in much less time.

First, we need to check if the binomial distribution is symmetrical enough to use the normal distribution. We know that the binomial for this problem is skewed because the probability of success, 0.1, is not the same as the probability of failure, 0.9. Nevertheless, both np = 10 and n(I–p)= 90 are larger than 5, the cutoff for using the normal distribution to estimate the binomial.

Figure 6.11 below shows the binomial distribution and marks the area we wish to know. The mean of the binomial, 10, is also marked, and the standard deviation is written on the side of the graph: $\sigma = \sqrt{npq} = 3$.

The area under the distribution from zero to 16 is the probability requested, and has been shaded in. Below the binomial distribution is a normal distribution to be used to estimate this probability. That probability has also been shaded.

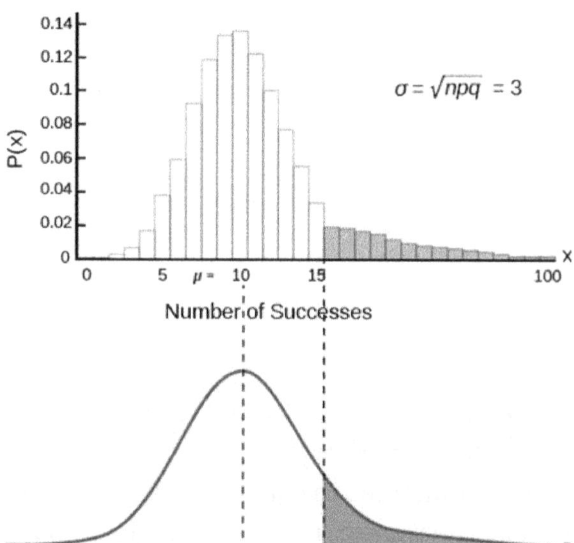

$$z_1 = \frac{X - \mu}{\sigma}$$

$$z_1 = \frac{16 - 10}{3} = 2$$

Figure 6.12

Standardizing from the binomial to the normal distribution as done in the past shows where we are asking for the probability from 16 to positive infinity, or 100 in this case. We need to calculate the number of standard deviations 16 is away from the mean: 10.

$$Z = \frac{x - \mu}{\sigma} = \frac{16 - 10}{3} = 2$$

We are asking for the probability beyond two standard deviations, a very unlikely event. We look up two standard deviations in the standard normal table and find the area from zero to two standard deviations is 0.4772. We are interested in the tail, however, so we subtract 0.4772 from 0.5 and thus find the area in the tail. Our conclusion is the probability of a kennel having 16 dogs with "perfect symmetry" is 0.0228. Dundee Kennels has an extraordinary record in this regard.

Mathematically, we write this as:

$$1 - [p(X = 0) + p(X = 1) + p(X = 2) + \ \dots \ + p(X = 16)] = p(X > 16) = p(Z > 2) = 0.0228$$

Key Terms

Normal Distribution a continuous random variable (RV) with $f(x) = \frac{1}{\sigma\sqrt{2\pi}} e^{\frac{-(x-\mu)^2}{2\sigma^2}}$, where μ is the mean of the distribution and σ is the standard deviation; notation: $X \sim N(\mu, \sigma)$. If $\mu = 0$ and $\sigma = 1$, the RV, Z, is called the standard normal distribution.

Standard Normal Distribution a continuous random variable (RV) $X \sim N(0, 1)$; when X follows the standard normal distribution, it is often noted as $Z \sim N(0, 1)$.

z-score the linear transformation of the form $z = \frac{x - \mu}{\sigma}$ or written as $z = \frac{|x - \mu|}{\sigma}$; if this transformation is applied to any normal distribution $X \sim N(\mu, \sigma)$ the result is the standard normal distribution $Z \sim N(0,1)$. If this transformation is applied to any specific value x of the RV with mean μ and standard deviation σ, the result is called the z-score of x. The z-score allows us to compare data that are normally distributed but scaled differently. A z-score is the number of standard deviations a particular x is away from its mean value.

CHAPTER REVIEW

6.1 The Standard Normal Distribution

A z-score is a standardized value. Its distribution is the standard normal, $Z \sim N(0, 1)$. The mean of the z-scores is zero and the standard deviation is one. If z is the z-score for a value x from the normal distribution $N(\mu, \sigma)$ then z tells you how many standard deviations x is above (greater than) or below (less than) μ.

6.3 Estimating the Binomial with the Normal Distribution

The normal distribution, which is continuous, is the most important of all the probability distributions. Its graph is bell-shaped. This bell-shaped curve is used in almost all disciplines. Since it is a continuous distribution, the total area under the curve is one. The parameters of the normal are the mean μ and the standard deviation σ. A special normal distribution, called the standard normal distribution is the distribution of *z*-scores. Its mean is zero, and its standard deviation is one.

FORMULA REVIEW

6.0 Introduction

$X \sim N(\mu, \sigma)$

μ = the mean; σ = the standard deviation

6.1 The Standard Normal Distribution

$Z \sim N(0, 1)$

z = a standardized value (*z*-score)

mean = 0; standard deviation = 1

To find the kth percentile of X when the *z*-scores is known: $x = \mu + (z)\sigma$

z-score: $z = \frac{x - \mu}{\sigma}$ or $z = \frac{|x - \mu|}{\sigma}$

Z = the random variable for *z*-scores

$Z \sim N(0, 1)$

6.3 Estimating the Binomial with the Normal Distribution

Normal Distribution: $X \sim N(\mu, \sigma)$ where μ is the mean and σ is the standard deviation.

Standard Normal Distribution: $Z \sim N(0, 1)$.

PRACTICE

6.1 The Standard Normal Distribution

1. A bottle of water contains 12.05 fluid ounces with a standard deviation of 0.01 ounces. Define the random variable X in words. X = ____________.
2. A normal distribution has a mean of 61 and a standard deviation of 15. What is the median?
3. $X \sim N(1, 2)$

   σ = _______
4. A company manufactures rubber balls. The mean diameter of a ball is 12 cm with a standard deviation of 0.2 cm. Define the random variable X in words. X = ______________.
5. $X \sim N(-4, 1)$

What is the median?

6. $X \sim N(3, 5)$

   σ = _______

7. $X \sim N(-2, 1)$

   μ = _______

8. What does a z-score measure?
9. What does standardizing a normal distribution do to the mean?
10. Is $X \sim N(0, 1)$ a standardized normal distribution? Why or why not?
11. What is the z-score of $x = 12$, if it is two standard deviations to the right of the mean?
12. What is the z-score of $x = 9$, if it is 1.5 standard deviations to the left of the mean?
13. What is the z-score of $x = -2$, if it is 2.78 standard deviations to the right of the mean?
14. What is the z-score of $x = 7$, if it is 0.133 standard deviations to the left of the mean?
15. Suppose $X \sim N(2, 6)$. What value of x has a z-score of three?
16. Suppose $X \sim N(8, 1)$. What value of x has a z-score of –2.25?
17. Suppose $X \sim N(9, 5)$. What value of x has a z-score of –0.5?
18. Suppose $X \sim N(2, 3)$. What value of x has a z-score of –0.67?
19. Suppose $X \sim N(4, 2)$. What value of x is 1.5 standard deviations to the left of the mean?
20. Suppose $X \sim N(4, 2)$. What value of x is two standard deviations to the right of the mean?
21. Suppose $X \sim N(8, 9)$. What value of x is 0.67 standard deviations to the left of the mean?
22. Suppose $X \sim N(-1, 2)$. What is the z-score of $x = 2$?
23. Suppose $X \sim N(12, 6)$. What is the z-score of $x = 2$?
24. Suppose $X \sim N(9, 3)$. What is the z-score of $x = 9$?
25. Suppose a normal distribution has a mean of six and a standard deviation of 1.5. What is the z-score of $x = 5.5$?
26. In a normal distribution, $x = 5$ and $z = -1.25$. This tells you that $x = 5$ is ____ standard deviations to the ____ (right or left) of the mean.
27. In a normal distribution, $x = 3$ and $z = 0.67$. This tells you that $x = 3$ is ____ standard deviations to the ____ (right or left) of the mean.

28. In a normal distribution, $x = -2$ and $z = 6$. This tells you that $x = -2$ is ____ standard deviations to the ____ (right or left) of the mean.

29. In a normal distribution, $x = -5$ and $z = -3.14$. This tells you that $x = -5$ is ____ standard deviations to the ____ (right or left) of the mean.

30. In a normal distribution, $x = 6$ and $z = -1.7$. This tells you that $x = 6$ is ____ standard deviations to the ____ (right or left) of the mean.

31. About what percent of x values from a normal distribution lie within one standard deviation (left and right) of the mean of that distribution?

32. About what percent of the x values from a normal distribution lie within two standard deviations (left and right) of the mean of that distribution?

33. About what percent of x values lie between the second and third standard deviations (both sides)?

34. Suppose $X \sim N(15, 3)$. Between what x values does 68.27% of the data lie? The range of x values is centered at the mean of the distribution (i.e., 15).

35. Suppose $X \sim N(-3, 1)$. Between what x values does 95.45% of the data lie? The range of x values is centered at the mean of the distribution(i.e., –3).

36. Suppose $X \sim N(-3, 1)$. Between what x values does 34.14% of the data lie?

37. About what percent of x values lie between the mean and three standard deviations?

38. About what percent of x values lie between the mean and one standard deviation?

39. About what percent of x values lie between the first and second standard deviations from the mean (both sides)?

40. About what percent of x values lie between the first and third standard deviations(both sides)?

Use the following information to answer the next two exercises: The life of Sunshine CD players is normally distributed with mean of 4.1 years and a standard deviation of 1.3 years. A CD player is guaranteed for three years. We are interested in the length of time a CD player lasts.

41. Define the random variable X in words. X = _______________.

42. $X \sim$ _____(_____,_____)

6.3 Estimating the Binomial with the Normal Distribution

43. How would you represent the area to the left of one in a probability statement?

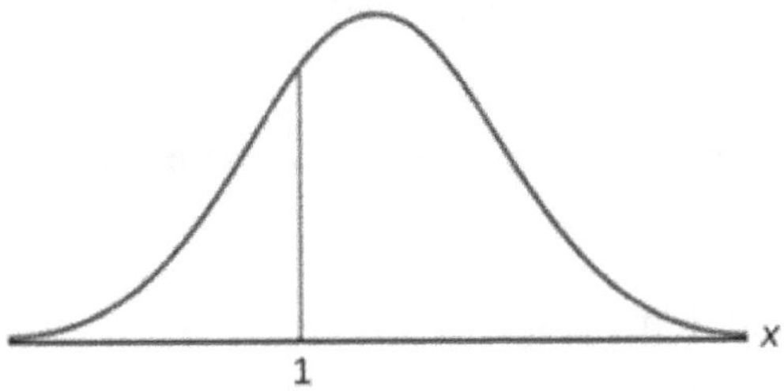

Figure 6.13

44. What is the area to the right of one?

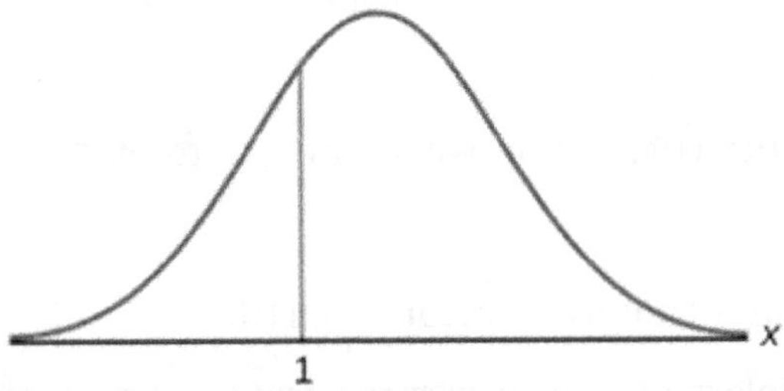

Figure 6.14

45. Is $P(x < 1)$ equal to $P(x \leq 1)$? Why?

46. How would you represent the area to the left of three in a probability statement?

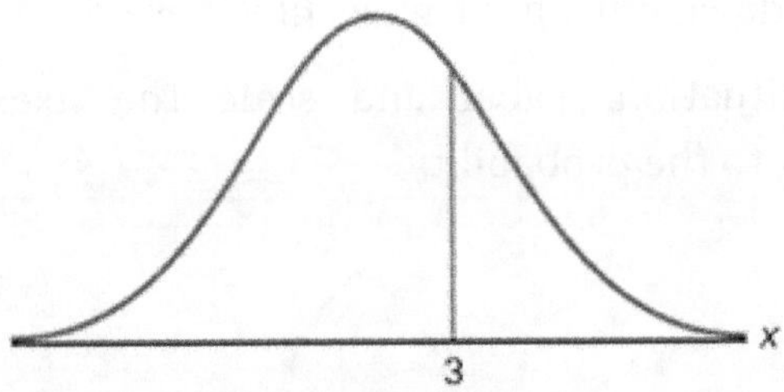

Figure 6.15

47. What is the area to the right of three?

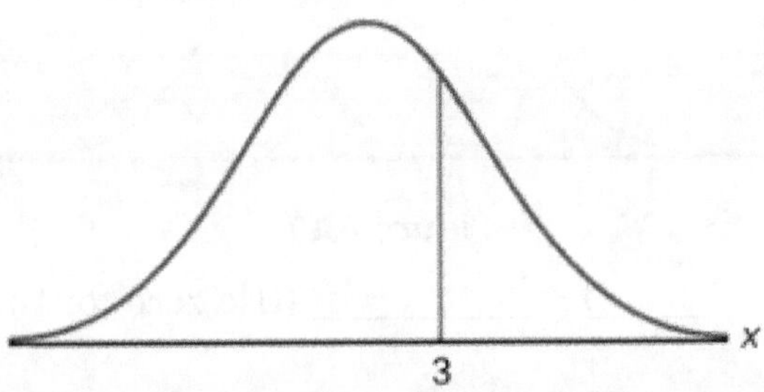

Figure 6.16

48. If the area to the left of x in a normal distribution is 0.123, what is the area to the right of x?

49. If the area to the right of x in a normal distribution is 0.543, what is the area to the left of x?

Use the following information to answer the next four exercises:

$X \sim N(54, 8)$

50. Find the probability that $x > 56$.

51. Find the probability that $x < 30$.

52. $X \sim N(6, 2)$

Find the probability that x is between three and nine.

53. $X \sim N(-3, 4)$

Find the probability that x is between one and four.

54. $X \sim N(4, 5)$

Find the maximum of x in the bottom quartile.

55. *Use the following information to answer the next three exercise:* The life of Sunshine CD players is normally distributed with a mean of 4.1 years and a standard deviation of 1.3 years. A CD player is guaranteed for three years. We are interested in the length of time a CD player lasts. Find the probability that a CD player will break down during the guarantee period.

a. Sketch the situation. Label and scale the axes. Shade the region corresponding to the probability.

Figure 6.17

b. $P(0 < x <$ ____________$) =$ ____________ (Use zero for the minimum value of x.)

56. Find the probability that a CD player will last between 2.8 and six years.

a. Sketch the situation. Label and scale the axes. Shade the region corresponding to the probability.

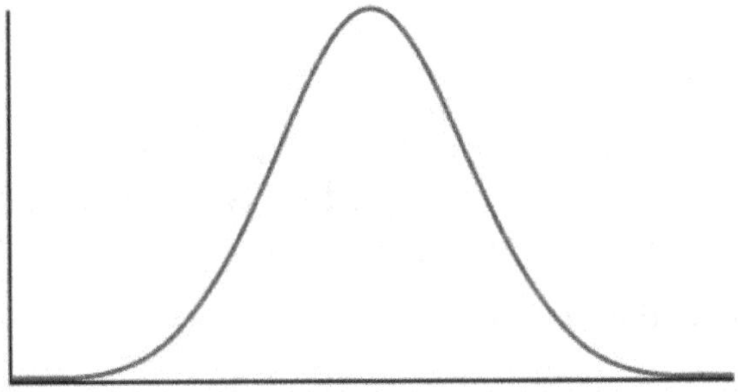

Figure 6.18

b. $P(\_\_\_\_\_\_\_\_\_\_ < x < \_\_\_\_\_\_\_\_\_\_) = \_\_\_\_\_\_\_\_\_\_$

57. An experiment with a probability of success given as 0.40 is repeated 100 times. Use the normal distribution to approximate the binomial distribution, and find the probability the experiment will have at least 45 successes.

58. An experiment with a probability of success given as 0.30 is repeated 90 times. Use the normal distribution to approximate the binomial distribution, and find the probability the experiment will have at least 22 successes.

59. An experiment with a probability of success given as 0.40 is repeated 100 times. Use the normal distribution to approximate the binomial distribution, and find the probability the experiment will have from 35 to 45 successes.

60. An experiment with a probability of success given as 0.30 is repeated 90 times. Use the normal distribution to approximate the binomial distribution, and find the probability the experiment will have from 26 to 30 successes.

61. An experiment with a probability of success given as 0.40 is repeated 100 times. Use the normal distribution to approximate the binomial distribution, and find the probability the experiment will have at most 34 successes.

62. An experiment with a probability of success given as 0.30 is repeated 90 times. Use the normal distribution to approximate the binomial distribution, and find the probability the experiment will have at most 34 successes.

63. A multiple choice test has a probability any question will be guesses correctly of 0.25. There are 100 questions, and a student guesses at all of them. Use the normal distribution to approximate the binomial distribution, and determine the probability at least 30, but no more than 32, questions will be guessed correctly.

64. A multiple choice test has a probability any question will be guesses correctly of 0.25. There are 100 questions, and a student guesses at all of them. Use the normal distribution to approximate the binomial distribution, and determine the probability at least 24, but no more than 28, questions will be guessed correctly.

HOMEWORK

6.1 The Standard Normal Distribution

Use the following information to answer the next two exercises: The patient recovery time from a particular surgical procedure is normally distributed with a mean of 5.3 days and a standard deviation of 2.1 days.

65. What is the median recovery time?

a. 2.7 b. 5.3
c. 7.4 d. 2.1

66. What is the *z*-score for a patient who takes ten days to recover?

a. 1.5 b. 0.2
c. 2.2 d. 7.3

67. The length of time to find a parking space at 9 A.M. follows a normal distribution with a mean of five minutes and a standard deviation of two minutes. If the mean is significantly greater than the standard deviation, which of the following statements is true?

i. The data cannot follow the uniform distribution.
ii. The data cannot follow the exponential distribution..
iii. The data cannot follow the normal distribution.

a. I only b. II only
c. III only d. I, II, and III

68. The heights of the 430 National Basketball Association players were listed on team rosters at the start of the 2005–2006 season. The heights of basketball players have an approximate normal distribution with mean, $\mu = 79$ inches and a standard deviation, $\sigma = 3.89$ inches. For each of the following heights, calculate the *z*-score and interpret it using complete sentences.

a. 77 inches
b. 85 inches
c. If an NBA player reported his height had a *z*-score of 3.5, would you believe him? Explain your answer.

69. The systolic blood pressure (given in millimeters) of males has an approximately normal distribution with mean $\mu = 125$ and standard deviation $\sigma = 14$. Systolic blood pressure for males follows a normal distribution.

a. Calculate the *z*-scores for the male systolic blood pressures 100 and 150 millimeters.
b. If a male friend of yours said he thought his systolic blood pressure was 2.5 standard deviations below the mean, but that he believed his blood pressure was between 100 and 150 millimeters, what would you say to him?

70. Kyle's doctor told him that the z-score for his systolic blood pressure is 1.75. Which of the following is the best interpretation of this standardized score? The systolic blood pressure (given in millimeters) of males has an approximately normal distribution with mean $\mu = 125$ and standard deviation $\sigma = 14$. If X = a systolic blood pressure score then $X \sim N(125, 14)$.

   a. Which answer(s) is/are correct?
      i. Kyle's systolic blood pressure is 175.
      ii. Kyle's systolic blood pressure is 1.75 times the average blood pressure of men his age.
      iii. Kyle's systolic blood pressure is 1.75 above the average systolic blood pressure of men his age.
      iv. Kyles's systolic blood pressure is 1.75 standard deviations above the average systolic blood pressure for men.
   b. Calculate Kyle's blood pressure.

71. Height and weight are two measurements used to track a child's development. The World Health Organization measures child development by comparing the weights of children who are the same height and the same gender. In 2009, weights for all 80 cm girls in the reference population had a mean $\mu = 10.2$ kg and standard deviation $\sigma = 0.8$ kg. Weights are normally distributed. $X \sim N(10.2, 0.8)$. Calculate the z-scores that correspond to the following weights and interpret them.

   a. 11 kg
   b. 7.9 kg
   c. 12.2 kg

72. In 2005, 1,475,623 students heading to college took the SAT. The distribution of scores in the math section of the SAT follows a normal distribution with mean $\mu = 520$ and standard deviation $\sigma = 115$.

   a. Calculate the z-score for an SAT score of 720. Interpret it using a complete sentence.
   b. What math SAT score is 1.5 standard deviations above the mean? What can you say about this SAT score?
   c. For 2012, the SAT math test had a mean of 514 and standard deviation 117. The ACT math test is an alternate to the SAT and is approximately normally distributed with mean 21 and standard deviation 5.3. If one person took the SAT math test and scored 700 and a second person took the ACT math test and scored 30, who did better with respect to the test they took?

6.3 Estimating the Binomial with the Normal Distribution

Use the following information to answer the next two exercises: The patient recovery time from a particular surgical procedure is normally distributed with a mean of 5.3 days and a standard deviation of 2.1 days.

73. What is the probability of spending more than two days in recovery?

a. 0.0580
b. 0.8447
c. 0.0553
d. 0.9420

Use the following information to answer the next three exercises: The length of time it takes to find a parking space at 9 A.M. follows a normal distribution with a mean of five minutes and a standard deviation of two minutes.

74. Based upon the given information and numerically justified, would you be surprised if it took less than one minute to find a parking space?

a. Yes
b. No
c. Unable to determine

75. Find the probability that it takes at least eight minutes to find a parking space.

a. 0.0001
b. 0.9270
c. 0.1862
d. 0.0668

76. Seventy percent of the time, it takes more than how many minutes to find a parking space?

a. 1.24
b. 2.41
c. 3.95
d. 6.05

77. According to a study done by De Anza students, the height for Asian adult males is normally distributed with an average of 66 inches and a standard deviation of 2.5 inches. Suppose one Asian adult male is randomly chosen. Let *X* = height of the individual.

a. *X* ~ _____(_____,_____)
b. Find the probability that the person is between 65 and 69 inches. Include a sketch of the graph, and write a probability statement.
c. Would you expect to meet many Asian adult males over 72 inches? Explain why or why not, and justify your answer numerically.
d. The middle 40% of heights fall between what two values? Sketch the graph, and write the probability statement.

78. IQ is normally distributed with a mean of 100 and a standard deviation of 15. Suppose one individual is randomly chosen. Let X = IQ of an individual.

a. X ~ _____(_____,_____)
b. Find the probability that the person has an IQ greater than 120. Include a sketch of the graph, and write a probability statement.
c. MENSA is an organization whose members have the top 2% of all IQs. Find the minimum IQ needed to qualify for the MENSA organization. Sketch the graph, and write the probability statement.

79. The percent of fat calories that a person in America consumes each day is normally distributed with a mean of about 36 and a standard deviation of 10. Suppose that one individual is randomly chosen. Let X = percent of fat calories.

a. X ~ _____(_____,_____)
b. Find the probability that the percent of fat calories a person consumes is more than 40. Graph the situation. Shade in the area to be determined.
c. Find the maximum number for the lower quarter of percent of fat calories. Sketch the graph and write the probability statement.

80. Suppose that the distance of fly balls hit to the outfield (in baseball) is normally distributed with a mean of 250 feet and a standard deviation of 50 feet.

a. If X = distance in feet for a fly ball, then X ~ _____(_____,_____)
b. If one fly ball is randomly chosen from this distribution, what is the probability that this ball traveled fewer than 220 feet? Sketch the graph. Scale the horizontal axis X. Shade the region corresponding to the probability. Find the probability.

81. In China, four-year-olds average three hours a day unsupervised. Most of the unsupervised children live in rural areas, considered safe. Suppose that the standard deviation is 1.5 hours and the amount of time spent alone is normally distributed. We randomly select one Chinese four-year-old living in a rural area. We are interested in the amount of time the child spends alone per day.

a. In words, define the random variable X.
b. X ~ _____(_____,_____)
c. Find the probability that the child spends less than one hour per day unsupervised. Sketch the graph, and write the probability statement.
d. What percent of the children spend over ten hours per day unsupervised?
e. Seventy percent of the children spend at least how long per day unsupervised?

82. In the 1992 presidential election, Alaska's 40 election districts averaged 1,956.8 votes per district for President Clinton. The standard deviation was 572.3. (There are only 40 election districts in Alaska.) The distribution of the votes per district for President Clinton was bell-shaped. Let X = number of votes for President Clinton for an election district.

a. State the approximate distribution of X.
b. Is 1,956.8 a population mean or a sample mean? How do you know?
c. Find the probability that a randomly selected district had fewer than 1,600 votes for President Clinton. Sketch the graph and write the probability statement.
d. Find the probability that a randomly selected district had between 1,800 and 2,000 votes for President Clinton.
e. Find the third quartile for votes for President Clinton.

83. Suppose that the duration of a particular type of criminal trial is known to be normally distributed with a mean of 21 days and a standard deviation of seven days.

a. In words, define the random variable X.
b. X ~ _____(_____,_____)
c. If one of the trials is randomly chosen, find the probability that it lasted at least 24 days. Sketch the graph and write the probability statement.
d. Sixty percent of all trials of this type are completed within how many days?

84. Terri Vogel, an amateur motorcycle racer, averages 129.71 seconds per 2.5 mile lap (in a seven-lap race) with a standard deviation of 2.28 seconds. The distribution of her race times is normally distributed. We are interested in one of her randomly selected laps.

a. In words, define the random variable X.
b. X ~ _____(_____,_____)
c. Find the percent of her laps that are completed in less than 130 seconds.
d. The fastest 3% of her laps are under _____.
e. The middle 80% of her laps are from _______ seconds to _______ seconds.

85. Thuy Dau, Ngoc Bui, Sam Su, and Lan Voung conducted a survey as to how long customers at Lucky claimed to wait in the checkout line until their turn. Let X = time in line. **Table 6.1** displays the ordered real data (in minutes):

Table 6.1

0.50	4.25	5	6	7.25
1.75	4.25	5.25	6	7.25
2	4.25	5.25	6.25	7.25
2.25	4.25	5.5	6.25	7.75
2.25	4.5	5.5	6.5	8
2.5	4.75	5.5	6.5	8.25
2.75	4.75	5.75	6.5	9.5
3.25	4.75	5.75	6.75	9.5
3.75	5	6	6.75	9.75
3.75	5	6	6.75	10.75

a. Calculate the sample mean and the sample standard deviation.
b. Construct a histogram.
c. Draw a smooth curve through the midpoints of the tops of the bars.
d. In words, describe the shape of your histogram and smooth curve.
e. Let the sample mean approximate μ and the sample standard deviation approximate σ. The distribution of X can then be approximated by X ~ _____(_____,_____)
f. Use the distribution in part e to calculate the probability that a person will wait fewer than 6.1 minutes.
g. Determine the cumulative relative frequency for waiting less than 6.1 minutes.
h. Why aren't the answers to part f and part g exactly the same?
i. Why are the answers to part f and part g as close as they are?
j. If only ten customers has been surveyed rather than 50, do you think the answers to part f and part g would have been closer together or farther apart? Explain your conclusion.

86. Suppose that Ricardo and Anita attend different colleges. Ricardo's GPA is the same as the average GPA at his school. Anita's GPA is 0.70 standard deviations above her school average. In complete sentences, explain why each of the following statements may be false.

a. Ricardo's actual GPA is lower than Anita's actual GPA.
b. Ricardo is not passing because his *z*-score is zero.
c. Anita is in the 70th percentile of students at her college.

87. An expert witness for a paternity lawsuit testifies that the length of a pregnancy is normally distributed with a mean of 280 days and a standard deviation of 13 days. An alleged father was out of the country from 240 to 306 days before the birth of the child, so the pregnancy would have been less than 240 days or more than 306 days long if he was the father. The birth was uncomplicated, and the child needed no medical intervention. What is the probability that he was NOT the father? What is the probability that he could be the father? Calculate the *z*-scores first, and then use those to calculate the probability.

88. A NUMMI assembly line, which has been operating since 1984, has built an average of 6,000 cars and trucks a week. Generally, 10% of the cars were defective coming off the assembly line. Suppose we draw a random sample of n = 100 cars. Let X represent the number of defective cars in the sample. What can we say about X in regard to the 68-95-99.7 empirical rule (one standard deviation, two standard deviations and three standard deviations from the mean are being referred to)? Assume a normal distribution for the defective cars in the sample.

89. We flip a coin 100 times (n = 100) and note that it only comes up heads 20% (p = 0.20) of the time. The mean and standard deviation for the number of times the coin lands on heads is $\mu = 20$ and $\sigma = 4$ (verify the mean and standard deviation). Solve the following:

a. There is about a 68% chance that the number of heads will be somewhere between ___ and ___.
b. There is about a ____chance that the number of heads will be somewhere between 12 and 28.
c. There is about a ____ chance that the number of heads will be somewhere between eight and 32.

90. A $1 scratch off lotto ticket will be a winner one out of five times. Out of a shipment of n = 190 lotto tickets, find the probability for the lotto tickets that there are

a. somewhere between 34 and 54 prizes.
b. somewhere between 54 and 64 prizes.
c. more than 64 prizes.

91. Facebook provides a variety of statistics on its Web site that detail the growth and popularity of the site.

On average, 28 percent of 18 to 34 year olds check their Facebook profiles before getting out of bed in the morning. Suppose this percentage follows a normal distribution with a standard deviation of five percent.

92. A hospital has 49 births in a year. It is considered equally likely that a birth be a boy as it is the birth be a girl.

a. What is the mean?
b. What is the standard deviation?
c. Can this binomial distribution be approximated with a normal distribution?
d. If so, use the normal distribution to find the probability that at least 23 of the 49 births were boys.

93. Historically, a final exam in a course is passed with a probability of 0.9. The exam is given to a group of 70 students.

a. What is the mean of the binomial distribution?
b. What is the standard deviation?
c. Can this binomial distribution be approximate with a normal distribution?
d. If so, use the normal distribution to find the probability that at least 60 of the students pass the exam?

94. A tree in an orchard has 200 oranges. Of the oranges, 40 are not ripe. Use the normal distribution to approximate the binomial distribution, and determine the probability a box containing 35 oranges has at most two oranges that are not ripe.

95. In a large city one in ten fire hydrants are in need of repair. If a crew examines 100 fire hydrants in a week, what is the probability they will find nine of fewer fire hydrants that need repair? Use the normal distribution to approximate the binomial distribution.

96. On an assembly line it is determined 85% of the assembled products have no defects. If one day 50 items are assembled, what is the probability at least 4 and no more than 8 are defective. Use the normal distribution to approximate the binomial distribution.

REFERENCES

6.1 The Standard Normal Distribution

"Blood Pressure of Males and Females." StatCruch, 2013. Available online at http://www.statcrunch.com/5.0/viewreport.php?reportid=11960 (accessed May 14, 2013).

"The Use of Epidemiological Tools in Conflict-affected populations: Open-access educational resources for policy-makers: Calculation of z-scores." London School of Hygiene and Tropical Medicine, 2009. Available online at http://conflict.lshtm.ac.uk/page_125.htm (accessed May 14, 2013).

"2012 College-Bound Seniors Total Group Profile Report." CollegeBoard, 2012. Available online at http://media.collegeboard.com/digitalServices/pdf/research/TotalGroup-2012.pdf (accessed May 14, 2013).

"Digest of Education Statistics: ACT score average and standard deviations by sex and race/ethnicity and percentage of ACT test takers, by selected composite score ranges and planned fields of study: Selected years, 1995 through 2009." National Center for Education Statistics. Available online at http://nces.ed.gov/programs/digest/d09/tables/dt09_147.asp (accessed May 14, 2013).

Data from the *San Jose Mercury News*.

Data from *The World Almanac and Book of Facts*.

"List of stadiums by capacity." Wikipedia. Available online at https://en.wikipedia.org/wiki/List_of_stadiums_by_capacity (accessed May 14, 2013).

Data from the National Basketball Association. Available online at www.nba.com (accessed May 14, 2013).

6.2 Using the Normal Distribution

"Naegele's rule." Wikipedia. Available online at http://en.wikipedia.org/wiki/Naegele's_rule (accessed May 14, 2013).

"403: NUMMI." Chicago Public Media & Ira Glass, 2013. Available online at http://www.thisamericanlife.org/radio-archives/episode/403/nummi (accessed May 14, 2013).

"Scratch-Off Lottery Ticket Playing Tips." WinAtTheLottery.com, 2013. Available online at http://www.winatthelottery.com/public/department40.cfm (accessed May 14, 2013).

"Smart Phone Users, By The Numbers." Visual.ly, 2013. Available online at http://visual.ly/smart-phone-users-numbers (accessed May 14, 2013).

"Facebook Statistics." Statistics Brain. Available online at http://www.statisticbrain.com/facebook-statistics/(accessed May 14, 2013).

SOLUTIONS

1. ounces of water in a bottle

3. 2

5. –4

7. –2

9. The mean becomes zero.

11. $z = 2$

13. $z = 2.78$

15. $x = 20$

17. $x = 6.5$

19. $x = 1$

21. $x = 1.97$

23. $z = -1.67$

25. $z \approx -0.33$

27. 0.67, right

29. 3.14, left

31. about 68%

33. about 4%

35. between –5 and –1

37. about 50%

39. about 27%

41. The lifetime of a Sunshine CD player measured in years.

43. $P(x < 1)$

45. Yes, because they are the same in a continuous distribution: $P(x = 1) = 0$

47. $1 - P(x < 3)$ or $P(x > 3)$

49. $1 - 0.543 = 0.457$

51. 0.0013

53. 0.1186

55. Check student's solution.

a. 3, 0.1979

57. 0.154

58. 0.874

59. 0.693

60. 0.346

61. 0.110

62. 0.946

63. 0.071

64. 0.347

66. c

68.

a. Use the z-score formula. $z = -0.5141$. The height of 77 inches is 0.5141 standard deviations below the mean. An NBA player whose height is 77 inches is shorter than average.

b. Use the z-score formula. $z = 1.5424$. The height 85 inches is 1.5424 standard deviations above the mean. An NBA player whose height is 85 inches is taller than average.

c. Height = 79 + 3.5(3.89) = 92.615 inches, which is taller than 7 feet, 8 inches. There are very few NBA players this tall so the answer is no, not likely.

70. a. iv

b. Kyle's blood pressure is equal to 125 + (1.75)(14) = 149.5.

72. Let X = an SAT math score and Y = an ACT math score.

a. $X = 720\ \frac{720-520}{15} = 1.74$ The exam score of 720 is 1.74 standard deviations above the mean of 520.

b. $z = 1.5$

The math SAT score is 520 + 1.5(115) ≈ 692.5. The exam score of 692.5 is 1.5 standard deviations above the mean of 520.

c. $\frac{X-\mu}{\sigma} = \frac{700-514}{117} \approx 1.59$, the z-score for the SAT. $\frac{Y-\mu}{\sigma} = \frac{30-21}{5.3} \approx 1.70$, the z-scores for the ACT. With respect to the test they took, the person who took the ACT did better (has the higher z-score).

75. d

77. a. $X \sim N(66, 2.5)$

b. 0.5404

c. No, the probability that an Asian male is over 72 inches tall is 0.0082

79. a. X ~ N(36, 10)

b. The probability that a person consumes more than 40% of their calories as fat is 0.3446.

c. Approximately 25% of people consume less than 29.26% of their calories as fat.

81. *a.* X = number of hours that a Chinese four-year-old in a rural area is unsupervised during the day.

b. $X \sim N(3, 1.5)$

c. The probability that the child spends less than one hour a day unsupervised is 0.0918.

d. The probability that a child spends over ten hours a day unsupervised is less than 0.0001.

e. 2.21 hours

83. *a.* X = the distribution of the number of days a particular type of criminal trial will take

b. $X \sim N(21, 7)$

c. The probability that a randomly selected trial will last more than 24 days is 0.3336.

d. 22.77

85. a. mean = 5.51, s = 2.15

b. Check student's solution.

c. Check student's solution.

d. Check student's solution.

e. $X \sim N(5.51, 2.15)$

f. 0.6029

g. The cumulative frequency for less than 6.1 minutes is 0.64.

h. The answers to part f and part g are not exactly the same, because the normal distribution is only an approximation to the real one.

i. The answers to part f and part g are close, because a normal distribution is an excellent approximation when the sample size is greater than 30.

j. The approximation would have been less accurate, because the smaller sample size means that the data does not fit normal curve as well.

88. $n = 100; p = 0.1; q = 0.9$

$\mu = np = (100)(0.10) = 10$

$\sigma = \sqrt{npq} = \sqrt{(100)(0.1)(0.9)} = 3$

i. $z=\pm1: x_1=\mu+z\sigma=10+1(3)=13$ and $x^2=\mu-z\sigma=10-1(3)=7.68$ of the defective cars will fall between seven and 13.

ii. $z=\pm2: x_1=\mu+z\sigma=10+2(3)=16$ and $x^2=\mu-z\sigma=10-2(3)=4$. 95 of the defective cars will fall between four and 16

iii. $z=\pm3: x_1=\mu+z\sigma=10+3(3)=19$ and $x^2=\mu-z\sigma=10-3(3)=1$. 99.7% of the defective cars will fall between one and 19.

90. $n = 190; p = \frac{1}{5} = 0.2; q = 0.8$

$\mu = np = (190)(0.2) = 38$

$\sigma = \sqrt{npq} = \sqrt{(190)(0.2)(0.8)} = 5.5136$

a. For this problem: $P(34 < x < 54) = 0.7641$

b. For this problem: $P(54 < x < 64) = 0.0018$

c. For this problem: $P(x > 64) = 0.0000012$ (approximately 0)

92. a. 24.5 b. 3.5

c. Yes d. 0.67

93. a. 63 b. 2.5

c. Yes d. 0.88

94. 0.02

95. 0.37

96. 0.50

7 Chapter THE CENTRAL LIMIT THEOREM: MOST POWERFUL AND USEFUL IDEAS IN ALL OF STATISTICS

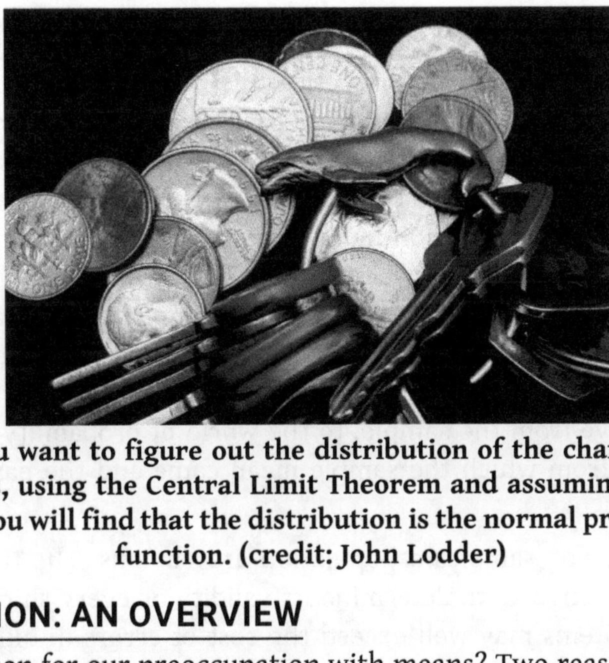

Figure 7.1 If you want to figure out the distribution of the change people carry in their pockets, using the Central Limit Theorem and assuming your sample is large enough, you will find that the distribution is the normal probability density function. (credit: John Lodder)

INTRODUCTION: AN OVERVIEW

What is the reason for our preoccupation with means? Two reasons are that they provide a basis for comparison and are simple to compute. This chapter will cover means and the Central Limit Theorem. The Central Limit Theorem is a fundamental and useful concept in statistics. The Central Limit Theorem is a theorem, indicating that it is not only a theory or an individual's conjecture regarding the functioning of phenomena. It is comparable to the Pythagorean Theorem and the theorem stating that the sum of the angles in a triangle equals 180 degrees. These are truths of the world's mechanisms meticulously validated by mathematical accuracy and reasoning. This significant theorem will delineate the permissible and impermissible assertions in inferential statistics. The Central Limit Theorem is concerned with drawing finite samples of size *n* from a population with a known mean, μ, and a known standard deviation, σ. The conclusion is that if we collect samples of size *n* with a "large enough *n*," calculate each sample's mean, and create a histogram (distribution) of those means, then the resulting distribution will tend to have an approximate normal distribution.

The astounding result is that it does not matter what the distribution of the original population is, or whether you even need to know it. The important fact is that the distribution of sample means tend to follow the normal distribution.

The size of the sample, *n*, that is required in order to be "large enough" depends on the original population from which the samples are drawn (the sample size should be at least 30 or the data should come from a normal distribution). If the original population is far from normal, then more observations are needed for the sample means. Sampling is done randomly and with replacement in the theoretical model.

7.1 THE CENTRAL LIMIT THEOREM FOR SAMPLE MEANS

The sampling distribution is a theoretical distribution. It is created by taking many samples of size n from a population. Each sample mean is then treated like a single observation of this new distribution, the sampling distribution. The genius of thinking this way is that it recognizes that when we sample we are creating an observation and that observation must come from some particular distribution. The Central Limit Theorem answers the question: from what distribution did a sample mean come? If this is discovered, then we can treat a sample mean just like any other observation and calculate probabilities about what values it might take on. We have effectively moved from the world of statistics where we know only what we have from the sample, to the world of probability where we know the distribution from which the sample mean came and the parameters of that distribution.

The reasons that one samples a population are obvious. The time and expense of checking every invoice to determine its validity or every shipment to see if it contains all the items may well exceed the cost of errors in billing or shipping. For some products, sampling would require destroying them, called destructive sampling. One such example is measuring the ability of a metal to withstand saltwater corrosion for parts on ocean going vessels.

Sampling thus raises an important question; just which sample was drawn. Even if the sample were randomly drawn, there are theoretically an almost infinite number of samples. With just 100 items, there are more than 75 million unique samples of size five that can be drawn. If six are in the sample, the number of possible samples increases to just more than one billion. Of the 75 million possible samples, then, which one did you get? If there is variation in the items to be sampled, there will be variation in the samples. One could draw an "unlucky" sample and make very wrong conclusions concerning the population. This recognition that any sample we draw is really only one from a distribution of samples provides us with what is probably the single most important theorem is statistics: the Central Limit Theorem. Without the Central Limit Theorem it would be impossible to proceed to inferential statistics from simple probability theory. In its most basic form, the Central Limit Theorem states that regardless of the underlying probability density function of the population data, the theoretical distribution of the means of samples from the population will be normally distributed. In essence, this says that the mean of a sample should be treated like an observation drawn from a normal distribution. The Central Limit Theorem only holds if the sample size is "large enough" which has been shown to be only 30 observations or more.

Figure 7.2 graphically displays this very important proposition.

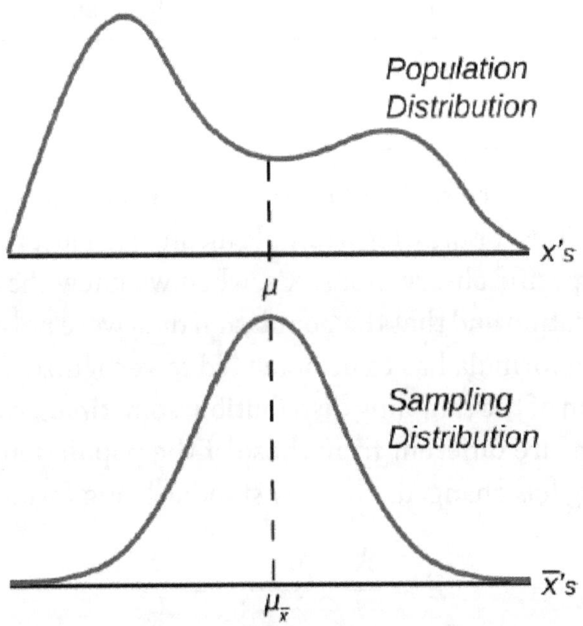

Figure 7.2

Notice that the horizontal axis in the top panel is labeled X. These are the individual observations of the population. This is the unknown distribution of the population values. The graph is purposefully drawn all squiggly to show that it does not matter just how odd ball it really is. Remember, we will never know what this distribution looks like, or its mean or standard deviation for that matter.

The horizontal axis in the bottom panel is labeled X–*X*–'s. This is the theoretical distribution called the sampling distribution of the means. Each observation on this distribution is a sample mean. All these sample means were calculated from individual samples with the same sample size. The theoretical sampling distribution contains all of the sample mean values from all the possible samples that could have been taken from the population. Of course, no one would ever actually take all of these samples, but if they did this is how they would look. And the Central Limit Theorem says that they will be normally distributed.

The Central Limit Theorem goes even further and tells us the mean and standard deviation of this theoretical distribution.

Table 7.1

Parameter	Population distribution	Sample	Sampling distribution of X–X–'s
Mean	μ	$\bar{X}$	$\mu_{\bar{x}}$ and $E(\mu_{\bar{x}}) = \mu$
Standard deviation	σ	s	$\sigma_{\bar{x}} = \frac{\sigma}{\sqrt{n}}$

The practical significance of The Central Limit Theorem is that now we can compute probabilities for drawing a sample mean, $\bar{x}$, in just the same way as we did for drawing specific observations, X's, when we knew the population mean and standard deviation and that the population data were normally distributed. The standardizing formula has to be amended to recognize that the mean and standard deviation of the sampling distribution, sometimes, called the standard error of the mean, are different from those of the population distribution, but otherwise nothing has changed. The new standardizing formula is

$$Z = \frac{\bar{X} - \mu_{\bar{X}}}{\sigma_{\bar{X}}} = \frac{\bar{X} - \mu}{\frac{\sigma}{\sqrt{n}}}$$

Notice that $\mu_{\bar{X}}$ in the first formula has been changed to simply μ in the second version. The reason is that mathematically it can be shown that the expected value of $\mu_{\bar{X}}$ is equal to μ. This was stated in Table 7.1 above. Mathematically, the E(x) symbol read the "expected value of x". This formula will be used in the next unit to provide estimates of the unknown population parameter μ.

7.2 USING THE CENTRAL LIMIT THEOREM

Examples of the Central Limit Theorem

Law of Large Numbers

The law of large numbers says that if you take samples of larger and larger size from any population, then the mean of the sampling distribution, $\mu_{\bar{x}}$ tends to get closer and closer to the true population mean, μ. From the Central Limit Theorem, we know that as n gets larger and larger, the sample means follow a normal distribution. The larger n gets, the smaller the standard deviation of the sampling distribution gets. (Remember that the standard deviation for the sampling distribution of $\bar{X}$ is $\frac{\sigma}{\sqrt{n}}$.) This means that the sample mean $\bar{x}$ must be closer to the population mean μ as n increases. We can say that μ is the value that the sample means approach as n gets larger. The Central Limit Theorem illustrates the law of large numbers.

This concept is so important and plays such a critical role in what follows it deserves to be developed further. Indeed, there are two critical issues that flow from the Central Limit Theorem and the application of the Law of Large numbers to it. These are

1. The probability density function of the sampling distribution of means is normally distributed regardless of the underlying distribution of the population observations and
2. standard deviation of the sampling distribution decreases as the size of the samples that were used to calculate the means for the sampling distribution increases.

Taking these in order. It would seem counterintuitive that the population may have **any** distribution and the distribution of means coming from it would be normally distributed. With the use of computers, experiments can be simulated that show the process by which the sampling distribution changes as the sample size is increased. These simulations show visually the results of the mathematical proof of the Central Limit Theorem.

Here are three examples of very different population distributions and the evolution of the sampling distribution to a normal distribution as the sample size increases. The top panel in these cases represents the histogram for the original data. The three panels show the histograms for 1,000 randomly drawn samples for different sample sizes: n=10, n= 25 and n=50. As the sample size increases, and the number of samples taken remains constant, the distribution of the 1,000 sample means becomes closer to the smooth line that represents the normal distribution.

Figure 7.3 is for a normal distribution of individual observations and we would expect the sampling distribution to converge on the normal quickly. The results show this and show that even at a very small sample size the distribution is close to the normal distribution.

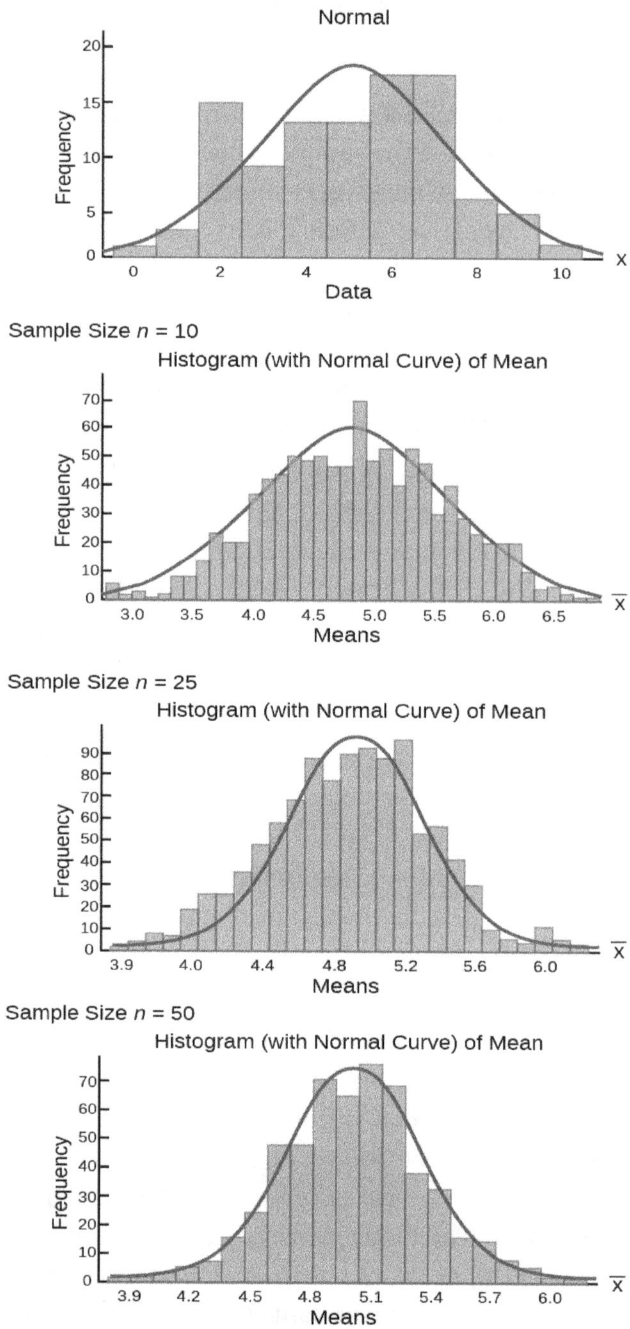

Figure 7.3

Figure 7.4 is a uniform distribution which, a bit amazingly, quickly approached the normal distribution even with only a sample of 10.

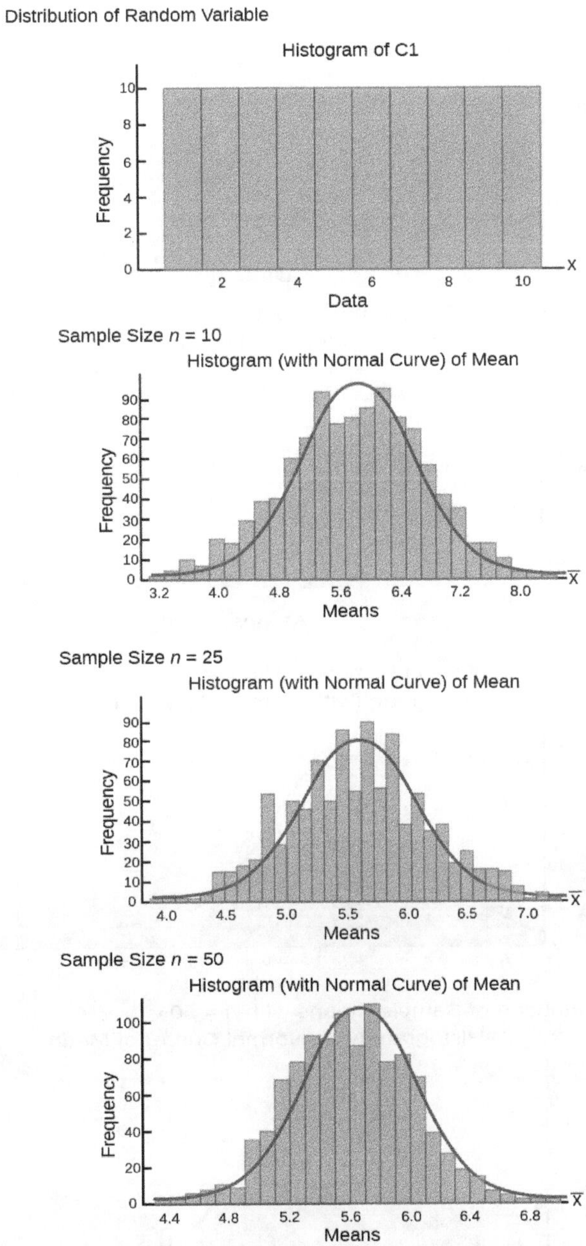

Figure 7.4

Figure 7.5 is a skewed distribution. This last one could be an exponential, geometric, or binomial with a small probability of success creating the skew in the distribution. For skewed distributions our intuition would say that this will take larger sample sizes to move to a normal distribution and indeed that is what we observe from the simulation. Nevertheless, at a sample size of 50, not considered a very large sample, the distribution of sample means has very decidedly gained the shape of the normal distribution.

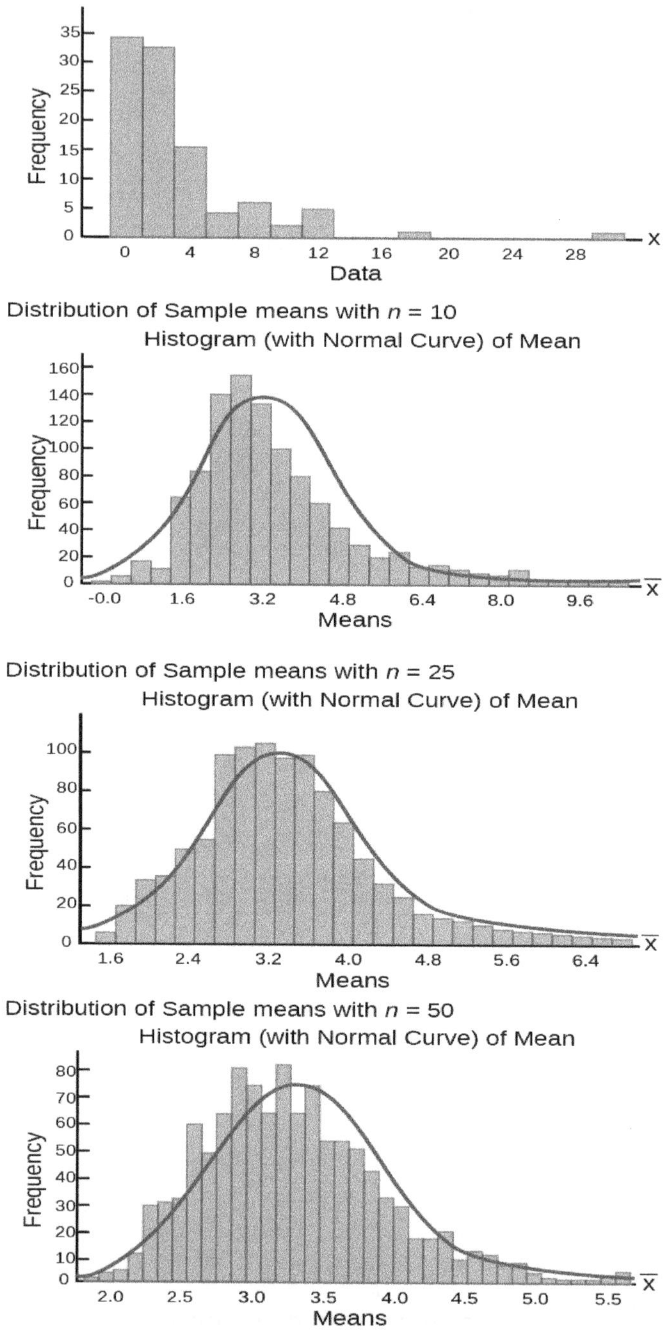

Figure 7.5

The Central Limit Theorem provides more than the proof that the sampling distribution of means is normally distributed. It also provides us with the mean and standard deviation of this distribution. Further, as discussed above, the expected value of the mean, $\mu_{\overline{x}}$, is equal to the mean of the population of the original data which is what we are interested in estimating from the sample we took. We have already inserted this conclusion of the Central Limit Theorem into the formula we use for standardizing from the sampling distribution to the standard normal distribution. And finally, the Central Limit Theorem has also provided the standard deviation of the sampling distribution, $\sigma_{\bar{x}} = \frac{\sigma}{\sqrt{n}}$, and this is critical to have to calculate probabilities of values of the new random variable, $\bar{x}$.

Figure 7.6 shows a sampling distribution. The mean has been marked on the horizontal axis of the $\bar{x}$'s and the standard deviation has been written to the right above the distribution. Notice that the standard deviation of the sampling distribution is the original standard deviation of the population, divided by the sample size. We have already seen that as the sample size increases the sampling distribution becomes closer and closer to the normal distribution. As this happens, the standard deviation of the sampling distribution changes in another way; the standard deviation decreases as n increases. At very large n, the standard deviation of the sampling distribution becomes very small and at infinity it collapses on top of the population mean. This is what it means that the expected value of $\mu_{\bar{x}}$ is the population mean, μ.

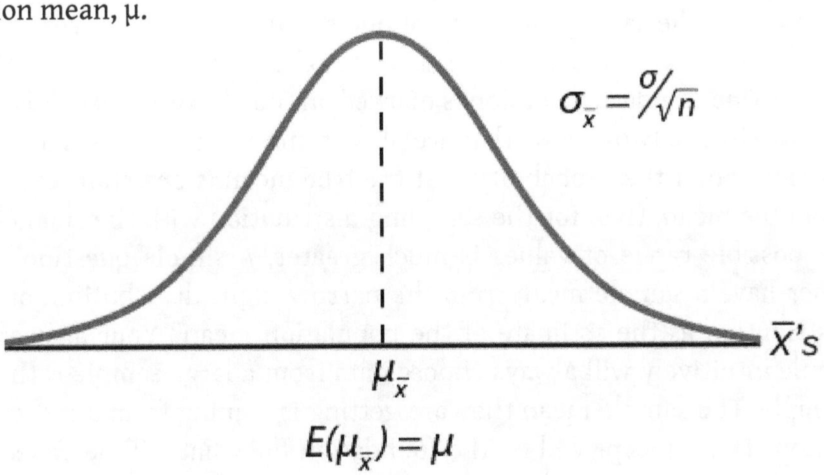

Figure 7.6

At non-extreme values of n, this relationship between the standard deviation of the sampling distribution and the sample size plays a very important part in our ability to estimate the parameters we are interested in.

Figure 7.7 shows three sampling distributions. The only change that was made is the sample size that was used to get the sample means for each distribution. As the sample size increases, n goes from 10 to 30 to 50, the standard deviations of the respective sampling distributions decrease because the sample size is in the denominator of the standard deviations of the sampling distributions.

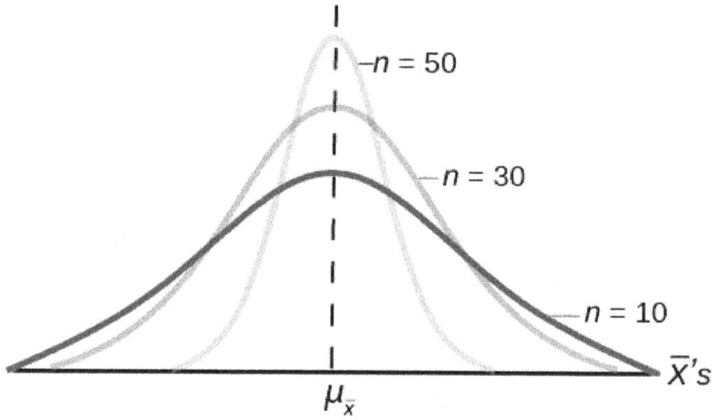

Figure 7.7

The implications for this are very important. Figure 7.8 shows the effect of the sample size on the confidence we will have in our estimates. These are two sampling distributions from the same population. One sampling distribution was created with samples of size 10 and the other with samples of size 50. All other things constant, the sampling distribution with sample size 50 has a smaller standard deviation that causes the graph to be higher and narrower. The important effect of this is that for the same probability of one standard deviation from the mean, this distribution covers much less of a range of possible values than the other distribution. One standard deviation is marked on the $\bar{X}$ axis for each distribution. This is shown by the two arrows that are plus or minus one standard deviation for each distribution. If the probability that the true mean is one standard deviation away from the mean, then for the sampling distribution with the smaller sample size, the possible range of values is much greater. A simple question is, would you rather have a sample mean from the narrow, tight distribution, or the flat, wide distribution as the estimate of the population mean? Your answer tells us why people intuitively will always choose data from a large sample rather than a small sample. The sample mean they are getting is coming from a more compact distribution. This concept will be the foundation for what will be called level of confidence in the next unit.

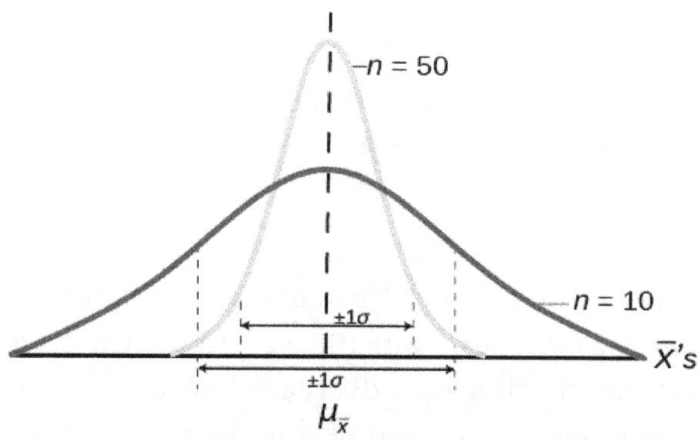

Figure 7.8

7.3 THE CENTRAL LIMIT THEOREM FOR PROPORTIONS

The Central Limit Theorem tells us that the point estimate for the sample mean, $\bar{x}$, comes from a normal distribution of $\bar{x}$'s. This theoretical distribution is called the sampling distribution of $\bar{x}$'s. We now investigate the sampling distribution for another important parameter we wish to estimate; p from the binomial probability density function.

If the random variable is discrete, such as for categorical data, then the parameter we wish to estimate is the population proportion. This is, of course, the probability of drawing a success in any one random draw. Unlike the case just discussed for a continuous random variable where we did not know the population distribution of X's, here we actually know the underlying probability density function for these data; it is the binomial. The random variable is X = the number of successes and the parameter we wish to know is p, the probability of drawing a success which is of course the proportion of successes in the population. The question at issue is: from what distribution was the sample proportion, $p' = \frac{x}{n}$ drawn? The sample size is n and X is the number of successes found in that sample. This is a parallel question that was just answered by the Central Limit Theorem: from what distribution was the sample mean, x¯x¯, drawn? We saw that once we knew that the distribution was the Normal distribution then we were able to create confidence intervals for the population parameter, μ. We will also use this same information to test hypotheses about the population mean later. We wish now to be able to develop confidence intervals for the population parameter "p" from the binomial probability density function.

In order to find the distribution from which sample proportions come we need to develop the sampling distribution of sample proportions just as we did for sample means. So again imagine that we randomly sample say 50 people and ask them if they support the new school bond issue. From this we find a sample proportion, p', and graph it on the axis of p's. We do this again and again etc., etc. until we have the theoretical distribution of p's. Some sample proportions will show high favorability toward the bond issue and others will show low favorability because random sampling will reflect the variation of views within the population. What we have done can be seen in Figure 7.9. The top panel is the population distributions of probabilities for each possible value of the random variable X. While we do not know what the specific distribution looks like because we do not know p, the population parameter, we do know that it must look something like this. In reality, we do not know either the mean or the standard deviation of this population distribution, the same difficulty we faced when analyzing the X's previously.

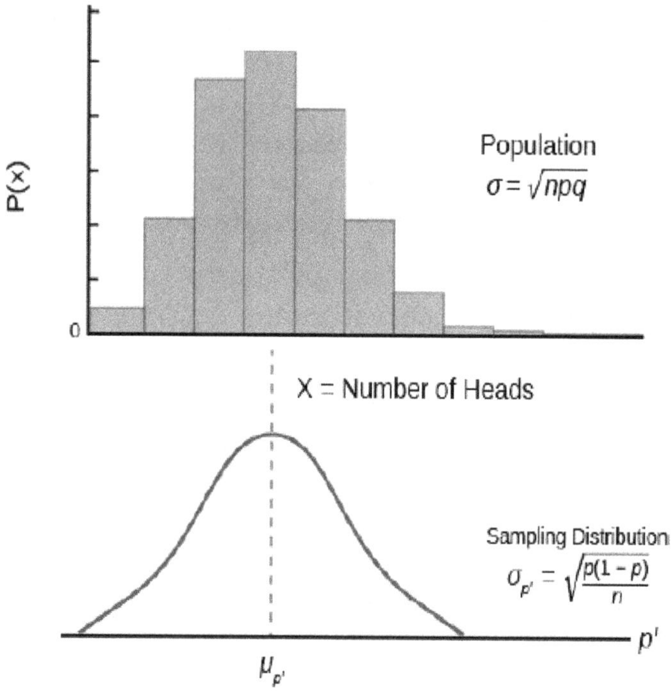

Figure 7.9

Figure 7.9 places the mean on the distribution of population probabilities as $\mu = np$ but of course we do not actually know the population mean because we do not know the population probability of success, p. Below the distribution of the population values is the sampling distribution of p's. Again the Central Limit Theorem tells us that this distribution is normally distributed just like the case of the sampling distribution for $\bar{x}$'s. This sampling distribution also has a mean, the mean of the pp's, and a standard deviation, $\sigma_{p'}$.

Importantly, in the case of the analysis of the distribution of sample means, the Central Limit Theorem told us the expected value of the mean of the sample means in the sampling distribution, and the standard deviation of the sampling distribution. Again the Central Limit Theorem provides this information for the sampling distribution for proportions. The answers are:

1. The expected value of the mean of sampling distribution of sample proportions, μp', is the population proportion, p.
2. The standard deviation of the sampling distribution of sample proportions,σp', is the population standard deviation divided by the square root of the sample size, n.

Both these conclusions are the same as we found for the sampling distribution for sample means. However in this case, because the mean and standard deviation of the binomial distribution both rely upon pp, the formula for the standard deviation of the sampling distribution requires algebraic manipulation to be useful. We will take that up in the next chapter. The proof of these important conclusions from the Central Limit Theorem is provided below.

$$E(p') = E\left(\frac{x}{n}\right) = \left(\frac{1}{n}\right)E(x) = \left(\frac{1}{n}\right)np = p$$

(The expected value of X, E(x), is simply the mean of the binomial distribution which we know to be np.)

$$\sigma_{p'}^{2} = \text{Var}(p') = \text{Var}\left(\frac{x}{n}\right) = \frac{1}{n^2}(\text{Var}(x)) = \frac{1}{n^2}(np(1-p)) = \frac{p(1-p)}{n}$$

The standard deviation of the sampling distribution for proportions is thus:

$$\sigma_{p'} = \sqrt{\frac{p(1-P)}{n}}$$

Table 7.2

Parameter	Population distribution	Sample	Sampling distribution of p's
Mean	µ = np	$p' = \frac{x}{n}$	p' and E(p') = p
Standard Deviation	$\sigma = \sqrt{npq}$		$\sigma_{p'} = \sqrt{\frac{p(1-p)}{n}}$

Table 7.2 summarizes these results and shows the relationship between the population, sample and sampling distribution. Notice the parallel between this Table and Table 7.1 for the case where the random variable is continuous and we were developing the sampling distribution for means.

Reviewing the formula for the standard deviation of the sampling distribution for proportions we see that as n increases the standard deviation decreases. This is the same observation we made for the standard deviation for the sampling distribution for means. Again, as the sample size increases, the point estimate for either μ or p is found to come from a distribution with a narrower and narrower distribution. We concluded that with a given level of probability, the range from which the point estimate comes is smaller as the sample size, n, increases. Figure 7.8 shows this result for the case of sample means. Simply substitute p' for $\bar{x}$ and we can see the impact of the sample size on the estimate of the sample proportion.

7.4 FINITE POPULATION CORRECTION FACTOR

We saw that the sample size has an important effect on the variance and thus the standard deviation of the sampling distribution. Also of interest is the proportion of the total population that has been sampled. We have assumed that the population is extremely large and that we have sampled a small part of the population. As the population becomes smaller and we sample a larger number of observations the sample observations are not independent of each other. To correct for the impact of this, the Finite Correction Factor can be used to adjust the variance of the sampling distribution. It is appropriate when more than 5% of the population is being sampled and the population has a known population size. There are cases when the population is known, and therefore the correction factor must be applied. The issue arises for both the sampling distribution of the means and the sampling distribution of proportions. The Finite Population Correction Factor for the variance of the means shown in the standardizing formula is:

$$Z = \frac{\bar{x} - \mu}{\frac{\sigma}{\sqrt{n}} * \sqrt{\frac{N-n}{N-1}}}$$

and for the variance of proportions is:

$$\sigma_{p'} = \sqrt{\frac{p(1-p)}{n}} \times \sqrt{\frac{N-n}{N-1}}$$

The following examples show how to apply the factor. Sampling variances get adjusted using the above formula.

EXAMPLE 7.1

It is learned that the population of White German Shepherds in the USA is 4,000 dogs, and the mean weight for German Shepherds is 75.45 pounds. It is also learned that the population standard deviation is 10.37 pounds.

If the sample size is 100 dogs, then find the probability that a sample will have a mean that differs from the true probability mean by less than 2 pounds.

Solution 7.1

$$N = 4000, \quad n = 100, \quad \sigma = 10.37, \quad \mu = 75.45, \quad \left(x - \mu\right) = \pm 2$$

$$Z = \frac{\bar{x} - \mu}{\frac{\sigma}{\sqrt{n}} * \sqrt{\frac{N-n}{N-1}}} = \frac{\pm 2}{\frac{10.37}{\sqrt{100}} * \sqrt{\frac{4000-100}{4000-1}}} = \pm 1.95$$

$$f(Z) = 0.4744 * 2 = 0.9488$$

Note that "differs by less" references the area on both sides of the mean within 2 pounds right or left.

EXAMPLE 7.2

When a customer places an order with Rudy's On-Line Office Supplies, a computerized accounting information system (AIS) automatically checks to see if the customer has exceeded his or her credit limit. Past records indicate that the probability of customers exceeding their credit limit is .06.

Suppose that on a given day, 3,000 orders are placed in total. If we randomly select 360 orders, what is the probability that between 10 and 20 customers will exceed their credit limit?

Solution 7.2

$N = 3000, \quad n = 360, \quad p = 0.06$

$$\sigma_{p'} = \sqrt{\frac{p(1-p)}{n}} \times \sqrt{\frac{N-n}{N-1}} = \sqrt{\frac{0.06(1-0.06)}{360}} \times \sqrt{\frac{3000-360}{3000-1}} = 0.0117$$

$$p_1 = \frac{10}{360} = 0.0278, \quad p_2 = \frac{20}{360} = 0.0556$$

$$Z = \frac{p' - p}{\sqrt{\frac{p(1-p)}{n}} * \sqrt{\frac{N-n}{N-1}}} = \frac{0.0278 - 0.06}{0.011744} = -2.74$$

$$Z = \frac{p' - p}{\sqrt{\frac{p(1-p)}{n}} * \sqrt{\frac{N-n}{N-1}}} = \frac{0.0556 - 0.06}{0.011744} = -0.38$$

$$p\left(\frac{0.0278 - 0.06}{0.011744} \angle z \angle \frac{0.0556 - 0.06}{0.011744}\right) = p\left(-2.74 \angle z \angle -0.38\right) = 0.4969 - 0.1480 = 0.3489$$

Terminology

Average a number that describes the central tendency of the data; there are a number of specialized averages, including the arithmetic mean, weighted mean, median, mode, and geometric mean.

Central Limit Theorem Given a random variable with known mean μ and known standard deviation, σ, we are sampling with size n, and we are interested in two new RVs: the sample mean, $\bar{X}$ If the size (n) of the sample is sufficiently large, then $\bar{X} \sim N(\mu, \frac{\sigma}{\sqrt{n}})$. If the size ($n$) of the sample is sufficiently large, then the distribution of the sample means will approximate a normal distributions regardless of the shape of the population. The mean of the sample means will equal the population mean. The standard deviation of the distribution of the sample means, $\frac{\sigma}{\sqrt{n}}$, is called the standard error of the mean.

Finite Population Correction Factor adjusts the variance of the sampling distribution if the population is known and more than 5% of the population is being sampled.

Mean a number that measures the central tendency; a common name for mean is "average." The term "mean" is a shortened form of "arithmetic mean." By definition, the mean for a sample (denoted by $\overline{x}$) is $\overline{x} = \frac{\text{Sum of all values in the sample}}{\text{Number of values in the sample}}$, and the mean for a population (denoted by μ) is $\mu = \frac{\text{Sum of all values in the population}}{\text{Number of values in the population}}$.

Normal Distribution a continuous random variable with $f(x) = \frac{1}{\sigma\sqrt{2\pi}} e^{\frac{-(x-\mu)^2}{2\sigma^2}}$, where μ is the mean of the distribution and σ is the standard deviation.; notation: $X \sim N(\mu, \sigma)$. If $\mu = 0$ and $\sigma = 1$, the random variable, Z, is called the standard normal distribution.

Sampling Distribution Given simple random samples of size n from a given population with a measured characteristic such as mean, proportion, or standard deviation for each sample, the probability distribution of all the measured characteristics is called a sampling distribution.

Standard Error of the Mean the standard deviation of the distribution of the sample means, or $\frac{\sigma}{\sqrt{n}}$.

Standard Error of the Proportion the standard deviation of the sampling distribution of proportions

CHAPTER REVIEW

7.1 The Central Limit Theorem for Sample Means

In a population whose distribution may be known or unknown, if the size (n) of samples is sufficiently large, the distribution of the sample means will be approximately normal. The mean of the sample means will equal the population mean. The standard deviation of the distribution of the sample means, called the standard error of the mean, is equal to the population standard deviation divided by the square root of the sample size (n).

7.2 Using the Central Limit Theorem

The Central Limit Theorem can be used to illustrate the law of large numbers. The law of large numbers states that the larger the sample size you take from a population, the closer the sample mean $\overline{x}$ gets to μ.

7.3 The Central Limit Theorem for Proportions

The Central Limit Theorem can also be used to illustrate that the sampling distribution of sample proportions is normally distributed with the expected value of p and a standard deviation of $\sigma_{p'} = \sqrt{\frac{p(1-p)}{n}}$

FORMULA REVIEW

7.1 The Central Limit Theorem for Sample Means

The Central Limit Theorem for Sample Means:

$$\bar{X} \sim N\left(\mu_{\bar{x}} \cdot \frac{\sigma}{\sqrt{n}}\right)$$

$$Z = \frac{\bar{X} - \mu_{\bar{X}}}{\sigma_{\bar{X}}} = \frac{\bar{X} - \mu}{\sigma / \sqrt{n}}$$

The Mean $\bar{X} : \mu_{\bar{x}}$

Central Limit Theorem for Sample Means z-score $z = \frac{\bar{x} - \mu_{\bar{x}}}{\left(\frac{\sigma}{\sqrt{n}}\right)}$

Standard Error of the Mean (Standard Deviation ($\bar{x}$)): $\frac{\sigma}{\sqrt{n}}$

Finite Population Correction Factor for the sampling distribution of means:

$$Z = \frac{\bar{x} - \mu}{\frac{\sigma}{\sqrt{n}} * \sqrt{\frac{N-n}{N-1}}}$$

Finite Population Correction Factor for the sampling distribution of proportions:

$$\sigma_{p'} = \sqrt{\frac{p(1-p)}{n}} \times \sqrt{\frac{N-n}{N-1}}$$

PRACTICE

7.2 Using the Central Limit Theorem

Use the following information to answer the next ten exercises: A manufacturer produces 25-pound lifting weights. The lowest actual weight is 24 pounds, and the highest is 26 pounds. Each weight is equally likely so the distribution of weights is uniform. A sample of 100 weights is taken.

1. a. What is the distribution for the weights of one 25-pound lifting weight? What is the mean and standard deviation?

b. What is the distribution for the mean weight of 100 25-pound lifting weights?

c. Find the probability that the mean actual weight for the 100 weights is less than 24.9.

2. Draw the graph from Exercise 7.1

3. Find the probability that the mean actual weight for the 100 weights is greater than 25.2.

4. Draw the graph from Exercise 7.3
5. Find the 90th percentile for the mean weight for the 100 weights.
6. Draw the graph from Exercise 7.5
7. a. What is the distribution for the sum of the weights of 100 25-pound lifting weights?
   b. Find $P(\Sigma x < 2{,}450)$.
8. Draw the graph from Exercise 7.7
9. Find the 90th percentile for the total weight of the 100 weights.
10. Draw the graph from Exercise 7.9

Use the following information to answer the next five exercises: The length of time a particular smartphone's battery lasts follows an exponential distribution with a mean of ten months. A sample of 64 of these smartphones is taken.

11. a. What is the standard deviation?
    b. What is the parameter m?
12. What is the distribution for the length of time one battery lasts?
13. What is the distribution for the mean length of time 64 batteries last?
14. What is the distribution for the total length of time 64 batteries last?
15. Find the probability that the sample mean is between seven and 11.
16. Find the 80th percentile for the total length of time 64 batteries last.
17. Find the IQR for the mean amount of time 64 batteries last.
18. Find the middle 80% for the total amount of time 64 batteries last.

Use the following information to answer the next eight exercises: A uniform distribution has a minimum of six and a maximum of ten. A sample of 50 is taken.

19. Find $P(\Sigma x > 420)$.
20. Find the 90th percentile for the sums.
21. Find the 15th percentile for the sums.
22. Find the first quartile for the sums.
23. Find the third quartile for the sums.
24. Find the 80th percentile for the sums.
25. A population has a mean of 25 and a standard deviation of 2. If it is sampled repeatedly with samples of size 49, what is the mean and standard deviation of the sample means?

26. A population has a mean of 48 and a standard deviation of 5. If it is sampled repeatedly with samples of size 36, what is the mean and standard deviation of the sample means?

27. A population has a mean of 90 and a standard deviation of 6. If it is sampled repeatedly with samples of size 64, what is the mean and standard deviation of the sample means?

28. A population has a mean of 120 and a standard deviation of 2.4. If it is sampled repeatedly with samples of size 40, what is the mean and standard deviation of the sample means?

29. A population has a mean of 17 and a standard deviation of 1.2. If it is sampled repeatedly with samples of size 50, what is the mean and standard deviation of the sample means?

30. A population has a mean of 17 and a standard deviation of 0.2. If it is sampled repeatedly with samples of size 16, what is the expected value and standard deviation of the sample means?

31. A population has a mean of 38 and a standard deviation of 3. If it is sampled repeatedly with samples of size 48, what is the expected value and standard deviation of the sample means?

32. A population has a mean of 14 and a standard deviation of 5. If it is sampled repeatedly with samples of size 60, what is the expected value and standard deviation of the sample means?

7.3 The Central Limit Theorem for Proportions

33. A question is asked of a class of 200 freshmen, and 23% of the students know the correct answer. If a sample of 50 students is taken repeatedly, what is the expected value of the mean of the sampling distribution of sample proportions?

34. A question is asked of a class of 200 freshmen, and 23% of the students know the correct answer. If a sample of 50 students is taken repeatedly, what is the standard deviation of the mean of the sampling distribution of sample proportions?

35. A game is played repeatedly. A player wins one-fifth of the time. If samples of 40 times the game is played are taken repeatedly, what is the expected value of the mean of the sampling distribution of sample proportions?

36. A game is played repeatedly. A player wins one-fifth of the time. If samples of 40 times the game is played are taken repeatedly, what is the standard deviation of the mean of the sampling distribution of sample proportions?

37. A virus attacks one in three of the people exposed to it. An entire large city is exposed. If samples of 70 people are taken, what is the expected value of the mean of the sampling distribution of sample proportions?

38. A virus attacks one in three of the people exposed to it. An entire large city is exposed. If samples of 70 people are taken, what is the standard deviation of the mean of the sampling distribution of sample proportions?

39. A company inspects products coming through its production process, and rejects detected products. One-tenth of the items are rejected. If samples of 50 items are taken, what is the expected value of the mean of the sampling distribution of sample proportions?

40. A company inspects products coming through its production process, and rejects detected products. One-tenth of the items are rejected. If samples of 50 items are taken, what is the standard deviation of the mean of the sampling distribution of sample proportions?

7.4 Finite Population Correction Factor

41. A fishing boat has 1,000 fish on board, with an average weight of 120 pounds and a standard deviation of 6.0 pounds. If sample sizes of 50 fish are checked, what is the probability the fish in a sample will have mean weight within 2.8 pounds the true mean of the population?

42. An experimental garden has 500 sunflowers plants. The plants are being treated so they grow to unusual heights. The average height is 9.3 feet with a standard deviation of 0.5 foot. If sample sizes of 60 plants are taken, what is the probability the plants in a given sample will have an average height within 0.1 foot of the true mean of the population?

43. A company has 800 employees. The average number of workdays between absence for illness is 123 with a standard deviation of 14 days. Samples of 50 employees are examined. What is the probability a sample has a mean of workdays with no absence for illness of at least 124 days?

44. Cars pass an automatic speed check device that monitors 2,000 cars on a given day. This population of cars has an average speed of 67 miles per hour with a standard deviation of 2 miles per hour. If samples of 30 cars are taken, what is the probability a given sample will have an average speed within 0.50 mile per hour of the population mean?

45. A town keeps weather records. From these records it has been determined that it rains on an average of 37% of the days each year. If 30 days are selected at random from one year, what is the probability that at least 5 and at most 11 days had rain?

46. A maker of yardsticks has an ink problem that causes the markings to smear on 4% of the yardsticks. The daily production run is 2,000 yardsticks. What is the

probability if a sample of 100 yardsticks is checked, there will be ink smeared on at most 4 yardsticks?

47. A school has 300 students. Usually, there are an average of 21 students who are absent. If a sample of 30 students is taken on a certain day, what is the probability that at most 2 students in the sample will be absent?

48. A college gives a placement test to 5,000 incoming students each year. On the average 1,213 place in one or more developmental courses. If a sample of 50 is taken from the 5,000, what is the probability at most 12 of those sampled will have to take at least one developmental course?

HOMEWORK

7.1 The Central Limit Theorem for Sample Means

49. Previously, De Anza statistics students estimated that the amount of change daytime statistics students carry is exponentially distributed with a mean of $0.88. Suppose that we randomly pick 25 daytime statistics students.

a. In words, X = ____________
b. $X \sim$ _____(_____,_____)
c. In words, $\bar{X}$ = ____________
d. $\bar{X} \sim$ ______ (______, ______)
e. Find the probability that an individual had between $0.80 and $1.00. Graph the situation, and shade in the area to be determined.
f. Find the probability that the average of the 25 students was between $0.80 and $1.00. Graph the situation, and shade in the area to be determined.
g. Explain why there is a difference in part e and part f.

50. Suppose that the distance of fly balls hit to the outfield (in baseball) is normally distributed with a mean of 250 feet and a standard deviation of 50 feet. We randomly sample 49 fly balls.

a. If $\bar{X}$ = average distance in feet for 49 fly balls, then $\bar{X} \sim$ _______ (_______,_______)
b. What is the probability that the 49 balls traveled an average of less than 240 feet? Sketch the graph. Scale the horizontal axis for $\bar{X}$. Shade the region corresponding to the probability. Find the probability.
c. Find the 80th percentile of the distribution of the average of 49 fly balls.

51. According to the Internal Revenue Service, the average length of time for an individual to complete (keep records for, learn, prepare, copy, assemble, and send) IRS Form 1040 is 10.53 hours (without any attached schedules). The distribution is unknown. Let us assume that the standard deviation is two hours. Suppose we randomly sample 36 taxpayers.

a. In words, X = _____________
b. In words, $\bar{X}$ = _____________
c. $\bar{X}$ ~ _____(_____,_____)
d. Would you be surprised if the 36 taxpayers finished their Form 1040s in an average of more than 12 hours? Explain why or why not in complete sentences.
e. Would you be surprised if one taxpayer finished his or her Form 1040 in more than 12 hours? In a complete sentence, explain why.

52. Suppose that a category of world-class runners are known to run a marathon (26 miles) in an average of 145 minutes with a standard deviation of 14 minutes. Consider 49 of the races. Let $\bar{X}$ the average of the 49 races.

a. $\bar{X}$ ~ _____(_____,_____)
b. Find the probability that the runner will average between 142 and 146 minutes in these 49 marathons.
c. Find the 80th percentile for the average of these 49 marathons.
d. Find the median of the average running times.

53. The length of songs in a collector's iTunes album collection is uniformly distributed from two to 3.5 minutes. Suppose we randomly pick five albums from the collection. There are a total of 43 songs on the five albums.

a. In words, X = _________
b. X ~ _____________
c. In words, $\bar{X}$ = _____________
d. $\bar{X}$ ~ _____(_____,_____)
e. Find the first quartile for the average song length.
f. The IQR(interquartile range) for the average song length is from _______-_______.

54. In 1940 the average size of a U.S. farm was 174 acres. Let's say that the standard deviation was 55 acres. Suppose we randomly survey 38 farmers from 1940.

a. In words, X = _____________
b. In words, $\bar{X}$ = _____________
c. $\bar{X}$ ~ _____(_____,_____)
d. The IQR for X–X– is from _______ acres to _______ acres.

55. Determine which of the following are true and which are false. Then, in complete sentences, justify your answers.

a. When the sample size is large, the mean of $\bar{X}$ is approximately equal to the mean of X.
b. When the sample size is large, $\bar{X}$ is approximately normally distributed.
c. When the sample size is large, the standard deviation of $\bar{X}$ is approximately the same as the standard deviation of X.

56. The percent of fat calories that a person in America consumes each day is normally distributed with a mean of about 36 and a standard deviation of about ten. Suppose that 16 individuals are randomly chosen. Let $\bar{X}$ = average percent of fat calories.

a. $\bar{X}$ ~ ______(______, ______)
b. For the group of 16, find the probability that the average percent of fat calories consumed is more than five. Graph the situation and shade in the area to be determined.
c. Find the first quartile for the average percent of fat calories.

57. The distribution of income in some Third World countries is considered wedge shaped (many very poor people, very few middle income people, and even fewer wealthy people). Suppose we pick a country with a wedge shaped distribution. Let the average salary be $2,000 per year with a standard deviation of $8,000. We randomly survey 1,000 residents of that country.

a. In words, X = ____________
b. In words, $\bar{X}$ = ____________
c. $\bar{X}$ ~ _____(_____,_____)
d. How is it possible for the standard deviation to be greater than the average?
e. Why is it more likely that the average of the 1,000 residents will be from $2,000 to $2,100 than from $2,100 to $2,200?

58. Which of the following is NOT TRUE about the distribution for averages?

a. The mean, median, and mode are equal.
b. The area under the curve is one.
c. The curve never touches the x-axis.
d. The curve is skewed to the right.

59. The cost of unleaded gasoline in the Bay Area once followed an unknown distribution with a mean of $4.59 and a standard deviation of $0.10. Sixteen gas stations from the Bay Area are randomly chosen. We are interested in the average cost of gasoline for the 16 gas stations. The distribution to use for the average cost of gasoline for the 16 gas stations is:

a. $\bar{X} \sim N(4.59, 0.10)$

b. $\bar{X} \sim N\left(4.59, \frac{0.10}{\sqrt{16}}\right)$

c. $\bar{X} \sim N\left(4.59, \frac{16}{0.10}\right)$

d. $\bar{X} \sim N\left(4.59, \frac{\sqrt{16}}{0.10}\right)$

7.2 Using the Central Limit Theorem

60. A large population of 5,000 students take a practice test to prepare for a standardized test. The population mean is 140 questions correct, and the standard deviation is 80. What size samples should a researcher take to get a distribution of means of the samples with a standard deviation of 10?

61. A large population has skewed data with a mean of 70 and a standard deviation of 6. Samples of size 100 are taken, and the distribution of the means of these samples is analyzed.

a. Will the distribution of the means be closer to a normal distribution than the distribution of the population?
b. Will the mean of the means of the samples remain close to 70?
c. Will the distribution of the means have a smaller standard deviation?
d. What is that standard deviation?

62. A researcher is looking at data from a large population with a standard deviation that is much too large. In order to concentrate the information, the researcher decides to repeatedly sample the data and use the distribution of the means of the samples? The first effort used sample sized of 100. But the standard deviation was about double the value the researcher wanted. What is the smallest size samples the researcher can use to remedy the problem?

63. A researcher looks at a large set of data, and concludes the population has a standard deviation of 40. Using sample sizes of 64, the researcher is able to focus the mean of the means of the sample to a narrower distribution where the standard deviation is 5. Then, the researcher realizes there was an error in the original calculations, and the initial standard deviation is really 20. Since the standard deviation of the means of the samples was obtained using the original standard deviation, this value is also impacted by the discovery of the error. What is the correct value of the standard deviation of the means of the samples?

64. A population has a standard deviation of 50. It is sampled with samples of size 100. What is the variance of the means of the samples?

7.3 The Central Limit Theorem for Proportions

65. A farmer picks pumpkins from a large field. The farmer makes samples of 260 pumpkins and inspects them. If one in fifty pumpkins are not fit to market and will be saved for seeds, what is the standard deviation of the mean of the sampling distribution of sample proportions?

66. A store surveys customers to see if they are satisfied with the service they received. Samples of 25 surveys are taken. One in five people are unsatisfied.

What is the variance of the mean of the sampling distribution of sample proportions for the number of unsatisfied customers? What is the variance for satisfied customers?

67. A company gives an anonymous survey to its employees to see what percent of its employees are happy. The company is too large to check each response, so samples of 50 are taken, and the tendency is that three-fourths of the employees are happy. For the mean of the sampling distribution of sample proportions, answer the following questions, if the sample size is doubled.

a. How does this affect the mean?
b. How does this affect the standard deviation?
c. How does this affect the variance?

68. A pollster asks a single question with only yes and no as answer possibilities. The poll is conducted nationwide, so samples of 100 responses are taken. There are four yes answers for each no answer overall. For the mean of the sampling distribution of sample proportions, find the following for yes answers.

a. The expected value.
b. The standard deviation.
c. The variance.

69. The mean of the sampling distribution of sample proportions has a value of p of 0.3, and sample size of 40.

a. Is there a difference in the expected value if p and q reverse roles?
b. Is there a difference in the calculation of the standard deviation with the same reversal?

7.4 Finite Population Correction Factor

70. A company has 1,000 employees. The average number of workdays between absence for illness is 80 with a standard deviation of 11 days. Samples of 80 employees are examined. What is the probability a sample has a mean of workdays with no absence for illness of at least 78 days and at most 84 days?

71. Trucks pass an automatic scale that monitors 2,000 trucks. This population of trucks has an average weight of 20 tons with a standard deviation of 2 tons. If a sample of 50 trucks is taken, what is the probability the sample will have an average weight within one-half ton of the population mean?

72. A town keeps weather records. From these records it has been determined that it rains on an average of 12% of the days each year. If 30 days are selected at random from one year, what is the probability that at most 3 days had rain?

73. A maker of greeting cards has an ink problem that causes the ink to smear on 7% of the cards. The daily production run is 500 cards. What is the probability that if a sample of 35 cards is checked, there will be ink smeared on at most 5 cards?

74. A school has 500 students. Usually, there are an average of 20 students who are absent. If a sample of 30 students is taken on a certain day, what is the probability that at least 2 students in the sample will be absent?

REFERENCES

7.1 The Central Limit Theorem for Sample Means

Baran, Daya. "20 Percent of Americans Have Never Used Email."WebGuild, 2010. Available online at http://www.webguild.org/20080519/20-percent-of-americans-have-never-used-email (accessed May 17, 2013).

Data from The Flurry Blog, 2013. Available online at http://blog.flurry.com (accessed May 17, 2013).

Data from the United States Department of Agriculture.

Solutions

1. a. $U(24, 26)$, 25, 0.5774

b. $N(25, 0.0577)$

c. 0.0416

3. 0.0003

5. 25.07

7. a. $N(2{,}500, 5.7735)$

b. 0

9. 2,507.40

11. a. 10

b. $\frac{1}{10}$

13. $N\left(10, \frac{10}{8}\right)$

15. 0.7799

17. 1.69

19. 0.0072

21. 391.54

23. 405.51

25. Mean = 25, standard deviation = 2/7

26. Mean = 48, standard deviation = 5/6

27. Mean = 90, standard deviation = 3/4

28. Mean = 120, standard deviation = 0.38

29. Mean = 17, standard deviation = 0.17

30. Expected value = 17, standard deviation = 0.05

31. Expected value = 38, standard deviation = 0.43

32. Expected value = 14, standard deviation = 0.65

33. 0.23

34. 0.060

35. 1/5

36. 0.063

37. 1/3

38. 0.056

39. 1/10

40. 0.042

41. 0.999

42. 0.901

43. 0.301

44. 0.832

45. 0.483

46. 0.500

47. 0.502

48. 0.519

49. a. X = amount of change students carry

b. $X \sim E(0.88, 0.88)$

c. $\bar{X}$ = average amount of change carried by a sample of 25 students.

d. $\bar{X} \sim N(0.88, 0.176)$

e. 0.0819

f. 0.1882

g. The distributions are different. Part a is exponential and part b is normal.

51. a. length of time for an individual to complete IRS form 1040, in hours.

b. mean length of time for a sample of 36 taxpayers to complete IRS form 1040, in hours.

c. $N\left(10.53, \frac{1}{3}\right)$

d. Yes. I would be surprised, because the probability is almost 0.

e. No. I would not be totally surprised because the probability is 0.2312

53. a. the length of a song, in minutes, in the collection

b. $U(2, 3.5)$

c. the average length, in minutes, of the songs from a sample of five albums from the collection

d. $N(2.75, 0.066)$

e. 2.74 minutes

f. 0.03 minutes

55. a. True. The mean of a sampling distribution of the means is approximately the mean of the data distribution.

b. True. According to the Central Limit Theorem, the larger the sample, the closer the sampling distribution of the means becomes normal.

c. The standard deviation of the sampling distribution of the means will decrease making it approximately the same as the standard deviation of X as the sample size increases.

57. a. X = the yearly income of someone in a third world country

b. the average salary from samples of 1,000 residents of a third world country

c. $\bar{X} \sim N\left(2000, \frac{8000}{\sqrt{1000}}\right)$

d. Very wide differences in data values can have averages smaller than standard deviations.

e. The distribution of the sample mean will have higher probabilities closer to the population mean. $P(2000 < \bar{X} < 2100) = 0.1537$

f. $P(2100 < \bar{X} < 2200) = 0.1317$

59. b

60. 64

61. a. Yes b. Yes

c. Yes d. 0.6

62. 400

63. 2.5

64. 25

65. 0.0087

66. 0.0064, 0.0064

67. a. It has no effect.

b. It is divided by $\sqrt{2}$.

c. It is divided by 2.

68. a. 4/5 b. 0.04

c. 0.0016

69. a. Yes

b. No

70. 0.955

71. 0.927

72. 0.648

73. 0.101

74. 0.273

www.ingramcontent.com/pod-product-compliance
Lightning Source LLC
LaVergne TN
LVHW062111200926
848484LV00007B/1639